History of U.S. Space Surveillance and Satellite Cataloging

A Long and Winding Road

History of U.S. Space Surveillance and Satellite Cataloging

A Long and Winding Road

Paul W. Schumacher Jr., Felix R. Hoots, and Kyle T. Alfriend

LIBRARY OF FLIGHT

Paul Park, Editor-in-Chief
University of Texas at Arlington
Arlington, Texas

Published by
American Institute of Aeronautics and Astronautics, Inc.
12700 Sunrise Valley Drive, Reston, VA 20191-5807

The images on the cover are courtesy of The Aerospace Corporation. They show realistic views of the original Sputnik orbit versus more than 45,00 satellites in orbit today.

American Institute of Aeronautics and Astronautics, Inc., Reston, Virginia

1 2 3 4 5

Library of Congress Cataloging-in-Publication Data

ISBN: 9781624107436

Ebook ISBN: 9781624107443

CONTENTS

PREFACE

Creation of a fundamentally new technical capability ideally begins with the recognition of some future need, a careful description of requirements, development, testing, and operational implementation. In contrast to this ideal situation, the launch of the Sputnik 1 satellite on 4 October 1957 created an immediate need within the United States to track and maintain a catalog of artificial Earth satellites. This required inventing a solution to a problem that did not even exist before 1957. Thus, the initial effort was driven by both external circumstances and highly constrained solution options such as computing and data communications capacities. This external event-driven paradigm has continued to be the case to this day. A good example is the unexpected explosion of an orbiting object in 1961 and the need to quickly catalog a huge number of new objects. Describing the system evolution based on a small mix of planning and far larger mix of reaction to external circumstance, in fact, makes a very interesting story.

We endeavored to make our historical account accessible to a general reader, while still including details of interest to engineers and other specialists in the field of space surveillance. To that end, this book begins with an introduction that provides a snapshot collection of episodes from the first half century of space surveillance, illustrating how events drove invention. The narrative describes how external problems arose, demanding near-real-time adaptation and invention to maintain the satellite catalog and provide timely space situational awareness. The discussion of these events is enhanced with several anecdotes and recollections from key practitioners, which provide a unique view into those times.

By design, many details underlying the stories in the Introduction are omitted, so that the reader will finish the introductory narrative with a good historical overview and an appetite for more in-depth information. It is hoped that this arrangement provides the reader with a sense of the importance and scope of space surveillance, right up front, to serve as motivation for the rest of the story. Of course, the underlying reasons why many elements of this history developed as they did certainly do have intrinsic interest, and those reasons are explained in more detail in the remaining chapters of the book.

The numbered chapters are organized by topic rather than by simple overall chronology, although the organization within each chapter is roughly chronological. The main elements of space surveillance are discussed separately, such as the observing sites, the orbit models, the computing systems, and so forth, as outlined in the Table of Contents. This topical arrangement allows a sufficient level of detail to be included in the appropriate context, so that, by the end, there is a reasonably clear record of each aspect of the systems and organizations involved, besides an overall chronology. The book concludes with an Epilogue briefly discussing the prospects and problems facing the future space surveillance mission.

For our story, the natural stopping place is not a single discrete date but rather sometime just before the advent of the most recent generation of tracking sensors, mainly the Space Fence radar and the Space Surveillance Telescope, and the organization of a new branch of military service, the Space Force. If a date had to be given, it would be no later than 2019. The coincidence of these new sensor developments and the national-level reorganization of military space surveillance, with the simultaneous emergence of commercial interest in space surveillance, guarantees that this mission area has now entered a completely new phase, whose history will have to be assessed in the next generation.

We realized years ago that space surveillance, despite its national importance, has always been poorly understood outside the specific community tasked to perform the mission. This book is intended to help remedy that situation. However, as just noted, the entire mission area is now entering a new and much more complex phase, with new types of sensors, a much larger population of space objects, a greatly expanded domain of surveillance throughout the Earth–Moon system, and a new military chain of command tasked with it all. Hence, part of our success criteria must be that others will eventually find the inspiration to write the next chapters of this evolving history.

ACKNOWLEDGEMENTS

Although we had not begun our professional careers at the time of the Sputnik launch, we feel extremely fortunate to have worked directly with many of those original space pioneers. We are very grateful for those firsthand experiences and knowledge that enabled us to present this history from the start of the Space Age. Those pioneers include Emory Bales, Ken Cranford, Roger Easton, Hubert Edge, Philip Fitzpatrick, John Gabbard, Carroll Hayden, Geoff Hilton, Max Lane, Fred Lipp, Paul Major, Neil Melson, Bob Morris, Ralph Schinnerer, Dick Smith, and Beat Wackernagel.

We also want to recognize the many colleagues we have worked with and learned from over the decades of our careers. We are especially grateful to those who have helped gather and compile information for this book: Gary Van Horn, Bob Morris, and Taft DeVere.

Most of all, we are grateful to our families for sharing us with the challenging exploration of astrodynamics that has been so fulfilling and rewarding. We especially mention our respective wives, Cindy, Judy, and Bonnie.

Introduction: How Events Drove Invention

I. How Will We Track and Keep Up with This New Satellite?

> ***On 13 April 1970, two days into the Apollo 13 lunar mission, an oxygen tank ruptured jeopardizing the mission and the lives of the three astronauts. Mission Control quickly devised improvisations including using the lunar module as a lifeboat for the 96-hour return to Earth. However, the CO2 circular shaped scrubbers would need replacement, and only square replacements were on board. Mission Control invented a circle to square adapter using random items like a flight manual cover, suit parts, a bungee cord, socks, and duct tape. Thus, the astronauts were saved using only materials on hand at the time of the event*** [1].

This is like the situation experienced by the United States when the Soviet Union launched the Sputnik satellite on 4 October 1957. Tracking this satellite and determining how to maintain a catalog of other satellites sure to follow was an exercise in figuring out what resources were available and how to best organize them to support this new mission.

Getting the Data

Tracking came from a surprising mix of resources. The Minitrack system that was developed in the late 1950s for the Naval Research Laboratory's Vanguard satellite program was designed to track satellites using signals emitted by onboard beacons to determine their positions and orbits. Sputnik and other Soviet satellites failed to conform to the beacon standard, illuminating a need to track nonradiating or noncooperative satellites.

Consequently, a high-priority program was initiated to acquire the capability to track satellites by signals reflected off them. An experiment using a transmitter in Fort Monmouth, New Jersey, and Minitrack receivers demonstrated the concept, and from this experiment a larger and more elaborate system was developed by NRL. The new system was to be called the Naval Space Surveillance System (NAVSPASUR), and the first two stations were operating in July 1958, less than six weeks after NRL was tasked to construct the system. At declaration of initial operational capability in 1961, it consisted of six field sites: two transmitter sites (Jordan Lake, Alabama, and Gila River, Arizona) plus four receiver sites (San Diego, California; Elephant Butte, New Mexico; Silver Lake, Mississippi; and Fort Stewart, Georgia).

Although construction of another important radar was not yet complete when Sputnik was launched, the Millstone Hill UHF radar was manually steered and was able to detect Sputnik on 5 October 1957, just one day after its launch. Another source of data was the Baker-Nunn cameras, which were developed in the mid-1950s to provide surveillance data. Baker-Nunn operations were labor intensive and required operators to develop film and perform manual data reductions to create observations.

Space enthusiasts and amateur astronomers also contributed as volunteer Moonwatch observers. These dedicated individuals took observations on satellites by visual means and stopwatches. At the time of the Sputnik launch, members of Moonwatch were the only organized worldwide network that was prepared to spot and help track satellites.

Besides operation Moonwatch, the Air Force funded the Western Satellite Research Network (WSRN) to obtain position and SOI (Space Object Identification) observations on satellites during the early 1960s. WSRN was a network of amateur satellite observers throughout the U.S., some of whom also operated Moonwatch stations in their backyards. Observations of new launches and difficult to track satellites were regularly telegraphed to North American Aviation, who had the WSRN contract and subsequently telegraphed the observations to Spacetrack. WSRN teams regularly received the Spacetrack satellite bulletins to generate their own look angles. Some objects were only maintained by the WSRN stations. I operated one of these WSRN stations in my backyard while I was in high school.

Bruce Bowman, Orbital Analyst, 1st Aerospace Control Squadron

In the 1960's the Space Surveillance Network (SSN) did not have the large phased-array radars that exist today. Many of the sensors were pencil beam dish radars. Only the Ballistic Missile Early Warning System (BMEWS) fences in the polar regions and the Navy Fence across the mid-U.S. could perform autonomous detection of new launches. The dish radars had to search laboriously when a new launch was underway.

An enterprising British crew of a dish radar (no longer in operation) designed an ingenious device to aid their acquisition attempts. They hung a piece of corrugated metal roofing material over the feed horn of the antenna to spread their beam size. However, when the object was finally acquired, the track data was not of good quality. So, they attached a rope to the metal sheet and at the appropriate time yanked it free and began precision tracking.

Bob Morris, Orbital Analyst, 1st Aerospace Control Squadron

MAKING A CATALOG

Data must be processed to obtain orbits. Those orbits must be sent to observers to get subsequent updates; the process flows in a closed loop. Space surveillance became a military mission almost as soon as the first Sputnik satellite was launched. In addition to the intense civilian and scientific interest in knowing the locations of space objects, the Air Force needed a practical way to prevent false missile-warning alarms as satellites transited through the coverage of warning systems, whereas the Navy needed a way to alert fleet units against possible overhead reconnaissance by satellites.

Both needs led to the creation of a complete catalog of detectable space objects, with satellite tracking data forwarded continually to a central processing facility and updated orbital data distributed routinely to defense users. Objects were categorized as payload, rocket body, or debris (payload covers, shrouds, fragments produced by wear and tear or collisions, etc.). Naturally, the catalog also served, and still serves, a variety of civilian and scientific purposes.

The first formalized effort to catalog satellites occurred at the National Space Surveillance Control Center located at Hanscom Field in Bedford, Massachusetts. The observations were transferred to the NSSCC by teletype, telephone, mail, and personal messenger. There a duty analyst reduced the data, determining corrections that should be made to the orbital elements before they were used for further prediction. Three different products were computed and sent back to the observing stations for their planning of future observing opportunities: 1) a table of revolution number, time of passage through the ascending node, and longitude of ascending node for three to seven days in advance; 2) tabulation of all satellite passes observable by a specific ground site; and 3) time and point of intersection of the satellite orbit with the vertical plane containing the sensing beam.

> *The Naval Proving Ground (pre-NAVSPASUR) started their catalog maintenance effort by using numerical integration (about 1959). That didn't last long because it took approximately 3 hours per satellite. They quickly transitioned to using Kepler's laws with two improvements, motion of node and motion of perigee. Early advisors included Dr. Paul Herget, Dr. Boris Garfinkle, and Dr. Kermet. The first orbital model was based on the Herget model and was used in the early sixties until the Brouwer model was adopted by about 1964. Herget taught at the University of Cincinnati but worked at the Proving Ground in the summers. During the '60–'64 period, a sequence of programs that identified, corrected, and predicted satellite locations called General Orbit Perturbation Improvements (GOPI) was put together by Emory Bales. Brouwer never came to SPASUR, but beginning in about 1960 Cox, Smith, and Bales spent part of their summers at Yale listening to Brouwer lecture from his hand-written notes which he later published in his book. Dick*

Smith made improvements to the Brouwer model and made it fully operational at SPASUR by about 1964.

Fred Lipp, Bob Cote, Emory Bales, Bob Cox

(Taken from a series of interviews of these men, all retired, by Gary Van Horn in 1998. Bob Cote had worked in the Operations Department of NAVSPASUR, Fred Lipp had been Head of the Analysis and Software Department at NAVSPASUR, and Bob Cox had previously served in that role. Emory Bales was Head of the Applications Section under Lipp and served at NAVSPASUR and later Naval Space Command, besides the pre-NAVSPASUR organization mentioned here.)

Thus, we see there were two independent efforts (Air Force and Navy) to create and maintain a complete satellite catalog driven by the tracking resources available to each and the need to support their respective missions. Information was regularly exchanged to assure the most complete and up-to-date catalog possible. This continued until 2004 when the NAVSPASUR Fence and support organization became part of Air Force Space Command.

From the mid-1960's to the mid-1980's, NAVSPASUR was deeply involved in providing reports to fleet units to alert them of overflights of Soviet reconnaissance satellites (Satellite Vulnerability Reports). This was initially provided to static locations based on their known geographic locations. As ship commanders became more concerned about space-based threats, NAVSPASUR was requested to provide similar reports to ships at sea. This required the ships to notify NAVSPASUR of their position and intended movements (PIMs). This was done either as a series of times and positions or as a starting time/position and a direction/speed of travel.

The first operational use of the "PIM" software occurred when the battleship U.S.S. New Jersey sailed from San Diego to the South China Sea during the Vietnam war to be used as an offshore artillery platform. The New Jersey provided NAVSPASUR with their planned transit via a series of times and positions, and the vulnerability reports were provided without a hitch. Shortly afterwards, another ship (don't recall its name) was planning a transit from Norfolk to Rota, Spain. Its PIM was provided in the form of a starting position and direction/speed. Vulnerability reports were then generated and transmitted (apparently) without a hitch. Several days into its transit, however, NAVSPASUR analysts were rechecking the PIM projections and found that the ship's location was predicted to be slightly west of Kansas City, MO! There was a "minor" bug in the PIM logic that created a 180-degree azimuth error. Corrected vulnerability reports were hastily produced and transmitted, and the ship made it safely to Rota without knowing about the snafu.

Hubert Edge

(From an interview by Gary Van Horn, circa 1998. Hubert Edge was retired from NAVSPASUR at the time. He had served as Deputy Head of the Analysis and Software Department at NAVSPASUR, under Fred Lipp, cited earlier; besides several previous roles in the same organization.)

During the early years of Navy involvement in Spacetrack, Bob Cox and Dick Smith worked for the Naval Proving Ground Facilities Department because there was no money for the Spacetrack effort. Bob and Dick used to have to cut grass for a couple of hours each pay period to justify their paycheck. That problem was solved by 1960, and so when Emory Bales joined them, he didn't have to cut grass.

Fred Lipp, Bob Cote, Emory Bales, Bob Cox

II. Oops! A Satellite Broke Up

Between 1957 and 1961, the process of tracking satellites and maintaining an up-to-date catalog had matured into a routine process. Predictions of future satellite locations were provided to tracking sites using the Bulletins. The sites performed tracking and sent their observations to the central processing center at Hanscom Field. Here the analysts did the data reduction and sent updated predictions using the Bulletins. There was a routine closed-loop process that maintained an ever-growing satellite catalog.

Figures 1a through 1d show a Satellite Situation Report from January 2, 1962 [2]. At that time, the convention of sequentially numbering the catalog entries had not yet been adopted. The entries are denoted by letters of the Greek alphabet. These early pioneers would indeed be surprised to learn that today's catalog contains over 60,000 entries.

On 29 June 1961, there were 115 objects that had been launched and cataloged. Of those, 50 were still in orbit. Suddenly our trackers detected extra objects in their scans. The Thor-Ablestar rocket upper stage had exploded, creating more than 300 fragments. Instantaneously, our well-tracked and well-maintained catalog of 50 on-orbit objects now included over 300 new objects whose trajectories were unknown. All we knew about the objects was their source; they were Thor-Ablestar debris. This was the first satellite breakup.

Previously, any new launch alerted us to the impending addition of a new satellite to our catalog, and so the catalog grew incrementally a few objects at a time: a payload, a rocket body, and perhaps a few pieces of launch debris. Adding these to the catalog was straightforward, and determining the category of each object was aided by radar cross-sectional data. But what do you do when there are 300 new pieces all at once? This is the first of many examples in the history of space surveillance where events drove innovation. This innovation is best described in the words of those from that era.

N63 22000

CODE-1

SPACE OPERATIONS CONTROL CENTER
GODDARD SPACE FLIGHT CENTER
NATIONAL AERONAUTICS AND SPACE ADMINISTRATION
GREENBELT, MARYLAND

Volume 2, No 1 SATELLITE SITUATION REPORT, January 2, 1962

The following report reflects data computed and compiled by Goddard Space Flight Center, NORAD, and the Smithsonian Astrophysical Observatory as of 1200Z on January 2, 1962.

(NASA TM X-50205) OTS PRICE

XEROX $ 1.10 ph

MICROFILM $ 0.80 mf.

RALPH R. STROBLE
HEAD, Operations Branch

TM X-50,205

Fig. 1a Satellite situation report – title page.

OBJECTS IN ORBIT

OBJECT	CODE NAME	SOURCE	LAUNCH	NODAL PERIOD	INCLI- NATION	APOGEE S. MILES	PERIGEE S. MILES	TRANSMITTING FREQ. (MC/S)
1958 ALPHA 1	EXPLORER I	US	1 FEB 58	106.0	33.19	1084	217	
1958 BETA 1	ROCKET BODY	US	17 MAR 58	138.3	34.26	2684	406	
1958 BETA 2	VANGUARD I	US	17 MAR 58	133.8	34.24	2449	404	108.022
1959 ALPHA 1	VANGUARD II	US	17 FEB 59	125.3	32.87	2079	313	
1959 ALPHA 2	ROCKET BODY	US	17 FEB 59	129.6	32.90	2254	370	
1959 ETA 1	VANGUARD III	US	18 SEP 59	129.7	33.30	2299	332	
1959 MU*	LUNIK I	USSR	2 JAN 59	450D	0.01	1.3177AU	0.9766AU	
1959 NU*	PIONEER IV	US	3 MAR 59	398D	0.127	1.1421AU	0.9871AU	
1959 IOTA 1	EXPLORER VII	US	13 OCT 59	101.1	50.31	669	344	
1959 IOTA 2	ROCKET BODY	US	13 OCT 59	100.9	50.30	656	345	
1960 ALPHA*	PIONEER V	US	11 MAR 60	311.6D	3.35	.9951AU	.8061AU	
1960 BETA 1	ROCKET BODY	US	1 APR 60	99.1	48.41	463	429	
1960 BETA 2	TIROS I	US	1 APR 60	99.1	48.39	467	429	107.997
1960 BETA 3	NONE	US	1 APR 60	97.8	48.46	445	375	
1960 BETA 4	NONE	US	1 APR 60	99.8	48.15	503	434	
1960 GAMMA 2	TRANSIT 1B	US	13 APR 60	94.7	51.28	401	229	
1960 GAMMA 4	NONE	US	13 APR 60	96.8	51.21	458	299	
1960 EPSILON 1	SPUTNIK IV	USSR	15 MAY 60	91.6	65.02	275	160	
1960 EPSILON 3	NONE	USSR	15 MAY 60	93.0	64.89	352	170	
1960 EPSILON 4	NONE	USSR	15 MAY 60	POSITION UNCERTAIN				
1960 ZETA 1	MIDAS II	US	24 MAY 60	94.2	33.01	310	299	
1960 ETA 1	TRANSIT 2A	US	22 JUN 60	101.6	66.77	649	389	162;216;
1960 ETA 2	GREB	US	22 JUN 60	101.6	66.77	657	381	
1960 ETA 3	ROCKET BODY	US	22 JUN 60	101.4	66.77	642	383	
1960 IOTA 1	ECHO I	US	12 AUG 60	116.1	47.30	1314	563	
1960 IOTA 2	ROCKET BODY	US	12 AUG 60	118.0	47.22	1049	932	
1960 IOTA 3	METAL OBJECT	US	12 AUG 60	118.2	47.20	1049	941	
1960 IOTA 4	METAL OBJECT	US	12 AUG 60	118.2	47.23	1071	922	
1960 IOTA 5	METAL OBJECT	US	12 AUG 60	118.3	47.20	1060	940	
1960 NU 1	COURIER 1B	US	4 OCT 60	106.8	28.20	777	577	107.9709
1960 NU 2	ROCKET BODY	US	4 OCT 60	106.4	28.29	756	570	
1960 XI 1	EXPLORER VIII	US	3 NOV 60	112.4	49.92	1409	258	
1960 XI 2	ROCKET BODY	US	3 NOV 60	112.2	49.98	1398	256	
1960 XI 3	NONE	US	3 NOV 60	110.8	49.39	1327	250	
1960 XI 4	NONE	US	3 NOV 60	111.7	50.52	1363	262	

- 1 -

Fig. 1b Satellite situation report – objects in orbit.

OBJECTS IN ORBIT (CONT'D)

OBJECT	CODE NAME	SOURCE	LAUNCH	NODAL PERIOD	INCLI-NATION	APOGEE S. MILES	PERIGEE S. MILES	TRANSMITTING FREQ. (MC/S)
1960 PI 1	TIROS II	US	23 NOV 60	98.2	48.57	461	378	
1960 PI 2	ROCKET BODY	US	23 NOV 60	98.1	48.57	455	378	
1960 PI 3	NONE	US	23 NOV 60	98.1	48.46	452	383	
1960 PI 4	NONE	US	23 NOV 60	98.3	48.51	455	389	
1961 ALPHA 1	SAMOS II	US	31 JAN 61	94.9	97.40	339	295	
1961 ALPHA 2	METAL OBJECT	US	31 JAN 61	94.9	97.40	338	295	
1961 GAMMA 1*	VENUS PROBE	USSR	12 FEB 61	300D	0.581	1.0190AU	0.7183AU	
1961 DELTA 1	EXPLORER IX	US	16 FEB 61	118.0	38.81	1519	466	
1961 DELTA 2	ROCKET BODY	US	16 FEB 61	118.4	38.77	1627	379	
1961 DELTA 3	NONE	US	16 FEB 61	118.1	38.87	1555	438	
1961 DELTA 4	NONE	US	16 FEB 61	POSITION UNCERTAIN				
1961 EPSILON 1	DISCOVERER XX	US	17 FEB 61	93.0	80.91	342	176	
1961 ZETA 1	DISCOVERER XXI	US	18 FEB 61	93.2	80.74	387	146	
1961 KAPPA 1	EXPLORER X	US	25 MAR 61	POSITION UNCERTAIN				
1961 LAMBDA 1	DISCOVERER XXIII	US	8 APR 61	91.8	82.31	281	166	
1961 LAMBDA 2	CAPSULE	US	8 APR 61	95.9	81.94	573	120	
1961 NU 1	EXPLORER XI	US	27 APR 61	107.8	28.78	1109	301	
1961 OMICRON 1	TRANSIT 4A	US	29 JUN 61	103.8	66.81	620	547	54;150;324;400
1961 OMICRON 2	INJUN-SR-3	US	29 JUN 61	103.8	66.82	619	548	136.5
1961 OMICRON 3-52**	METAL OBJECTS	US	29 JUN 61					
1961 RHO 1	TIROS III	US	12 JUL 61	100.3	47.90	511	457	108.0;108.03
1961 RHO 2	ROCKET BODY	US	12 JUL 61	100.3	47.85	503	462	
1961 RHO 3	METAL OBJECT	US	12 JUL 61	98.8	47.92	500	376	
1961 RHO 4	METAL OBJECT	US	12 JUL 61	101.9	47.89	563	498	
1961 SIGMA 1	MIDAS III	US	12 JUL 61	161.5	91.13	2197	2084	
1961 SIGMA 3	METAL OBJECT	US	12 JUL 61	161.2	91.13	2178	2085	
1961 SIGMA 4	METAL OBJECT	US	12 JUL 61	161.9	91.15	2205	2096	
1961 UPSILON 1	EXPLORER XII	US	16 AUG 61	1588.3	33.43	47651	481	
1961 ALPHA DELTA 1	MIDAS IV	US	21 OCT 61					
1961 ALPHA EPSILON 1	DISCOVERER XXXIV	US	5 NOV 61	96.4	82.52	583	140	

- 2 -

Fig. 1c Satellite situation report – objects in orbit (continued).

OBJECTS IN ORBIT (CONT'D)

OBJECT	CODE NAME	SOURCE	LAUNCH	NODAL PERIOD	INCLI-NATION	APOGEE S. MILES	PERIGEE S. MILES	TRANSMITTING FREQ. (MC/S)
1961 ALPHA ETA 1	TRANSIT IV-B	US	15 NOV 61	105.6	32.43	689	593	136.8;54;324;150;400
1961 ALPHA ETA 2	TRACC	US	15 NOV 61	105.6	32.43	686	597	136.65;54;324
1961 ALPHA ETA 3	ROCKET BODY	US	15 NOV 61	105.5	32.41	687	588	
1961 ALPHA KAPPA 1	DISCOVERER XXXVI	US	12 DEC 61	91.5	81.21	280	148	20.005;40.01
1961 ALPHA KAPPA 2	OSCAR	US	12 DEC 61	91.1	81.21	258	146	145

*APHELION, PERIHELION IN ASTRONOMICAL UNITS, INCLINATION TO ECLIPTIC.
**FIFTY METAL OBJECTS HAVE BEEN IDENTIFIED IN THE PLANE OF THE 1961 OMICRON 1 AND 1961 OMICRON 2 ORBITS.

PLEASE ADD THE FOLLOWING TO THE DECAYED OBJECTS LIST:

OBJECT	CODE NAME	SOURCE	LAUNCH	DECAY
1961 ALPHA KAPPA 3	METAL OBJECT	US	12 DEC 61	20 DEC 61

- 3 -

Fig. 1d Satellite situation report – objects in orbit (concluded).

A very real problem did occur during the operational acceptance check-out of USAF SPACETRACK on 29 June 1961. On that day there were 50 manmade objects in orbit, and we were preparing for the launch of the first multiple-payload launch – Transit 4A, Solar Radiation 3, and Injun. Launch occurred ("1961 Omicron") and seemed nominal. Suddenly there appeared to be dozens of ICBM "impacts" on North America. What was going on?

As we later found out, the upper stage of the launch, an Ablestar booster, which had gone into orbit with the triple-payload, had exploded in response to an accidental signal sent by launch control to the range safety detonator package. This added approximately 500 new satellites to the existing 50—a tenfold increase in the satellite population! These new satellites (explosion debris) were going through the ballistic missile early warning (BMEWS) fan radars so close together that vectors from different satellites were incorrectly combining to form "false" impacts! WWIII thus was averted!

Bill Craig, Orbital Analyst, 1st Aerospace Control Squadron

In the 1960's the first sensor to detect a breakup was usually the Navy Fence across the mid-U.S., or sometimes the BMEWS fences in the polar regions with their ability to autonomously detect uncorrelated targets (UCT's). A UCT was a track that did not match or "correlate" with any satellite in the known catalog. Once multiple UCT's had been collected in the orbital plane of the breakup orbit, IBM cards would be punched for each observation (time, azimuth, elevation, range).

The first break-up processing in the early 60's was a manual procedure using graph paper. The residuals between these observations and the parent orbit would be hand plotted on graph paper, but the amount of data and resolution needed required a 3-foot by 4-foot plotting surface. Individual pieces of 8.5 by 11-inch paper were taped together over the surface of a desk and each new track was plotted as it was processed (this continued over a period of days). Rulers and French curves would be used to "connect the dots" and string together a trending group of UCT's. (the analysts' favorite curves to use were long "ship's curves" used by naval architects).

The IBM punch cards for each observation on the trend curve would be pulled from the large drawer of suspected breakup data. The orbital analyst would then run each candidate group on the mainframe in a batch least squares differential correction program to obtain an element set (if one of the fits failed, he would remove any cards whose observation had high residuals and try again). After he had a good fit, an "analyst satellite" with a temporary number would be generated and sent out to the network for tracking. If the network was able to repeatedly acquire the candidate object, it would be cataloged as a new satellite. If it wasn't tracked, its punched cards would be returned to the drawer and tried in another candidate grouping. Once enough pieces of the breakup had been catalogued, the breakup time could be determined by flying the satellite orbits back to the common time when they all were closest.

Bob Morris, Orbital Analyst, 1st Aerospace Control Squadron

NAVSPASUR had a satellite catalog of less than 50 satellites. Observations of these satellites from the "fence" were displayed at Dahlgren on Sanborn recorders (as phase differences between the several antenna at each site), manually read and recorded on paper by operators, the elevation angle from the detecting site determined by a "slide rule" device that resolved the ambiguity between the various antenna pairs, and the observation was (sometimes) identified manually by comparing against a listing of predicted time/angles of fence penetrations. Unidentified observations were passed to an analyst for further examination. The observation was then converted to card form by keypunch operators, stored on computer tapes, and, at 8-hour intervals, loaded on the NORC (Naval Ordinance Research Computer). Analysts would then periodically run an early version of the manual differential correction program and generate predictions for future passes of this satellite.

During this time period, a computerized version of this process (called ADR, automatic data reduction) was being developed by NAVSPASUR software personnel. This program was designed to read the collected phase data from the sites, perform the elevation angle resolution, and identify the satellite based on pre-computed fence crossing predictions. There was considerable doubt among the NAVSPASUR staff as to the reliability of this program, however, and most were reluctant to give up the old reliable "slide rule" method. After all, 50 satellites produced only a few observations per hour, and these could easily be handled by the manual method.

On a Friday morning in June 1961, during the launch of a Transit satellite by the U.S., the final stage booster exploded after achieving orbit (at about 400–500 miles altitude) and instantly created several hundred new satellites. The ADR program was declared operational on the following Monday!

Hubert Edge

This was early in the Fence's operational life and, fortunately, very shortly after the first computer upgrade at Dahlgren which replaced the NORC for ops. Dahlgren dealt with all the unassociated Fence obs, without using any outside data. The job could not have been done at all with the NORC, and Dahlgren ops people were laying out strip chart traces of analog data from the receiver sites, in the basement hallway of the old building in Dahlgren. They had to do that to find all the time-coincident obs data from all the receiver sites so they could associate them in the orbit solutions.

Paul Schumacher

(Schumacher was recalling the story as he heard it as a new-hire in the Analysis and Software Department at NAVSPASUR.)

These efforts were the beginning and the foundation of subsequent innovations for breakup processing, which unfortunately has been frequently needed ever since. Figure 2 shows the actual count of satellites launched and still in orbit. It includes those satellites that were later cataloged but were in orbit at the time. The dramatic increase due to the Ablestar booster breakup is evident.

III. We Do Not Have a Big Enough Piece of Graph Paper

In 1961, a new mission emerged with the first launch of humans into orbit. The Soviet Union achieved this on 12 April 1961, and the United States followed on 5 May 1961. Then on 12 September 1962 President John F. Kennedy made the historical declaration:

I believe that this nation should commit itself to achieving the goal, before this decade is out, of landing a man on the moon and returning him safely to the Earth [4].

This set the tone for the decade. It was clear that space activity was going to grow dramatically, and we needed the tracking and cataloging infrastructure to support that national goal. At the time of the speech, there were 100

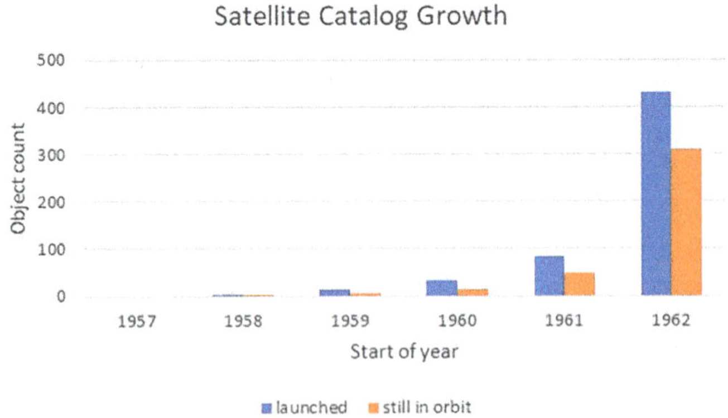

Fig. 2 Catalog growth in the early years [3].

satellites in orbit, and on 20 July 1969, when the Apollo lunar module touched down on the moon's surface, there were 1850.

TRACKING

The BMEWS consisted of three sites located at very high northern latitudes to provide the earliest possible warning of a ballistic missile attack from the Soviet Union.

The site at Clear, Alaska, became operational on 1 July 1961. It consisted of three fixed detection fan radars and one mechanical tracking radar. The site at Thule, Greenland, also became operational in early 1961. It consisted of four detection fan radar antennas driven by three transmitters, and one mechanical tracking radar. The triad of sites was completed in September 1963 when the site at Fylingdales in the North York Moors, England, became operational. It consisted of three mechanical tracking radars. Two of these were employed to scan the horizon, instead of the fixed detection fans used at the other two BMEWS sites.

NAVSPASUR had begun operations in July 1958. Between 1961 and 1965, a more powerful transmitter at Lake Kickapoo, Texas, and two more receiver sites at Red River, Arkansas, and Hawkinsville, Georgia, were added, completing the configuration.

A missile test tracking site located at Ascension Island in the Atlantic Ocean just south of the equator was upgraded during 1962 and 1963 as part of tracking improvements for the Gemini manned missions. This added an important capability to get early tracking on satellite launches originating from Cape Canaveral, Florida.

As the number of objects in orbit grew, it became apparent that we needed dedicated resources to track satellites. Getting the missile-warning data "for free" was nice, but newer technologies could enable more accurate and more extensive tracking data. Additionally, it was important to have tracking capability pointing in directions other than north.

Eglin Air Force Base located in Florida's northwest panhandle on the Gulf Coast was selected for a south-pointing radar. Its location and orientation give it the potential to track all low Earth satellites at least once per day. The FPS-85 radar was an early implementation of the phased array radar concept and was optimized for the space surveillance function. It employed separate transmitter and receiver systems and antenna arrays. Radar construction began in October 1962 at Eglin's test site. Radar testing was originally scheduled for 1 April 1965. However, just four months before, the building and all the equipment were destroyed in a fire caused by arcing electrical equipment (see Fig. 3).

But those satellite guys were serious about wanting to track satellites, and so the radar was rebuilt, and the Air Force took ownership of the site on 29 January 1969 with the 20th Surveillance Squadron (20 SURS) as the primary operator.

Fig. 3 FPS-85 fire. (From AFSPC/A8XY SSN history.)

CATALOGING

The original cataloging center was located at Hanscom Field in Massachusetts. But the center of satellite cataloging was destined to be in Colorado Springs, Colorado. It was not coincidental that construction had begun there in May 1961 of a huge underground complex for secure military operations.

> *By direction of Air Force Chief of Staff, Gen. Thomas D. White, we officially transferred operations on 1 July 1961. We (the orbital analysts) came to Colorado Springs in May 1961, and the remaining Hanscom cadre—representing the development community—gave us the official "turnover validation test case" (1961 Omicron) on 29 June 1961. As you will recall, the Ablestar orbital upper stage booster on Omicron exploded, and the space population went from around 50 to approximately 500.*
>
> *We used the old Hanscom methods in Colorado Springs—the primarily manual nodal period/ascending-node epoch procedures—until Aeronutronic introduced semi-automated anomalistic period procedures in early 1962.*
>
> *Bill Craig, Orbital Analyst, 1st Aerospace Control Squadron*

Besides the physical move, this was a time of major upgrades to the Air Force processing capability and the orbital prediction fidelity. Contractor support was provided by Aeronutronic, a defense-related division of Ford Motor Company. In addition to operational code, they published very thorough and tutorial-level documentation (which one of the authors still uses today). These pioneers are Geoff Hilton, Lou Walters, Jeanne Arsenault, and Ralph Schinnerer. Equally important was the small contingent of civil service scientists. The astrodynamics office was established by Max Lane in early 1965, and he was joined by Paul Major and Ken Cranford later that year.

In April 1960, the Naval Space Surveillance Facility was established at Dahlgren, Virginia, and responsibility for operating the system transferred from NRL to the Navy. In the next year, the system, now formally recognized as the Naval Space Surveillance System (NAVSPASUR), came under operational control of North American Aerospace Defense (NORAD) as part of its Space Detection and Tracking System (SPADATS).

The Air Force facility in Colorado Springs and the Navy facility in Dahlgren were now established as the permanent centers on which the cataloging maintenance mission would be carried out over the coming decades. Each had its primary function (missile-warning support and fleet warning support), but they had the common mission of "maintaining a catalog of all Earth orbiting objects in space." That amiable partnership worked exceedingly well (except for the Saturday each fall when the Navy and Air Force football games occurred).

IV. Brouwer and Kozai: The Great Astrodynamics Bifurcation

> ***NASA lost its Mars Climate Orbiter because spacecraft engineers failed to convert from English to metric measurements when exchanging vital data before the craft was launched*** [5].

Likewise, in the satellite prediction business when you exchange data you need to have a clear understanding of what the data mean and how they should be used. In 1609, Kepler published his first two laws of planetary motion. These same laws also apply to satellites orbiting the Earth. The first of these stated that the planets move in elliptical orbits around the Sun. The Sun is at one focus of the ellipse, and the planetary angular speed is fastest when the planet is closest to the Sun. Similarly, the angular speed is slowest when the planet is farthest from the Sun. The average angular speed over a complete revolution is called the mean motion of the planet and is usually denoted as n_0. That all seems simple enough and should be an amenable definition on which we can all agree. That was the case for 350 years until the publication in 1959 of Vol. 64, No. 1274 of *The Astronomical Journal*. That volume contained two papers that each provided analytical equations for predicting the motion of an artificial Earth satellite with significantly more accuracy than the simple Kepler elliptical motion. Pages 367–377 contained a paper by Yoshihide Kozai, and pages 378–397 contained a paper by Dirk Brouwer. These two important scientific papers led directly to *The Great Astrodynamics Bifurcation*.

Let's begin with a quick comparison of the Brouwer and Kozai solutions. The chief difference can be isolated to the equation for predicting the along-track motion of the satellite denoted by the angle M.

The Brouwer equation is

$$M = M_0 + n_0 t + O(J_2)t$$

and the Kozai equation is

$$M = M_0 + [n_0 + O(J_2)]t$$

where J_2 represents effects due to the oblate shape of the Earth, a subscript *o* denotes the initial value of a variable, and *t* is the time since that initial value.

The equations provide similar numerical results. The bifurcation comes in the language used to describe the equations. Those who adopted the Brouwer solution think of n_0 as the mean motion of the satellite, whereas those who adopted the Kozai solution think of the entire term in brackets as the mean motion of the satellite. Consequently, when we speak of the satellite mean motion, we must be careful to say the *Brouwer* mean motion or the *Kozai* mean motion.

The mean motion is the angular speed of the satellite. It is easy to see that use of a Kozai mean motion in a Brouwer-based model would cause a significant and rapidly growing difference. The same can be said of using a Brouwer mean motion in a Kozai-based theory. Both models are perfectly good when using properly compatible data. Both a gasoline engine and a diesel engine will propel an automobile just fine, but each must be provided the proper fuel.

Those centers responsible for tracking and predicting satellites were eager to incorporate a more accurate model whose data would be accurate for a longer time. Some adopted Brouwer, and some adopted Kozai. Thus began *The Great Astrodynamics Bifurcation*.

How do we assure everyone uses the correct fuel for their propagation engine? The U.S. Space Surveillance System provides a free catalog of information for predicting satellite locations. The information for each satellite is called a two-line element (TLE) set. The TLE set contains parameters that describe the ellipse, one of which is a mean motion. Notice that we said "a" mean motion. Which one is it? It is the Kozai mean motion. There is an easy formula to compute the Brouwer mean motion from the Kozai mean motion. The key to assuring good predictions is to know what model you have (Kozai or Brouwer) and to know whether a data conversion is required or not.

This all seems straightforward, but it may take another 350 years to make sure everyone gets it right.

When I was a young Air Force civil servant working for the Aerospace Defense Command one of my first business trips was to Dahlgren, VA to visit with our Navy counterpart to discuss satellite cataloging. I had enthusiastically learned all I could about the Navy orbit modeling before my trip but had found that the documentation was sparse. I was particularly interested in the details of the Navy implementation of the complete Brouwer solution.

It was with great awe and respect that I engaged in a delightful afternoon of discussion with Dick Smith in the NAVSPASUR headquarters building. Near the end of the day as the shadows grew long, I remember discussing how the critical inclination (potential zero divisor) was handled. Dick lit his pipe for the umpteenth time that afternoon and patiently went to the chalk board and derived his clever modification that "patched" the Brouwer solution with a function that smoothly went to zero as one approached the critical inclination. I was truly impressed and asked if we might "make up" some satellite orbits progressively approaching the critical inclination, so I could see how the model behaved. The test cases were generated and together we watched the results scroll by until the exactly critical inclination case promptly took an error exit because of a divide by zero! Dick could not explain the result since he had been diligent in his testing by running this very case. I left Dick later that evening scratching his head and still puffing on his pipe.

The next morning when I returned to the base, I found Dick in his office with fresh equations on the chalk board. I asked if he had figured out what the problem was, and he proceeded to carefully explain what had happened. The formulation of his "patched" solution smoothly went to zero but became zero divided by zero at exactly the critical inclination. The solution had been originally implemented on an IBM computer whose logic produced a result of zero when given a zero over zero division. However, recently the Navy had rehosted their system on new CDC computers which were intolerant of zero over zero division. But Dick was not to be deterred. He had reformulated his solution using a series expansion so that he could algebraically remove the offending divisor. I was quite impressed with his late evening miracle. The next obvious question to me was when the Navy might be able to implement his modification. Having worked in the Air Force system, I was accustomed to at least a year long process of implementation, testing, validation, and scheduling for a formal software release. To my surprise and amazement, Dick replied that he had implemented his modification before going home for the evening so that it might be used for the daily update of the catalog! . . . But all good things do come to an end. A few years later I heard that Dick had at last come under the shadow of configuration management and they had taken away his system privileges!

Felix Hoots, Mathematician, 14th Aerospace Force

(After he began his career in the first space surveillance activity in Dahlgren, Dick Smith went on to serve as Scientific Director at

NAVSPASUR for many years, retiring in 1986. He served alongside Harold Hasenfus, who was the NAVSPASUR Technical Director from 1961 to 1986. Hasenfus, an Army veteran of the Manhattan Project, continued in his civilian career after World War II as Head of the Rocket Branch at the Army Ballistics Research Laboratory at Aberdeen Proving Ground in Maryland. He came to NAVSPASUR as its first Technical Director about the time that the organization officially became a Navy command. Hasenfus was an engineer with wide-ranging interests and accomplishments when he moved to Dahlgren, and he took full advantage of Smith's interest in the emerging fields of astrodynamics and computer programming. Both Hasenfus and Smith had firsthand experience with some of the earliest computers, the former on the ENIAC at Aberdeen and the latter on the NORC and the IBM 7090 at Dahlgren.)

V. SATELLITE OR MISSILE? YOU DO NOT KNOW WHICH ONE UNTIL YOU COMPUTE THE VELOCITY

The precursor of placing satellites into orbit is missile technology. Once you have a missile with global reach, it is not a big stretch to give it a little more velocity and put a satellite into orbit. The technology for missile launches came from Germany and the von Braun team. Both the United States and the Soviet Union pursued this technology to achieve an offensive capability. At the same time, each pursued a defensive capability to detect and track incoming missiles. How fortunate for the satellite guys! If you have missile-warning radars constantly watching for missiles, then those same resources can also track satellites.

The trajectory of a missile launched from the Soviet Union to the United States would come over the North Pole. Thus, the United States constructed a family of BMEWS radars pointing toward the north. To provide detection of an incoming missile, the radars projected a broad beam of energy near the horizon. If a missile or satellite flies through the beam, then the radar detects the reflected energy and must determine whether this is an incoming missile or just a satellite passing through the neighborhood. The key discriminator is the velocity. To stay in orbit, a near-Earth satellite must travel at about 7 km/s. A missile trajectory is designed to fall back to Earth and hit a point on the surface. It therefore has substantially smaller velocity than an orbiting satellite. The missile-warning radar design included a second beam at a higher elevation. The object detected by the low-elevation beam would fly through the higher-elevation beam and produce a second detection. The pair of detections allow rapid determination of the object velocity and hence characterization of missile versus satellite.

In October 1960, Ballistic Missile Early Warning Site 1 (BMEWS 1) outside of Thule, Greenland was in final preparations for going into operations (IOC). BMEWS 1 was to be the U.S.'s first heavy radar to provide warning of an ICBM attack by the Soviet's newly developed capability. The BMEWS radar for IOC consisted of four AN/FPS-50 fan type radars with dual (over/under) 60-degree horizontal scanning beams. An ICBM fired from the Soviet Union to strike America normally would penetrate the lower beam and then the upper. State vectors would be generated from both penetrations, and then would be combined for a best prediction of impact time and location.

The test for IOC was a team from Aerospace Defense Command (1st Aerospace Control Squadron), Colorado Springs, CO. When they turned on the fan radars for the first part of the IOC test, thousands of radar returns came in! It looked like a mass ICBM raid, and there was initial panic! Then someone noted that the Doppler on the radar returns was near zero. This meant that the "ICBMs" weren't closing on North America. So, what were the BMEWS radars seeing? Why, the Moon of course!

It turned out that the radar processing software was set up for ICBMs at ranges up to several thousand kilometers (first "main bang range") but had no feature that would account for more distant targets—such as the Moon, a really good radar reflector at several hundred thousand kilometers!

Bill Craig, Orbital Analyst, 1st Aerospace Control Squadron

This was the beginning of an important and mutually beneficial marriage between missile warning and space surveillance. Following are the key elements:

1. We need to be watching for missiles 24 hours per day every day of the year and so we get space surveillance as a by-product.
2. Using the by-product data, we can maintain an up-to-date catalog of satellite orbits.
3. We can provide that catalog of satellite orbits to the missile-warning radars.
4. They can use that catalog to predict when satellites will penetrate their early warning beams.
5. This will give them an immediate indication that this penetration is just a satellite.
6. Of course, they would still compute the velocity to be sure.

An interesting fact is that a very important use of an up-to-date satellite catalog is to provide a critical filter for the missile-warning mission. Fortunately, most penetrations of our missile-warning radars are identified as satellites. This partnership continues today and is still the foundation of both the missile-warning and the space surveillance missions.

VI. Working in a Windowless Steel Building, Mounted on Springs, Deep Inside a Man-Made Cave

The Facility

As the Cold War materialized, it became apparent that aboveground facilities for attack warning were vulnerable to aircraft and missile attacks. The search for a more secure facility led to the small town of Colorado Springs, Colorado, roughly situated in the middle of the United States. Most importantly, the town was nestled at the base of Cheyenne Mountain, a granite massif. It is here that the United States began excavating 5 acres of caverns and tunnels at the core of the mountain, shielded above by 2000 ft of solid granite. Construction began on 18 May 1961, and the entire complex was fully operational on 6 February 1967.

Inside the caverns are 15 three-story steel buildings. The buildings rest on a system of giant springs to protect them from movement due to earthquakes or explosions. The system is designed to prevent any of the buildings from shifting by more than one inch. The complex has its own power plant, heating and cooling system, and water supply. The main access tunnel leads to a set of 25-ton blast doors designed to withstand a blast wave traveling down the entrance tunnel. There is a network of blast valves with specialized filters to capture chemical, biological, radiological, and nuclear contaminants. Support facilities include a cafeteria, recreation facilities, and medical facilities. Besides the extensive physical protection, the facility can button up and sustain operational manning for 30 days.

Life on a Crew

> *In 1966 as NORAD prepared to open the Cheyenne Mt Complex the Air Force personnel system was routing new 2nd Lt's with STEM degrees to Colorado Springs to work on the expanded crews in the new facility. When I reported in Oct the CMC was not yet ready, so I was assigned to Ent, AFB. I arrived at night and drove through town three times trying to find the Air Force base (no GPS back then). Finally, I found the postage stamp size base (an Air Force base without a runway) in downtown Colorado Springs. The Space Defense Center was in an old tuberculosis hospital building with bricked up windows (the base is now the site of the U.S. Olympic training Center).*
>
> *In Dec of 1966 we started dual operations at the old location and in the CMC. The drive up to the CMC was a winding road terminating in a large parking lot consisting of all the material dug out to make the underground complex. The signs around the fenced entry area read" Deadly Force Authorized" (the TV show "Star Gate" actually had a very accurate depiction of the entrance). Once through security we would ride the bus*

the considerable distance in the main tunnel to get to the side tunnel with the blast doors. There were two doors, one on each end of a small chamber able to fit a busload of people. When being used, only one door would be open at a time. There was a deep boom as the doors finally closed and I was always glad when the second door started to open.

After passing through the blast door area, we walked a fair distance to the entrance of the building where the SDC was located. Several days after a heavy rain the water would work its way through the granite mountain down to tunnel level and drip on us as we walked; sometimes there were puddles of water. As we walked, we passed several of the three story, metal buildings sitting on huge springs. There was only a freight elevator, so we did a lot of steps to move around the complex. The 3-story tunnel adjacent to the SDC building was empty in 1966 (a building was built there later), and we had a balcony off the 3rd floor location of the SDC with a view of a large cavern.

We had five crews who worked two days each of day, swing, and mid shifts with 3 days off between cycles. My body never adjusted to the constant change in routine. The CMC had a good mess hall, and we always took the appropriate meal there. The midnight chow was popular as it helped you stay awake. However, the Soviets liked to begin their space operations day at dawn there, which was early AM in Colorado, so we had to be done with breakfast in time to be back at the SDC in case any Soviet space launches occurred.

After I was more experienced, I became the launch officer for my crew. It was my job to process new foreign or domestic space launches (at the SDC we didn't worry about missiles; there were a bunch of people down the hall that did that). In those days the radar network was not as expansive as later and the ability of the radars to find something that they weren't quite sure of was limited. As the launch progressed, I would sift through the tracking data to ensure that what the radars had sent was really on the launch and when enough data was collected, I would determine what the orbits associated with the payload, rocket body, and associated debris (e.g., fairings) were. I could then publish "Elset 1". This was the first elset (an element set consisted of the Keplerian orbit parameters) for that object. We had relatively short timelines on how long we could wait until it was ready. The network needed help in knowing where to look as well as the need to know where a potentially hostile satellite was. The crew commander would constantly ask me if I was ready to publish as the time since launch progressed.

It was not easy to perform the orbital analysis rapidly. The incoming tracking data was punched onto IBM cards (thin carboard cards about the size of a dollar bill). Each azimuth, elevation, and range set was on one card. We used a key punch the size of a small desk to cut IBM control cards for the astrodynamics software to read which were then bundled into a deck with the tracking data cards, all to be sent off to the mainframe Philco-2000 computers some distance away within the CMC. To get the decks of cards to there, a large vacuum tube system was employed. The carriers placed into the tubes were the size of two footballs, end to end. Once at the computer room the deck would be read into the mainframe computer, the job run, the results written to a magnetic tape, the tape carried to an off-line computer to be printed, the printout folded to fit into the tube system and sent to the SDC for examination of the results. Later a card input device and online printer were added up in the SDC area to speed the turn around process.

Bob Morris, Orbital Analyst, 1st Aerospace Control Squadron

Cave temperature in Colorado is 45–55 deg. I remember that there was an exhaust fan near the building entrance. It conveyed the dining kitchen smells. That warm moist air and the smell of food were welcoming after the walk thru the chilly tunnel.

Felix Hoots, Mathematician, 14th Aerospace Force

VII. Now They Are Putting Satellites Way Out at the Geosynchronous Equatorial Orbit: What Took So Long?

In October 1945, long before he broke through as a fiction writer Arthur C. Clarke published a paper in the U.K. journal Wireless World that mapped out the potential of the geostationary orbit for satellite communications [6].

According to Kepler's laws, as we increase the altitude of a satellite, the angular speed of the satellite is reduced. This is to say, it takes longer for the satellite to complete a revolution around the Earth. If the altitude is high enough (about 22,236 miles) and the orbit is in the plane of the equator, then the motion in the orbit is the same as the daily rotation of the Earth. Thus, the satellite appears to hover over a fixed spot on the Earth. This is the "geostationary orbit" that Clarke proposed.

In 1960 the science of rocketry had not matured to the capability needed to reach geosynchronous altitude. But there was great interest in using satellites for communications. Project Echo was designed as an experiment to demonstrate the potential of satellite communications. The concept proposed to deploy a 98-ft aluminum-coated Mylar balloon at a high altitude. A ground station would direct a microwave signal toward the balloon that would reflect the signal back down toward the Earth where it would be detected by large antennas.

On 12 August 1960, Echo was successfully put into a 947-mile by 1046-mile orbit using a Thor-Delta rocket. A microwave transmission from the Jet

Propulsion Laboratory Goldstone facility in California was relayed by the satellite to Bell Laboratories in Holmdel, New Jersey, that same day. Although the message was a recording by President Dwight D. Eisenhower, Echo was best known to the public because it could be seen at night with the naked eye.

By 1963 rockets were able to reach geosynchronous altitude, and one more step was made toward achieving Clarke's vision. On 26 July 1963, the Syncom 2 satellite was launched from Cape Canaveral into a geosynchronous orbit. Because of the location of the launch site (33 deg latitude), the orbit plane was tilted by 33 deg from equatorial. Even though the satellite had an orbital period that matched the Earth rotation period, it did not hover over a fixed point on the Earth. Rather it moved in a figure eight north and south with a mean position over the equator. Thus, the orbit is geosynchronous (synchronized with the Earth rotation) but is not geostationary (remains over a fixed point).

To achieve a truly geostationary orbit, one must either launch from a site along the equator or perform a large maneuver to change the plane from that of the launch site to the equatorial plane. This was finally accomplished on 3 August 1964 with the launch of Syncom 3, the world's first geostationary satellite. The satellite was positioned above the International Date Line and was used to telecast the 1964 Summer Olympics held in Tokyo to the United States.

Today we enjoy the luxury of small satellite dishes for receipt of numerous television broadcasts. However, not everyone can benefit from geostationary satellites. If you live far enough north, your dish antenna does not have a line of sight to the geostationary satellites. This was a problem that the Soviet Union had to address in the mid-1960s.

Soviet scientists discovered a unique orbit that they named a Molniya orbit. The orbit plane is tilted 63.4 deg relative to the equator with a moderate perigee altitude and a very high apogee altitude. This stable highly elliptical orbit enabled the satellite to fly over the high latitudes of the Soviet Union. Furthermore, by orienting the ellipse so the portion farthest from the Earth is centered over the Northern Hemisphere, a Molniya orbit provides communications over a large portion of its 12-h period orbit. By deploying more than one Molniya suitably phased, the Soviets could obtain continuous communications. The first use of the Molniya-type orbit was launched on 14 October 1965.

Other than the scientific Explorer satellites, most orbits were either near circular or only moderately elliptical. The Molniya orbits caused prediction difficulties as well as challenges for observing. Because of the extremely high altitude reached at their highest point, the effect of the gravity of the Moon and Sun caused significant variations from the widely used orbit models. Furthermore, because of the exact half-day period, the orbits experience a resonance effect that tends to amplify normally small variations in the Earth gravity effect.

Once again, external circumstances (use of Molniya-type orbits) necessitate the creation of a new orbit model to maintain tracking and cataloging of these satellites.

> *In 1968 when I arrived at NORAD the current high eccentricity 12-hour satellites were being maintained by hand using such long observation spans that the orbit corrections were terrible. Also, only a few NAVSPASUR observations were being obtained on these satellites. I worked with the people at 14th Air Force to develop a new method of using the Baker-Nunn cameras to track these satellites at very long ranges. I then changed the tasking of the cameras from low earth objects to these deep space satellites. Many more very accurate observations were then obtained from the four Air Force Baker-Nunn sites world-wide, and the orbit corrections were run daily on a small IBM 1130 computer located in the Space Defense Center. However, no one understood the orbit perturbations that were occurring with these 12-hour resonance orbits, so work was started on developing a theory to maintain these satellites. I developed the theory and incorporated it on the IBM 1130 in the early 1970s to improve the orbit corrections, and these orbits were maintained on the IBM 1130 until the theory was incorporated into SGP4 in 1979.*
>
> *Bruce Bowman, Orbital Analyst, 1st Aerospace Control Squadron*

VIII. NEEDLES, GLOVES, AND PARTS, OH MY

Mishaps during space walks provide by far the more interesting and diverse assortment of orbital debris: 1) a spatula used for spreading heat-shield repair material; 2) a 30-lb tool bag, filled with grease guns, a scraper tool, and a couple of bags for debris; 3) a glove lost by the first American space walker; 4) Gene Roddenberry's ashes in a lipstick-size container; 5) a camera; and 6) a pair of needle-nose pliers.

A WELL-MEANING EXPERIMENT

Previously we discussed the accidental explosion of a spacecraft in June 1961. This added more than 300 pieces of debris to the space catalog and began a trend that continues to this day. There are far more pieces of debris in space than spacecraft. In May 1963, the West Ford Needles Experiment [7] was launched into an orbit 2200 by 2400 miles altitude. The experiment planned to disperse 480 million copper wires, each of length 0.7 in. These would form a metallic ring around the Earth that would reflect radio waves and provide a more reliable means of communications with overseas forces. This experiment received widespread criticism from the astronomy community as well as other nations. It was seen as dirtying space. Although the needles were expected to reenter within three years due to natural forces, some of the needles did not disperse properly but remained in clumps. Many of those clumps are still in orbit today.

NEVER EXPECTED THIS TO HAPPEN

Another source of orbital debris is sodium-potassium droplets. This substance was used as a coolant for nuclear reactors aboard spacecraft during the 1980s. It leaked from the reactors and self-formed into droplets up to several centimeters in size. These droplets are the second largest contributor to the space debris environment at 560-mile altitude range [8].

The Cosmic Background Explorer (COBE) satellite was launched on 18 November 1989 into a near circular 560-mile orbit. Two additional pieces of debris were cataloged and associated with the launch, which is not unusual. However, beginning in March 1993, a small piece of debris seems to have shed from the spacecraft. Subsequently more than 70 additional pieces were shed by the spacecraft, with the last occurring in April 1994. Various investigations were carried out to see if some other cataloged object had impacted the COBE spacecraft, but no evidence was found. Furthermore, the spacecraft never experienced any degradation in performance.

FRAGMENTATION EVENTS

The most significant source of space debris is the result of fragmentation events. We first consider fragmentation events not caused by a collision. The Delta rocket bodies sometimes would explode years after they had delivered their payload to orbit. After several such events, engineers eventually determined that the bulkhead between the fuel and oxidizer tanks would fail, and so unspent fuel and oxidizer would combine and explode, fragmenting the rocket body. The solution was to vent any remaining fuel after payload delivery. After putting this practice in place, the Delta rocket bodies no longer exploded.

> *John Gabbard did pioneering work in the areas of satellite breakups, sensor tasking and space debris. His work led to the discovery of a problem with Delta rockets that caused many fragmentation events, and he worked with NASA to resolve this by venting residual fuel, and the problem was essentially eliminated. A little-known fact about John is that he reached GS-15 without the benefit of any college courses; he was quite proud that his first college credits came in his late 60's when he took a metal working class at the local community college.*
>
> *G. Taft DeVere, Captain U.S. Air Force, AFSPACECOM/XPYS*

Another cause of fragmentation of defunct spacecraft is battery explosions. Even after the spacecraft had reached end of life, the batteries still retained a charge. In some cases, they developed a short, overheated, and exploded. The solution was to passivate all systems when the spacecraft reached end of life. Last, some spacecraft suffered structural failure and fragmented. Through the end of 2023, there have been 645 breakup events [9].

Next, we consider collisions between orbiting objects. There are millions of tiny particles orbiting the Earth that we do not track and are not in the satellite catalog, and sometimes they impact a spacecraft. Paint flecks have damaged the space shuttle window. Debris damage is also evident on the shuttle robotic arm. A small piece of debris caused a crack in a window of the International Space Station (ISS). In 2016 a piece of debris too small to track impacted the solar array of the European Space Agency Sentinel-1a satellite, visibly damaging the solar array. Decades ago, the Long Duration Exposure Facility satellite was deployed into orbit by the space shuttle specifically to collect small debris encountered in its orbital journey. After six years of exposure, the satellite was retrieved by the space shuttle, returned to Earth, and the collected debris counted and cataloged. This experiment provided important data to characterize the "background" debris in space.

The Orbital Debris Program Office was created at NASA/Johnson in 1979, but NASA was worried about debris even before then. The paint-fleck damage to the space shuttle window happened in June 1983 during the STS-7 mission, which happened to be Space Shuttle *Challenger's* second mission. That was the first external damage to a shuttle window that was severe enough (the pitting was 0.4-mm deep) to force the replacement of the outer pane, although microscopic pitting from micrometeoroids and presumably debris had been noticed after earlier shuttle missions.

ACCIDENTAL COLLISIONS

As far as accidental collisions between orbiting objects in the catalog, there have been only five confirmed events:

1. The French Cerise satellite and the Ariane rocket body collided on 24 July 1996. Cerise was launched on 17 July 1995 using the Ariane rocket (France handles the production). The collision on 24 July 1996 was with a piece of debris from an Ariane rocket that launched the SPOT 1 satellite on 22 February 1986. The collision tore off a portion of the Cerise gravity gradient boom. It is ironic that the first inadvertent orbital collision would be between a French satellite and a French launch vehicle.
2. The Chinese YunHai 1-02 satellite and Russian rocket debris collided in March 2021.

3. The DMSP FB F5 rocket body was hit by #26207, a piece of debris from CZ-4, in January 2005.
4. Cosmos 1934 was hit by #13475, a piece of debris from Cosmos 926, in December 1991.
5. Iridium 33 and Cosmos 2251 collided in February 2009. The collision of Iridium 33 and Cosmos 2251 is the most severe accidental fragmentation on record. More than 2300 pieces of debris have been cataloged. The altitude is about 490 miles, so that the debris field is likely to persist for decades.

Following is an event that could have been a collision. This illustrates the extent of creative analysis sometimes needed for the forensics following a breakup event.

- TVSAT-1 and something collided in November 1987 [10].

  This example is unique in NAVSPASUR records of breakup processing. Three satellites were involved in the analysis, and at the time, some suspicion was raised that an inadvertent on-orbit collision had occurred. The first player in this cosmic dance is Cosmos 1646 in an orbit of 65-deg inclination at an altitude of 249×269 miles. The satellite broke up at 0131UT on 20 November 1987 at lat 64.9° north, long. 60.3° west. Early piece counts were about 50, although later estimates ranged up to 150.

  And now the victim enters the picture. On 21 November, TVSAT-1 was launched due east from Kourou, French Guiana, aboard the Ariane V20 vehicle. Third-stage injection into geosynchronous transfer orbit commenced at 0235UT, and payload separation occurred on schedule at 0238UT. Thirty seconds later, the payload and the spent third stage crossed the Cosmos 1646 orbit plane near the west coast of Africa at approximately the altitude of the debris. At about 0244UT, it was discovered that one of the solar panels on the payload had failed to extend. Between 0530UT and 0726UT, the third stage was tracked from Kwajalein (by ALTAIR) and observed to have an anomalous low thrust. Launch plans had called for the third stage to remain in orbit, but instead the low thrust caused reentry on the first revolution at about 1249UT.

  The coincidence of these two malfunctions led debris scientists at NASA/ Johnson Space Center to speculate that collisions with small particles, even millimeter-scale ones, from the Cosmos breakup could have punctured the pressurized third stage and damaged the solar panel. (The relative velocity was about 9 km/s.) NAVSPASUR was asked to investigate the orbital conjunction. Calculation of Miss Between Orbits (COMBO) analysis indicated that TVSAT-1 did indeed penetrate the debris cloud but had approached no closer than 119 miles to any of the known pieces. There was some uncertainty in this value because of fairly high decay rates in the debris orbits. Meanwhile, contractor analysts at NASA/ Johnson Space Center (JSC) pursued a parallel study and found a close conjunction with the third stage, which showed the stage somewhat below and behind the known debris pieces rather than among them. Either COMBO result could have been used to argue for taking the collision risk seriously, but the actual verdict on collision was at most a weak "not proven." At NAVSPASUR, the collision hypothesis was considered very unlikely in view of the fact that the payload was later reported to be functioning normally, while the Ariane itself had not had a trouble-free history up to that date.

INTENTIONAL COLLISIONS

Intentional collisions are another source of debris, but the amount and lifetime of that debris can vary considerably. In 2008, the U.S. Navy used an SM-3 missile, meant to intercept ballistic missiles, to destroy an old reconnaissance satellite. The satellite contained a large quantity of dangerous hydrazine fuel that was predicted to survive reentry. The intercept intentionally took place at a very low altitude to limit orbital debris, and the majority of the 174 pieces of debris reentered within four months.

> *Everything about this incident was motivated by balancing risks: mitigating the risk from the hydrazine, without creating undue risk from the orbital debris. Hydrazine is extremely toxic and dangerous in its gaseous form, and the fuel was frozen solid in an internal tank. Frozen hydrazine is like water in density and melting temperatures. Our analysis showed that the tank and its contents would survive reentry all the way to ground impact, the tank would rupture, and the result would be a huge toxic Slurpee that would vaporize and expose anyone in the area to extreme risk. Worse, we couldn't predict or control where the tank might land. The whole intercept was designed to eliminate the risk to people on the ground, but at the same time avoid creating risk to other satellites from the orbital debris we would create. The best way to do that was to perform the intercept at the lowest possible altitude, while still ensuring the highest probability of a successful intercept. We were racing the clock, too, since the satellite was going to reenter on its own schedule. What if we missed? Would we get a second chance?*
>
> *Our team examined countless analyses of alternative plans and what-if exercises, and we developed new models and techniques on the fly. We internalized the meaning of the phrase "battle rhythm." The amount of planning, analysis, and exploration of alternatives was both exhilarating and inspiring to see all these brilliant people dedicated to a single end. I handled the interface with senior decision makers, and in the end, the decision chart that went to the White House compared the risks to people on the ground vs. satellites in orbit, for both successful and failed intercepts. It was probably the single most exciting few months of my career.*
>
> *Ted Muelhaupt, Principal Director, Center for Orbital and Reentry Debris Studies, The Aerospace Corporation*

The concept of an antisatellite weapon is as old as the Space Age. The use of satellites for important communications and reconnaissance functions leads to an inevitable quest by others to negate such a capability. The first U.S. antisatellite weapon was tested in October 1959. Only one test flight of the antisatellite mission was carried out, making a mock attack on the Explorer 6 at an altitude of 251 km (156 miles). The missile successfully passed within 6.4 km (4 miles) of the satellite [11]. This began significant efforts by the United States and other nations to develop an antisatellite capability. Unfortunately, any successful test will produce orbital debris. Depending on the location of the target, that debris may quickly fall out of orbit or may pollute space for hundreds of years.

The Soviet Union also began research into antisatellite weapons with the first test of an ASAT system conducted in 1963. Unlike the U.S. strategy of relying on missiles, the Soviet Union adopted a co-orbital method, which involved launching a weapon into the same orbit as its target. As it neared the target, the weapon would explode a shrapnel warhead close enough to destroy the target satellite. With more than 20 such tests, this made a considerable contribution to orbital debris.

In 1985, the United States successfully tested a miniature ASAT weapon launched from an F-15 fighter. The missile impacted the defunct U.S. Solwind satellite used for solar observation. The satellite was in low Earth orbit and produced 289 pieces of debris that all reentered within about four years.

> *The Space Surveillance Center in Cheyenne Mountain provided the predictions on Solwind's position for the intercept and provided verification that no other satellites (collision avoidance) would be in the vicinity of Solwind. The BMEWS radar at Clear AFS was the first radar able to confirm a successful intercept in real time (about 12-15 minutes after intercept). AF Space Command activated the Battle Staff in Cheyenne Mountain because the Solwind ground track, after flying north along the California coast, would pass over the artic and then pass from north to south over the Soviet Union.*
>
> *Earlier in the week I had coordinated with Clear and had the SOI Analyst who would be on duty during the intercept collect and look at the SOI signature for Solwind so they would recognize what they would be looking at. I had Clear on the line as Solwind broke the horizon. The SOI Analyst reported that it looked like there were three pieces, but it didn't look anything like the signature from earlier in the week. I reported a successful intercept to the Battle Staff and notified Pirinclik Radar of the success and requested they provide a headcount when the object came through their coverage. Pirinclik Radar counted about 35–40 objects before they lost count because there were so many objects coming through simultaneously.*
>
> *Ronald Berkley, Satellite Processing Officer, HQ SPACECMD/DO1SS*

That test was considered to be the last of the ASAT tests until China demonstrated their ASAT capability in January 2007. The target was the Chinese Fengyun-1C weather satellite at an altitude of 537 miles. Most of the more than 3500 pieces of debris are expected to remain in orbit for many decades. Understandably, the Chinese ASAT test triggered waves of protest from Western countries.

This event created so much debris in a highly populated orbital regime that there were immediate concerns about the risk posed by the debris cloud to operational satellites. Analysis was quickly assembled using two key mature tools to assess risk to resident satellites:

1. A physics-based collision model showed the distribution of particles in terms of number, size, mass, and spreading speed.
2. A tool was developed to propagate the particles and compute risk as they encounter resident satellites.

This tool has evolved to a state that is now the standard method used to evaluate the risk to resident satellites from any debris-producing event in space [12].

In 2019, India conducted a successful ASAT test by intercepting the Microsat-R satellite orbiting at an altitude of 176 miles. Most of this debris reentered within a year.

In November 2021, Russia conducted a successful ASAT test by intercepting the defunct Cosmos 1408 satellite at an altitude of 300 miles. More than 1500 pieces of trackable debris have been cataloged with more expected. At this altitude, the debris is likely to persist for about two years. Besides the long-lived debris, it is troubling that this test was near the nominal operating altitude of the ISS (250 miles). Decaying debris from the test will likely spiral down through the orbit of the ISS in the future.

IX. Reality Check: The Earth Is Fat

Now that we revealed that both intentional and accidental events have spread a lot of debris around space, there is one more culprit that may surprise you. In the ideal world, we think of the Earth shape as a nice round ball. Well, it is not. Because of the daily rotation and resulting centrifugal force, the Earth slightly bulges out at the equator. Technically we say the shape of the Earth is an oblate spheroid. It is hard to say which term sounds more offensive: "oblate" or "bulge." Regardless, it is only about 13 miles larger at the equator than at the poles. Mathematically, we refer to that oblate shape as J_2, which is not offensive at all.

But those 13 miles contain a lot of mass as compared to an average spherical Earth. That extra mass tugs at the satellite each time it crosses the equator and causes the satellite to veer slightly from its orbital path. Cumulatively over many orbits, this causes the orbit plane to gradually shift its orientation. We say that the orbital plane regresses. The amount of regression depends on the altitude of the satellite as well as other parameters.

Now let's see what this does to the debris. When a breakup occurs, it is like fireworks. Pieces go in every direction. For the pieces that get pushed forward, they have extra speed, and so they go into a higher orbit. For the pieces that get pushed backward, they have less speed, so that they go into a lower orbit. At this point, you can imagine a debris cloud that has some width because of the pieces that were pushed sideways, but it also has an altitude depth above and below the original orbit. Your imagination should now be picturing a 3-D cloud around the fragmented satellite. Okay, that still seems fairly confined to the vicinity of the breakup. But remember that some particles are now moving faster, and some are moving slower than the original satellite. After some time, we can imagine these particles spreading out along the orbit and beginning to look like a donut shape full of debris pieces. Well, that is a good bit more space polluted with debris, but it still seems confined.

Enter the fat Earth villain! Because those particles went up and down to new orbits at different altitudes, their orbits will regress at a different rate than the original orbit. Some will regress slower, and some will regress faster. This will eventually spread the donut of debris into a shell all around the Earth. Now we see that a single fragmentation event can have a widespread pollution of the space environment.

And let's not forget about the atmosphere. Even at satellite altitudes, there is air, and it ever so slightly slows down satellites. This will cause the debris cloud to slowly sink lower in altitude. Consequently, a satellite that resided in an orbit lower than a breakup event may be determined to be safe, for now. But some day, that debris from above will pass through the lower orbit and present a danger of collision.

Figure 4 shows the effective numbers of objects per 10-km altitude bin between 200 and 2000 km altitude [13]. The satellites currently tracked and in the catalog are based on the Space-Track.org catalog snapshot from 16 December 2021 and are depicted by the gray line in the figure. Currently operational constellations are highlighted by the green line. The orange line depicts satellites theoretically trackable by the new Space Fence radar site that can detect objects 5 cm and larger. Finally, the blue line depicts the theoretical number of objects 1 cm or greater in size. A collision with one of these objects is generally considered to be mission ending.

X. Bookkeeping and Inventorying the Garbage

When a satellite fragmentation occurs, we have many new objects that must be added to the catalog. We must not only determine the orbit of each piece,

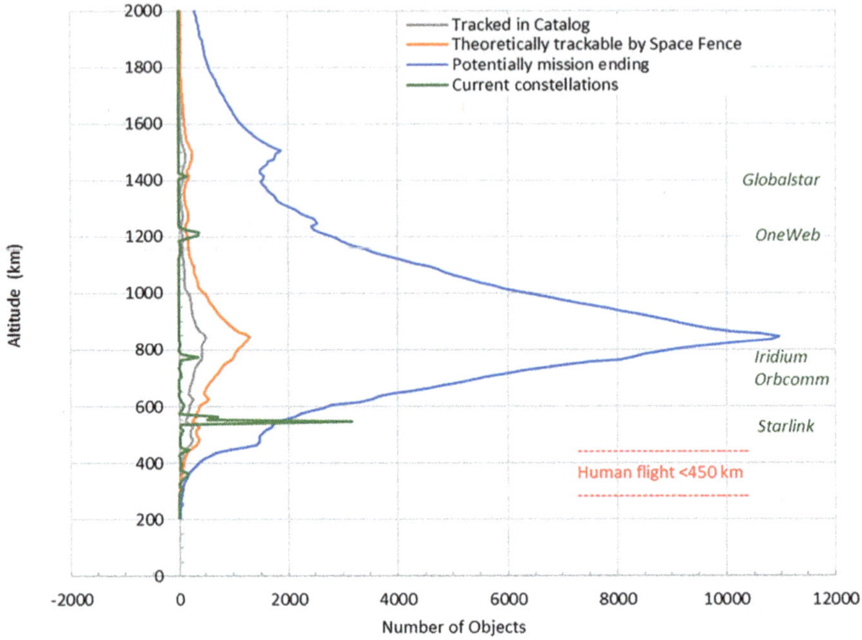

Fig. 4 Satellite population versus altitude.

but we must also attribute its identity to the parent that fragmented. Because the fragments follow orbits that are different from existing cataloged objects, the tracking data on the fragments will not correlate. The data are referred to as uncorrelated tracks (UCTs). At some time after the fragmentation, there will be a large collection of UCTs.

Early pioneers of space surveillance recognized the need to develop algorithms to examine the UCTs and find groupings that potentially belong to a single piece. In this way, a collection of new orbits could be developed that describe the trackable pieces from the fragmentation event. The algorithms were specifically designed to take advantage of the capabilities of the tracking resources.

The processing innovations depend highly on the type of tracking data available. For example, optical data report direction but no range, whereas radar data report all three. Hypothesizing orbits to fit tracks together will require very different creativity for 2-D data versus 3-D data.

NAVY SPACE SURVEILLANCE SYSTEM (NAVSPASUR)

The Navy Space Surveillance System (NAVSPASUR) consisted of three transmitters and six receivers spread along a great circle near the 33rd parallel reaching from Tattnal County, Georgia, to San Diego, California. The transmitters produced a thin wall of energy at very high frequency (VHF).

When a satellite passed through the Fence, energy was reflected and received by one of more of the receivers. The basic problem was to triangulate single-pass Fence angles-only observations into a single position estimate. That data-association problem was feasible because the Fence beam was very thin North–South, and so the observations on a single object were always close together in time. This fact made the triangulations very reliable, providing an accurate position at the time of passage. But one position and time do not describe a satellite orbit. However, if we take a pair of such positions and times, we can use a classical technique invented by Lambert and Gauss to determine an orbit passing through the two positions at the observed times. Some further modifications were made to account for the orbit perturbations occurring over the time span from the first observation to the second observation. Nevertheless, we now had a candidate orbit for a satellite that was observed twice as it flew through different Fence beam passages. We could try to find other Fence observations that correlate to this candidate orbit. These could be used in an orbit determination to create a refined candidate orbit.

SPACE SURVEILLANCE NETWORK

The SSN produces numerous radar tracks (3-D data) when a fragmentation occurs. Unlike the single observations from the NAVSPASUR Fence, the radars produce a sequence of multiple observations. These can be combined to form an estimated orbit. UCT processing for radar tracking data becomes a matter of finding similar orbits formed from the radar tracks.

The first attempt at automation was designed by Larry Cuthbert in 1966. It was called SPACE SWEEP and took 11 of the Philco computer's 15 tape drives to run. It tried to match up orbital planes and period trends followed by an orbit determination. It ran for many hours and at the end still required manual actions to evaluate results and properly refile the observation data.

In the 1970s, Bob Morris designed an automated breakup processing program called SEQ-50 that made use of the Philco computer drum storage devices (similar to a disk drive), and it ran a batch of observations in less than an hour. This was also a trend search process followed by an orbit determination. However, it minimized the amount of manual interaction required and automatically produced separate observation files for each candidate object. In the 1980s, the SEQ-50 program was updated to run on the new Honeywell computers and became the breakup processing method for the new system.

In the 1990s, Bob Morris worked with Larry Cuthbert to combine their methodologies, and this combined method (Cuthbert–Morris) is what was installed in the SPADOC system in the mid-1980s when the Honeywell computers were replaced by IBM mainframes.

Although breakups were not an issue for the geosynchronous regime, small pieces of launch-related debris may not get cataloged and later will be detected as UCTs. These need to be cataloged but need a different technique because the tracking data are from 2-D telescope tracking without range measurements. In the 1990s, Bob Morris developed a UCT processing technique for the GEO regime optical data called SST-5. The technique is based on orbital plane grouping and drift trend matching followed by an orbit determination.

Over the last 20 years, the UCT processing tools have continued to be improved (particularly for lower-altitude low-Earth-orbit (LEO) satellites experiencing significant atmospheric drag) and now include alternative methods based on multi-hypothesis techniques. To have a well-maintained catalog, UCT processing is run on a regular basis. It not only finds previously uncataloged objects but also helps recover satellites without recent tracking.

XI. The Sky Is Falling

Satellites with altitude below about 750 miles experience atmospheric drag. This small force gradually removes energy from the orbit causing a slow spiral to lower altitudes. When the orbit altitude reaches around 65 miles, the orbit will spiral into the Earth. Most small satellites and pieces of debris will break apart on their descent and burn up before impacting the Earth's surface. Rocket bodies and larger satellites will have sizable portions that will reach the Earth's surface.

Skylab was launched on 14 May 1973, having a length of 82 ft and a mass of 84 tons. The initial orbit was near circular at 273 miles altitude. After completion of the planned missions accumulating 172 days of habitation, NASA performed an extensive evaluation and determined that Skylab could be resupplied with expendables plus other technical enhancements and could be used for other missions.

But what about that pesky atmospheric drag? It will continue to gradually lower the Skylab orbit even as plans for reuse were being discussed. British astrodynamicist Desmond King-Hele had predicted in 1973 that the Skylab orbit would decay in 1979. It was clear that a mission would be needed to recover the altitude being lost due to atmospheric drag.

NASA considered several plans for boosting the Skylab orbit. But the most fully developed used a series of space shuttle flights. Would Skylab still be in orbit when the first Space Suttle was ready for flight? Believe it or not, that answer depends on the behavior of the Sun.

We have seen amazing pictures of solar flare eruptions from the Sun's surface. These send large amounts of charged particles into space. If a solar flare is directed toward the Earth, those charged particles will interact with the Earth's magnetic field, and this heats up the atmosphere. If you heat the

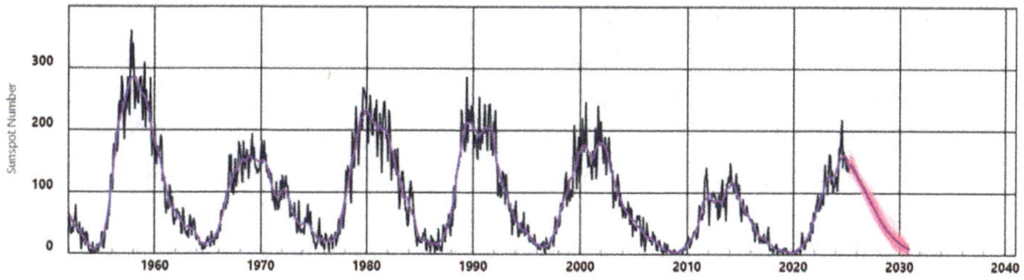

Fig. 5 Sunspot 11-year cycles.

atmosphere, then the density and atmospheric drag will change. This is a fact of nature. There is nothing we can do about it, but we can try to predict it. Figure 5 shows the sunspot history over the last 70 years. The red line shows the prediction for the current cycle [14].

Since 1755, scientists have been aware that the amount of solar activity was related to the number of sunspots, and the sunspot activity has been actively recorded for nearly 270 years. Scientists found that the level of activity follows an average 11-year cycle. The cycle spanning the years 1954–1965 had been one of the highest in history, and the cycle (1965–1976) in which Skylab was launched was one of the lowest in history. What would the next cycle be like? The activity level of the next cycle would determine whether there was time to rescue Skylab or not.

The best laid plans for a shuttle rescue hinged on the unknown future behavior of the Sun. Studies were undertaken by multiple agencies to predict the behavior of the next cycle. The National Oceanic and Atmospheric Administration (NOAA) produced a model, as did NASA and others. Scientists at the NORAD Command used the various models and predicted reentry in mid-1979.

The race was on between the Sun and the shuttle rescue. In March 1978, NASA contacted Skylab to recharge its batteries, and a teleoperator retrieval system was nearing completion. However, the shuttle would not be ready in time, and NASA abandoned boost plans for the Skylab in December 1978. The Sun won, and attention turned to predicting where and when Skylab would impact the Earth. Experts estimated that 25 tons of metal would survive reentry. Skylab reentered on 11 July 1979, impacting 300 miles east of Perth in Western Australia.

Maybe there is a lesson to be learned here. What goes up must come down, and so a plan must be in place before it goes up. The Soviet Union Space Station *Mir* was launched on 20 February 1986. A carefully planned reentry was carried out using a series of maneuver burns while the Progress vehicle was docked with the Mir. Most of the unburned fragments of Mir landed in the South Pacific Ocean on 23 March 2001.

The Soviet Union Cosmos 954 was launched on 18 September 1977 into a 161-mile by 172-mile orbit. The satellite was able to maintain this low-altitude orbit using thrusters powered by a nuclear reactor. By December 1977, it became apparent that the operators had lost control of the satellite. Normally, at the end of life, the satellite would eject the reactor core into a safe disposal orbit. That did not happen, and this reentry event would be much more concerning than Skylab for a different reason. This reentry would not just scatter chunks of metal along a reentry footprint, but that footprint would also contain chunks of radioactive materials. The satellite and its nuclear fuel reentered on 24 January 1978 into Canada, and debris was scattered over a 370-mile footprint extending from the Northwest Territories to Saskatchewan. A search was immediately begun to recover the radioactive material, successfully finding 12 large pieces of the spacecraft in nine months, but this only accounted for 1% of the radioactive material.

On 14 December 2006, the United States launched a classified satellite. The satellite achieved orbit but lost contact with ground stations. In late January 2008, it was reported that the satellite would reenter within weeks. This was particularly concerning because the satellite had a large supply of highly toxic hydrazine. This had likely become frozen and might survive reentry, where the toxic cloud could endanger human life if reentry were near an inhabited area. In response, the United States developed plans to use a missile to intercept and destroy the satellite shortly before its reentry. That successfully occurred on 21 February 2008.

The first components of the ISS were launched in 1998 with additional modules added over the next decade. Currently, the station is the size of a football field and has a mass of 460 tons. It is expected to be permanently inhabited and is periodically boosted to maintain operational altitude.

On 8 November 2011, Russia launched a sample return mission to Phobos, one of the moons of Mars. The mission was named Phobos-Grunt (ground). Unfortunately, subsequent burns to send it on its way to Mars failed, and the spacecraft remained in Earth orbit. The time to reentry was predicted to be two to three months and caused international concern because the spacecraft still carried large amounts of highly toxic hydrazine and nitrogen tetroxide. This was certainly not as planned, and there was no capability for a controlled reentry. The reentry occurred on 15 January 2012 in the Pacific Ocean just west of Chile.

China launched the first component of its Tiangong space station on 29 April 2021. As with all satellites, the rocket body also achieved orbit. The Long March rocket body is about one-third the mass of the space station. Two other modules were launched in 2022 using the Long March rocket. The reentry of the rocket bodies caused international concern because their reentries were uncontrolled. The Chinese have since reported improvements to dispose of the booster after launch.

Small satellites and related debris orbits contract due to drag and reenter quite frequently but generally burn up in the atmosphere. For larger satellites and rocket bodies, some parts of these may survive reentry. The space surveillance system routinely predicts such events and confirms an approximate time and location of the reentry. That has worked just fine for decades and is part of the routine duties of orbital analysts. But a proliferated constellation is one of those distinctly new uses of space that require changes in operations. On average, Starlink has a new launch every two weeks and deploys 60 satellites. That means that sometime in the future there will be an average of 30 Starlink satellites per week that decay in addition to other catalog objects. Starlink has developed a standardized procedure when satellites come to the end of their operational life. As the orbit spirals down to about 300-km altitude, the satellite is put into a tumbling mode that maximizes the effect of drag and hastens the altitude decay. These satellites impact the Earth in about 10 days. This is more predictable than other random uncontrolled reentries. Furthermore, the uncertainty spread of Starlink reentries is significantly smaller than for uncontrolled reentries.

XII. Do Satellites Have Dandruff?

You may have seen pictures of satellites covered with what appears to be shiny Christmas wrapping. That wrapping is called multilayer insulation (MLI). MLI is made up of layers of thin metal-coated plastic film. Its primary purpose is to reduce heat loss from thermal radiation. It does this job very well and is present on a wide variety of spacecraft. But space is a harsh environment, and the MLI begins to deteriorate, and pieces are shed from the spacecraft. Thomas Schildknecht of the Astronomical Institute of the University of Bern first reported detecting such objects in 2003.

Each shed piece of MLI becomes a new orbiting object with unique characteristics. A piece of MLI will not weigh very much but will appear to be large because of its reflectivity. We refer to these objects as high area-to-mass ratio (HAMR) objects. So now, in addition to astronauts losing track of a hammer while on a spacewalk, we have thousands of new HAMRs in orbit.

If it is in orbit, then we need to track and catalog it. This is where the hard part begins. When electromagnetic radiation (sunlight) strikes the surface of a spacecraft, it exerts a mechanical pressure on the spacecraft. This is called solar radiation pressure. This pressure has only a small effect on a weighty spacecraft but can be significant for a piece of MLI. You can think of it like the poor little MLI is trying its best to follow a nice elliptical orbit, but it is being blown around by a strong wind. The wind is what we call a perturbation of the orbit, and it is very strong relative to other perturbations at that altitude. What makes it especially difficult is that sometimes the MLI is oriented broadside to the sunlight and receives a strong push but later may be oriented sideways and receives very little push. This makes it very hard to maintain and predict these orbits.

In the two decades since the HAMR discovery, the telescopes used to track satellites have become more sensitive and are producing more tracks on MLI objects. With the recognition of such objects in space, our processing is now more attuned to cataloging and predicting such objects. There have been instances where operators of large commercial communications satellites have been alarmed by the approach of a bright object on a near collision course. Although it turned out to be a piece of wandering MLI, it would be nice if it were maintained as an object in the catalog, and the operators could be alerted ahead of time.

The space surveillance system is making progress to add such items to the catalog, but it really is like dandruff. It just does not quit, and there seems to be no cure for it.

XIII. A FUNNY THING HAPPENED ON THE WAY TO THE NUMBER STORE

Consider the problem of how to organize a catalog of satellites. Probably the initial idea would be to give each a number as it gets tracked and cataloged. This is what was done, and the honor of being satellite 1 belongs to the rocket that launched the Sputnik satellite. Satellite 2 is the Sputnik satellite. Along with the sequential number, we would probably have a name for the object. From the earliest records, the object identifier began with the year of the launch. A further identifier affixed a Greek letter after the launch year. The first satellite launched in 1958 would have the identifier 1958 ALPHA, the next would be 1958 BETA, etc. There are only 24 letters in the Greek alphabet. That worked well for 1957, 1958, and 1959. By 1960, satellite names were given a longer name such as 1960 GAMMA 3. This naming system continued until the end of 1962 when it was replaced by the current convention called the International Designator. The latter is formed by a four-digit year of launch, a three-digit increment of the launch number of the year, and a three-letter code designating the sequential identifier of the piece in the launch. For example, the ISS is designated 1998-067A and has the sequential catalog number 25544.

The satellite number has remained as a simple counter of the object number since the launch of Sputnik. This brings us to the subject of our story. When the cataloging system was moved from Hanscom Field, Massachusetts, to Colorado Springs, Colorado, in July 1961, the computer software was designed to handle a three-digit satellite number. How could there ever be more than 999 satellites? Little did we know that we would catalog satellite 1000 on 11 February 1965.

> *As an historical note, sometime in 1964 when it became apparent that the 496L SPO, MITRE, and SDC (still an FCRC at the time) had no realistic plans for converting the space surveillance-related programs (around 72 of them) to handle a four-digit field for satellite number (limit was 999, and we were fast approaching 1000), I formed up a team of six experts—three Air Force officers and three experts from our support contractor, Wolf R&D—and we "covertly" took on the project. The project (Project Update) expanded into a major end-to-end rework of all the mainstream surveillance programs to correct/improve the astrodynamic algorithms and procedures. When completed, Update was sprung on a horrified headquarters—which had no recourse other than to set up formal testing. The Update programs were used during the last months of operations in building P4 and formed the basis for the Delta system in the Mountain. Some of the routines are still running in SPADOC4C.*
>
> *Bill Craig, Orbital Analyst, 1st Aerospace Control Squadron*

Engineers always try to anticipate future needs, so that the next upgrade to the system allowed a five-digit number that would accommodate up to 99,999 satellites. That is so many numbers! Let's reserve some blocks of numbers for special uses. And so they did. The 70,000 range is for launches, the 80,000 range is for analyst satellites, and the 90,000 range is for tracking sites to designate UCTs. That means the usable range for the satellite catalog ends at 69,999. The current catalog numbering is now over 60,000. Satellite number assignment is permanent, and they are not reused. Thus, the highest catalog number will always be greater than the number of objects still in orbit because many low-altitude objects decay.

Well, why can't we just use bigger numbers with six or seven digits or more? That is where we get constrained by the communications system between the tracking sites and the processing center. The message only has space for a five-digit number. The message system has been in place for decades and is not planned to be changed anytime soon.

You now see, contrary to what they taught us in school, it is possible to run out of numbers! But do not lose sleep over this. This is just another one of those dilemmas that we have to solve within existing system constraints. The message actually allows five spaces for the satellite identifier. It does not have to be a five-digit number, just five characters. In anticipation of the impending doom when someone launches satellite 70,000, the Space Surveillance System is implementing a "numbering" system called alpha 5. This means that the first character is a letter of the alphabet or an integer in the range zero to nine, and the next four characters are integers. Thankfully the letters I and O are not used to avoid possible confusion with the integers one and zero. This will extend the sequential numbering to 340,000, which should last us for a while. Of course, that is what we thought when we went from four digits to five digits.

XIV. IT SURE IS GETTING CROWDED UP THERE

In October 2024, there were more than 45,000 objects in orbit around the Earth. That sounds like a lot of satellites. People often wonder if space is getting crowded and if they ever run into one another.

Let's first consider how crowded space is using a thought experiment. Imagine the surface of the Earth covered by land, no water anywhere. Now take 45,000 golf balls and randomly hide them around the Earth surface. Now ask your friend to try and find one. I will save you the calculation. That is an average of one golf ball every 4376 square miles. But wait! Satellites fly around above the surface of the Earth, and so we have to consider the altitude dimension. If we just consider the low-altitude satellite environment, that encompasses a volume between 190 miles and about 1250 miles. Your friend is searching for a golf ball somewhere in an average volume of 7×10^{20} cubic miles.

It would seem that, on average, space is not really crowded at all. Now let's make our example more realistic. First, we put the oceans back where they belong. Rather than tramping out through the Arctic, or barren deserts, or the backcountry of Alaska, let's only travel on highways to hide the golf balls. It turns out that space has popular highways, orbits that are good places to put communications satellites or Earth-observing satellites, for example.

Some highways are busy like an East Coast interstate, and some are not busy, like a county road in Montana. Now let's forget about hiding golf balls and imagine satellites moving along the highways. The most likely location for a collision is at the highway intersections. That is the same for satellites. Imagine two identical circular rings around the Earth. Those rings will cross each other at two places. Now put a bead on each ring representing the two satellites. There can only be a collision if the two beads arrive at the intersection point at the same time. The orbits are rarely the same size, so that the intersections are above or below each other.

Fairly early in the Space Age, analysts were interested in trying to determine what other satellites might be doing and why they were in the orbit they were in. Ralph Schinnerer wrote the first tool to determine where and when two satellites had points of closest approach. This was in the early 1960s, and the tool is called COMBO.

> *I used it as a young 2nd Lt in 1967 to do analysis of some Soviet testing activity. I also used COMBO in 1967 to do maneuver analysis on Soyuz orbits as the USSR started its manned spaceflight program. The Salyut space station was launched in 1971.*
>
> *Bob Morris, Orbital Analyst, 1st Aerospace Control Squadron*

Besides watching what others may be doing, the advent of manned spaceflight necessitated another use of COMBO: flight safety. The Soviet Union Salyut was launched in 1971. The United States started launching pieces of Skylab in 1973. Once it was completed and manned, the use of COMBO for orbital safety screening became routine.

Today, COMBO is routinely run for all objects in space. This supports collision alerting, whereby a satellite owner may be notified of the impending close approach with another active satellite or some inoperable rocket body or piece of debris.

XV. *Can We Make the Catalog More Accurate? The Astronauts Sure Would Appreciate It*

> ***On 24 August 2023, the International Space Station's Zvezda service module engines were fired for 21.5 seconds . . . to maneuver the complex away from the predicted track of an orbital debris fragment***[15].

Knowing when satellites come close to one another dates back to the early 1960s to answer questions like "What are those other guys doing, and how close are they to me?" In the early 1970s with the advent of long-term residence on space stations, the more important question became "Please tell me when anything might come close to my space station." The priority had shifted from computing proximity of a pair of operational spacecraft to computing proximity of my space station and anything else in orbit. By this time, the majority of the catalog was rocket bodies and debris. Given the large number of satellites and the limited tracking resources, priority was given to tracking operational spacecraft sufficiently often that maneuvers could be quickly recognized, and good positional knowledge could be maintained (still worried about what those other guys are doing). In contrast, rocket bodies and debris in higher orbits are well behaved and need only to be revisited every few days. Most objects in lower orbits experiencing more drag need frequent tracking.

Up until this time, space safety calculations were based on finding the time when the separation distance was smallest. Various users select different thresholds for deciding if something was too close for comfort. Predictions of close approaches were done a few days in advance. If the two objects were both operational spacecraft, they were already heavily tracked. But if one was a rocket body or piece of debris with infrequent tracking, the tracking network could be tasked to increase the revisit rate on this object. More data would help to refine the predictions and would often show that the dangerous close approach was in fact larger than the threshold of concern.

Through the 1990s, the orbits of almost all satellites were predicted using an analytic orbit model. This is called a general perturbation (GP) model. Such a model is similar to the formula we learned in science:

Distance = Rate × Time. Granted, analytic orbit models contain a lot more equations, but they simply require putting in the requested prediction time and out comes the predicted position. Another method called special perturbations (SP) uses numerical integration for prediction and is much more accurate than GP models. A simple analogy to understand SP is to imagine using the familiar Distance = Rate × Time but only using it to predict for one minute. Then do it again to predict for the next minute and continue this until you reach the desired prediction time three days later, for example. We can see that SP models will take a lot longer to run on the computer but can be more accurate because each 1-min step is tailored to the forces acting during that one minute rather than a single average force acting over the entire prediction time.

Here is where we face the quandary of needing more accuracy for safety of flight knowing there are more accurate prediction methods but not having sufficient computer resources to meet that goal. We might as well add another thing to our worry list. We really did not know how accurate our prediction models were. In general, we expected SP to be more accurate than GP. But prediction accuracy depends highly on the amount of tracking used, the accuracy of the trackers, and the geometric and time distribution around the orbit of those trackers. It so happens there is something called covariance that accounts for all those factors.

The good news is that covariance allows us to consider both miss distance and probability of collision in our determination of whether a maneuver might be required. Considering the increasing space population, this additional metric certainly helps to eliminate many potential close approach events. Wow! This is all coming together nicely.

The bad news is that covariance is most effective when used in conjunction with SP models. In theory it could be used with GP models, but the accuracy limitations of GP models themselves reduces the reliability of covariance error predictions. To summarize, we have a great solution to the problem, but we do not have sufficient computer resources to implement it.

By the late 1990s, innovations in parallel processing allowed networking of multiple dissimilar platforms rather than just multiple processors on a single machine. During 1997, Shannon Coffey from NRL implemented SP orbit determination and prediction using distributed parallel processing. This was demonstrated on the operational system in Dahlgren by maintaining the complete catalog using SP for the month of December 1997. At this time, there were approximately 9000 satellites in the catalog. Based on this successful demonstration, the decision was made to purchase multiprocessor IBM SP-2 computers with the goal of maintaining an SP catalog in parallel to the GP catalog. An interim capability was implemented on a network of workstations and began operational support to maintain a low-altitude catalog of about 2000 satellites in July 1999. After delivery and checkout of the IBM system, the process expanded to maintain the entire catalog in May 2005.

In a parallel effort, the Air Force was pursuing implementation of catalog maintenance using SP. The target processing environment was a multiprocessor workstation. In 1998, the Astrodynamics Support Workstation (ASW) became operational in Cheyenne Mountain and began supporting NASA ISS operations with collision assessment and covariance data. This operational support included maintaining a subset of the space catalog that includes about 700 satellites that pose a collision threat to the ISS. In September 1999, the ASW began operational processing of an SP space catalog of approximately 10,000 objects in Cheyenne Mountain.

Thus, to solve this operational need, the solution was a marriage of improved astrodynamics and improved computational capability.

XVI. Proliferated Constellations: Just When We Thought We Had Everything Working Smoothly

We have come a long way from the first experiment with satellite communications in 1960. The latest advancement is Internet connections using satellites. The Starlink constellation began launch and deployment in 2019 and currently contains over 6600 satellites. We already discussed the challenge of scaling up the processing capacity and the limitation of five-digit satellite numbers. But Starlink introduced a new problem to be solved.

Processing a new launch requires manual interaction. Some launches are announced ahead of time. In those cases, the analyst knows the planned time and location of the launch as well as the planned orbit for the satellite. Furthermore, information is provided on the number of payloads if more than one is planned. Historically, most satellite launches contained only a single large satellite or at most a few smaller satellites. As the upper stage deploys the payloads, they are initially close enough together that tracking radars gather data on all the objects, the upper stage and payloads. The analyst plots the tracks from multiple radars and "connects the dots" to identify the orbit of each satellite. This process is known as manual piece separation. Once the individual orbits are identified, they are added to the catalog, and normal catalog maintenance begins.

Starlink deploys up to 60 satellites with each launch. The satellites slowly drift apart, which means that radars may not detect 60 separate objects until they are sufficiently separated from one another. The combination of incomplete data and an order of magnitude more objects demands something other than a human-intensive piece separation method.

There are multiple proposals for what is known as proliferated or megaconstellations of satellites. Some of these are from countries that do not traditionally share planned launch information. That means an analyst will not only need to catalog multiple payloads but will not know how many there are to be cataloged.

This introduces one of the latest instances where an external event (megaconstellations) has created a need for processing innovation. Coincidentally, near simultaneous deployment of 60 or more satellites behaves much like a satellite fragmentation. The term "fragmentation" is not appropriate, but the dynamics of the deployed satellite motion presents a similar cataloging problem. After several orbits, the analyst has multiple radar tracks on the objects. A breakup processor will automatically tie together the tracks on each object and present the analyst with candidate states. The main limitation is the same as with manual piece separation: timeliness. The pieces cannot all be separated either manually or automatically until they have drifted far enough apart to produce unique radar detections. In this case, the solution to an external event is to apply a more than 50-year-old tool, the Cuthbert–Morris breakup processor, to the new problem.

Sometimes external events are solved by policy decisions rather than new tools. At the start of the Space Age, satellites were built and operated by the government. As communication satellites became feasible, commercial providers entered the picture. Later, commercial providers invested in satellites that gather information on weather and Earth resources. Now the megaconstellations overwhelmingly tip the balance toward commercial satellites versus government satellites. Consequently, in 2018, the U.S. government passed Space Policy Directive-3 that transfers responsibility from the U.S. Department of Defense (DoD) to the U.S. Department of Commerce for providing basic space situational awareness (SSA) to commercial and civil operators. This legislation recognizes the pressing need for more accurate and timelier SSA data and enables the DoD to concentrate resources on national security. The NOAA reorganized and expanded the Office of Space Commerce to address these challenges. OSC is developing an SSA capability that will blend government and commercial data to provide actionable SSA for civil and commercial operators.

The DoD will continue to provide the satellite tracking and catalog maintenance for national security purposes. In the initial phases, that catalog will be provided to the Department of Commerce that will perform a service to alert satellite owner-operators of future close approaches between their satellite and any other orbiting object. This is planned to go operational during Fiscal Year 2026.

XVII. Now They Want to Go to the Moon and Mars

It has been nearly 70 years since the first satellite launch, and the evolution of catalog processing requirements continues to be driven by distinctly new uses of space. These uses will continue to grow the number of objects in space, but the limitations of message formats previously discussed are not a worry at this time. But there are always new worries.

Consider the megaconstellations such as Starlink that plan to contain 12,000 satellites. Each satellite is maneuverable and contributes to the autonomously maintained constellation. Additionally, if a close approach is predicted for any other object in the satellite catalog, the constellation members execute small maneuvers to reduce the probability of collision. All the Starlink satellites know where they are and can successfully dodge an approaching satellite in the catalog.

Other companies such as OneWEB are deploying large constellations as well. This constellation has a 750-mile altitude, whereas the Starlink constellation is at a 342-mile altitude. These constellations are in two well-separated concentric shells and will not experience close approaches with each other. But several other companies are in line. Amazon is planning a constellation of over 3000 satellites situated in three concentric shells at altitudes between 367 and 391 miles. As the number of megaconstellations grows, it remains to be seen whether the existing COMBO tools simply need to be scaled up or will need some yet to be invented variation to handle the growing challenge of space safety support.

In recent years, there has been more and more interest in cislunar space. Cislunar space is the region of space from the Earth out to and including the region around the surface of the Moon. This interest is motivated by civil programs to return to the Moon and military concepts using the "high ground" of cislunar space.

Since the start of the Space Age, there have been agreed lanes of responsibility between NASA and the DoD. If a launch is headed to the Moon or another planet, this is the responsibility of NASA. DoD may help assure safe passage of the NASA mission as it passes the many satellites in the catalog. On the other hand, if any object completes at least one revolution of the Earth, then it will be entered in the satellite catalog. For example, this is the case for the NASA Mercury, Gemini, and Apollo missions.

The cislunar regime is not new to the satellite cataloging business. The earliest scientific satellite orbit well beyond geosynchronous altitude was the Soviet Union Luna 3 launched on 4 October 1959. The satellite had a highly eccentric orbit with perigee of four Earth radii and apogee more than 45 Earth radii, well beyond the Moon. It is object #21 in the catalog. Currently, we maintain and track over 700 satellites with period greater than one day. Of these, more than 40 have a period greater than two days. Our current system has been dealing with satellites in cislunar space for a long time. What is new is that we now have a special name for that domain, and we have the expectation that cislunar will become busy with traffic between the Earth and the Moon. This traffic will be not only from the United States (NASA) but from other nations such as China or Russia. A more important concern are military missions placed in unknown positions in cislunar space.

We currently have very good knowledge of the satellite population out to the geosynchronous region. This is largely due to a very extensive worldwide network of radars and telescopes that provide approximately 750,000 observations per day. The volume of space between geosynchronous altitude and the Moon is a factor of 760 larger than the volume from the Earth surface to geosynchronous altitude. That is a lot of volume to search for new unknown objects. In fact, it is impossible to do quickly. The most probable solution is to track the launch and follow the trajectory until it reaches final deployment. This is not easy, but at least it is feasible, although the difficulties of traditional tracking in the lunar exclusion zone must be considered. Maintaining track custody is one of the biggest challenges facing space surveillance and catalog maintenance in the cislunar region.

Besides establishing and tracking the new orbit, many other cataloging functions will have to be rethought. Consider UCT processing that involves finding groups of tracks that appear to have a common orbit. Such tracks are likely to be widely separated from one another, both physically and dynamically. Current techniques discussed earlier will not work, and so new techniques will be needed.

Cislunar catalog maintenance is reminiscent of 1957 when scientists had to figure out how to do it with the resources they already had. We did it before, and we are confident it can be done again. Creative use of space has fueled invention and innovation in the space surveillance and cataloging community for decades, and it will again.

REFERENCES

1 Diaz, J., "This Is the Actual Hack that Saved the Astronauts of the Apollo XIII," *Gizmodo*, 1 July 2014. https://gizmodo.com/this-is-the-actual-hack-that-saved-the-astronauts-of-th-1598385593.

2 Stroble, R. R., "Satellite Situation Report," NASA-TM-X-50205, 2 Jan. 1962. https://ntrs.nasa.gov/citations/19630012120.

3 Landry, P. M., "Satellite Population Characteristics and Projections (Amendment)," Landry 14th Aerospace Analysis Memorandum 71-18, 20 Oct. 1971.

4 Ostovar, M., "The Decision to Go to the Moon: President John F. Kennedy's May 25, 1961 Speech before a Joint Session of Congress," NASA, 22 Sep. 1998. https://www.nasa.gov/history/the-decision-to-go-to-the-moon/.

5 Hotz, R. L., "Mars Probe Lost Due to Simple Math Error," *Los Angeles Times*, 1 Oct. 1999. https://www.latimes.com/archives/la-xpm-1999-oct-01-mn-17288-story.html.

6 "Clarke Suggests Geosynchronous Orbit," ETHW, 23 Nov. 2017. https://ethw.org/Clarke_Suggests_Geosynchronous_Orbit.

7 "Project West Ford," *Wikipedia*. https://en.wikipedia.org/wiki/Project_West_Ford.

8 Wiedemann, C., Gamper, E., Horstmann, A., Braun, V., and Stoll, E., "The Contribution of NaK Droplets to the Space Debris Environment," *Proceedings of the 7th European Conference on Space Debris*, edited by T. Florer and F. Schmitz, Darmstadt, Germany, 18–21 April 2017. https://conference.sdo.esoc.esa.int/proceedings/sdc7/paper/367/SDC7-paper367.pdf.

9 "ESA Annual Space Environment Report," ESA Space Debris Office, GEN-DB-LOG-00288-OPS-SD, Issue 8.0, 19 July 2024.

10 Schumacher, P. W., Jr., "NAVSPASUR Orbital Processing for Satellite Break-up Events," *Fourth Annual Workshop on Space Operations, Applications and Research (SOAR 90)*, Albuquerque, NM, NASA Conf. Publ. 3101, Vol. II, 26–28 June 1990, pp. 718–723.

11 "Anti-Satellite Weapon," *Wikipedia*. https://en.wikipedia.org/wiki/Antisatellite_weapon#:~:text=It%20too%20was%20adapted%20for,under%20the%20DOMINIC%20I%20series.

12 Aerospace Corporation, "Understanding Space Debris: Causes, Mitigations, and Issues," *Crosslink*, Vol. 16, No. 1, 2015. https://aerospace.org/paper/crosslink-fall-2015.

13 Sorge, M. E., Henning, G. A., Bologna, D. G., Jenkin, A. B., Mains, D. L., and Peterson, G. E., "Aerospace Space Sustainability Studies," OTR 2024-01310, presented at the *International Astronautical Conference*, Milan, Italy, 15 Oct. 2024.

14 Space Weather Prediction Center, "Solar Cycle Progression," National Oceanic and Atmospheric Administration https://www.swpc.noaa.gov/products/solar-cycle-progression [Retrieved 9 May 2025]

15 Howell, E., "ISS Fires Thrusters to Avoid Oncoming Space Junk," *Space*, 24 Aug. 2023. https://www.space.com/international-space-station-debris-avoidance-maneuver-august-2023

Chapter 1

The Need for Space Surveillance

The reasons for the military's approach to keeping track of objects in space, or space surveillance, as that mission came to be called, are multifaceted. Understanding the rationale for the large effort that the job required (and that it still requires) is the common thread running throughout our account of this history. In this chapter, we describe the original motivations of the military in pursuing space surveillance and something about the situation in which military leaders found themselves at the beginning of the Space Age. In later chapters, we explain how the job expanded in many unforeseen ways, while becoming ever more complex and more necessary as time went on. By the end of our tale, it will be easier for the reader to understand the present-day convergence of military, civil, and commercial needs for unprecedented levels of global SSA and international space traffic coordination.

I. The First Satellites

The Soviet Union's launch of Sputnik 1 into Earth orbit on Friday, 4 October 1957, followed by Sputnik 2 a month later on 3 November, famously startled everyone to some degree at the time. Even those engaged in satellite-related research and engineering were impressed by how much the Soviets had accomplished without any fanfare. Yet, these two launches did not take scientists, engineers, government officials, or even the public at large, completely by surprise; the idea had been in the air for a few years. For example, beginning on 5 April 1950, ad hoc discussions about an International

Geophysical Year (IGY) had been carried on among geophysicists.[1] In May 1952, the International Council of Scientific Unions (ICSU) established a Special Committee for the IGY to promote scientific cooperation and data sharing, in part to counter the early Cold War chill on international collaboration. Then, in November 1952, the ICSU issued an open invitation for national governments to commit to the collaborative effort in Earth sciences to be mounted during the 18-month period 1 July 1957–31 December 1958, coinciding with the expected peak of Solar Activity Cycle 19. Not surprisingly, Cold War nations reacted to this invitation with diplomatic pause and thoughtful caution. However, the United States joined the effort in February 1953, and the Soviet Union followed in March 1954 after the death of Josef Stalin. It was not long before talk of launch vehicles and satellites, and their manifestly huge potential for scientific research, became part of IGY planning. The IGY Special Committee published a statement at its meeting in Rome on 4 October 1954, proposing that thought be given to the launching of small satellite vehicles. On 29 July 1955, the United States announced its intention to participate in the effort to launch an artificial Earth satellite as part of the IGY. The next day, the Soviet Union announced that it too would endeavor to launch a satellite during the IGY.[2] That each country actually did launch, not just one but two, satellites by March 1958, with several more later in the same year, was a marvel of advancing space technology of the kind that most of us in the 21st century have come to take for granted. At this writing, the last of those four original satellites, Vanguard 1, launched by the United States on 17 March 1958, is still in orbit, the oldest man-made object in space.

Despite the many scientific successes of the IGY, the Cold War context of these early launches brought a sudden urgency to several long-standing

[1] The date is certain. James Van Allen [1, 2] recounted the first mention of the idea of an IGY at a small private dinner party that he hosted at his home for five of his colleagues. Van Allen later [3] told the story that the actual basis of the IGY was the superb chocolate layer cake that his wife Abigail made for the occasion, because the suggestion of an IGY did not even come up until after she had served dessert! He added that Abigail participated in the conversation that followed. Both the eventual dates of the IGY and its basic content were agreed among these six colleagues that evening and later realized in the most extensive international scientific collaboration in history. In the years immediately after World War II, Van Allen had been involved in using sounding rockets for high-altitude magnetospheric research, working directly with some of the German rocket engineers who came to the United States after the war. In 1954, Van Allen was approached by Ernst Stuhlinger representing Wernher von Braun, both by that time working on the U.S. Army's Redstone missile project, to become the "real honest-to-goodness scientist" on their as-yet-unofficial effort to build and launch an artificial satellite ([4], Chapter VII; [5]). Van Allen did not join them. However, in November 1954, he proposed a detailed design for satellite instrumentation to measure cosmic ray intensity in space, a topic in which Stuhlinger had expressed interest. Van Allen went on to use data from the Explorer 1 satellite, launched by von Braun's team, to confirm the existence, already suspected from his sounding-rocket data, of the inner radiation belt that now bears his name. Data from two other American satellites, launched later in 1958, allowed him to make an initial map of the radiation intensity in the belt. He discovered the outer radiation belt shortly afterward, again using American satellite data.

[2] The original press release about the U.S. intention is available online from the Dwight D. Eisenhower Presidential Library [6]. See also Hughes [7] for a recent perspective on the situation at that time.

geopolitical issues. In both the United States and Russia, the propulsion technology needed for the IGY satellite programs had come from military ballistic missile projects started in the aftermath of World War II. Consequently, in both countries, the launch vehicles were developed by the respective defense establishments, partly to save time and meet the aggressive IGY schedule goals. Of course, such practicalities did nothing to lighten the heavy Cold War question of who had the better rockets. In fact, in the United States, concern over our national security completely dominated the news of these satellite launches. The American public began to wonder just how far ahead of us technologically the Russians were, given that America's first satellite in space, Explorer 1, would not be launched until 31 January 1958. On one hand, both the Army's Explorer program and the Navy's simultaneous Vanguard program that would launch its first satellite two months after Explorer 1 produced some spectacular launch failures among their epoch-making successes. On the other hand, thanks to the customary Soviet censorship in that period, news of Russian technical setbacks rarely surfaced in the United States.

Almost any citizen at the time could understand the political and military overtones of what was, by all measures, an astonishing achievement by Russia, something that in unalloyed peacetime would have caused celebration all around the world.[3] The basic reason for such ambivalence was that the specter of intercontinental ballistic missiles (ICBMs) capable of carrying nuclear warheads had already become part of the collective consciousness. By October 1957, it did not take much explanation to make even the least-informed American see that putting a satellite into orbit around the Earth also meant that a nuclear weapon could, in principle, be delivered, not merely from Russia to U.S. territory, or soon, *vice versa*, by a direct trajectory over the north polar region, but, at least in principle, from anywhere to anywhere on the Earth, within a few tens of minutes from launch. Moreover, a persistent question occurred to many persons, which no one knew how to answer: was Sputnik 1 really the first satellite launched by Russia? Perhaps it was only the first one that conveniently broadcast radio signals that we could detect. What if other satellites were already up there, or soon would be, not advertising their presence, but quietly photographing the Earth and listening to our radio signals? Some commentators pointed out that Sputnik's success proved that

[3]The seriousness of the American public's reaction to Sputnik, and even of the world's reaction, is conveyed succinctly in a White House internal confidential assessment dated 16 October 1957 and referenced later. That assessment as well as its final version dated the next day, 17 October 1957, both now declassified, are available online from the Dwight D. Eisenhower Presidential Library [8]. The unusual speed with which this document was produced and reviewed is itself evidence of the gravity of the issues that American decision-makers suddenly had to address as quickly and as confidently as possible. Even the need for radically improved technical education at all levels in the United States was recognized in this early assessment. Portree [9] cites and discusses the final version of the White House assessment and provides a scanned copy of that version in an online appendix.

nuclear bombs, disguised as peaceful satellites, could be put into orbit, kept there for long periods of time, and dropped, without any possible warning, on any location under the orbital path. Although this latter menace ultimately did not materialize as some had feared, plenty of cause for concern remained. The worries of the general public over news of satellites in orbit, serious as they were, concerned mostly a speculative, if risky, future. In contrast, the better-informed but less-publicized worries of the American military belonged to the dangerous present and called for immediate action.

II. MILITARY SPACE SURVEILLANCE

To appreciate the extent of the military's concerns with satellites, we should keep in mind a few fundamental questions. If satellites were first launched into Earth orbit for peaceful scientific reasons, albeit with military-derived rockets, why did the American military become immediately and permanently concerned with tracking those satellites? Why didn't the unprecedented collaborative spirit of the IGY lead to an international civilian or scientific effort to keep track of satellites in space? Why does the job of tracking satellites, whatever their purpose, remain a military mission to the present day and, in fact, for the foreseeable future? These questions improve as one examines them more closely. For example, early in the Space Age, using a satellite for research amounted to tuning into radio signals broadcast by the satellite and recording and interpreting the signals. Those same signals also allowed the researchers to monitor and predict the position of the satellite in its orbit by applying the laws of celestial mechanics as part of the interpretation of the data. The ground facilities needed for such data collection were, at least from our point of view at this writing, simple and low cost. When the satellite stopped broadcasting (or when the project funding ran out), it was of no further scientific interest and could be forgotten by the research community as they went on to other projects. However, the military, primarily the Air Force and the Navy, as we discuss, also needed to track that same satellite as soon as it was launched, even before the scientists began to collect data from it. The military typically did not rely on signals broadcast by the satellite but rather used comparatively elaborate and expensive multitarget radar systems, plus a few telescopes, to monitor and predict the satellite's position in orbit, a much more complicated job than merely collecting scientific data. Moreover, they continued to track it after it went silent, for as long as the satellite was still in orbit. In other words, the military could never forget about it unless it disappeared from its orbit around the Earth, whether by reentering the lower atmosphere or by being boosted into interplanetary space. Another aspect of the military tracking job, often poorly understood outside the small community that does it, is that it necessarily involves more than the functional payloads that are deployed in orbit. Typically, several objects go into orbit along with the payload.[4] These include the empty upper stage of the launch vehicle and several pieces of "launch debris" such as nosecone fairings and payload adapters that kept the payload intact during the ascent to orbit. The military tracking sensors detect these objects in addition to the payload. Substantial effort must be expended to identify them for what they are and to recognize them whenever they are detected in the future. The questions remain, why did the military immediately go to so much trouble to try to track every detectable man-made object in space, whether it was functioning or inert? Why do they continue to do it at considerable expense? Why is there still no end in sight for this military tracking mission?

The military's concern with satellites was not as sudden as it might have seemed to the average citizen in October 1957. Just as in the civilian domain, the idea of satellites had been in the air for some years. In a research paper for the Air Force issued in October 1946, James Everett Lipp, head of Project RAND in the Engineering Division at Douglas Aircraft Company in Southern California, had written:

> *Since mastery of the elements is a reliable index of material progress, the nation which first makes significant achievements in space travel will be acknowledged as the world leader in both military and scientific techniques. To visualize the impact on the world, one can imagine the consternation and admiration that would be felt here if the United States were to discover suddenly that some other nation had already put up a successful satellite!*[5]

It is not recorded whether Lipp ever received an award for this remarkable prescience. In the months after Sputnik 1, as the U.S. government struggled to formulate an adequate response and some viable policy for the future, a senior Air Force officer admitted the following in writing:

> *As you would expect, the Russian launching of an earth satellite has caused considerable alarm not only in the Air Force but also in the Department of Defense.*[6]

---

[4]Sputnik 1 was the first case in point. Many persons still alive can remember going out at night to "see Sputnik" as it passed overhead. In fact, the small 58-cm-diam payload flying more than 300 miles high was almost always too dim to be seen with the unaided eye: it was never brighter than visual magnitude 6. What most observers remember seeing was the much larger upper-stage rocket body traveling in a similar orbit. The Sputnik 1 upper stage remained in orbit for almost two months, reentering on 1 December 1957. Sputnik 1 remained in orbit until 4 January 1958. At this writing, the goal of having a completely reusable upper stage for every launch, one that does not have to remain in orbit, empty of fuel and useless, is still many years away.

[5]Lipp [10], p. 8; Lipp [11], p. 48. Quoted and discussed by Campbell [12], p. 53.

[6]Lt. Gen. Donald L. Putt, Air Force Deputy Chief of Staff for Development, as quoted by Neufeld [13], p. 169. Neufeld's Chapter VI describes the context of this remark. The Air Force had been trying to mount an ICBM program at least since 1954. However, in 1957, budget negotiations with Congress appeared likely to further delay such an effort because the statutory national debt limit was about to be exceeded. Even Russia's open announcement in late August 1957 that it had successfully tested an ICBM, seemed to have no effect on this stalemate. However, the news of Sputnik quickly dissolved the impasse on American ICBM development, thereby, in effect, also guaranteeing several more years of satellite launch vehicle development.

Lipp's statement in October 1946 had been part of follow-on work inspired by another analysis already completed at Project RAND earlier in the year, regarding the possibilities inherent with rockets and Earth-orbiting satellites. The report on that previous study, dated 2 May 1946, had been, in fact, RAND's first report, entitled "Preliminary Design of an Experimental World-Circling Spaceship." This forward-looking analysis had concluded that useful 20-cubic-ft satellites weighing 500 lb, and the rockets needed to launch them into orbit above 300 miles altitude, were entirely feasible with then-current technology. Simultaneously with the RAND work and following its own December 1945 publication of a Request for Study,[7] the Navy had contracted with three other companies for the engineering design of satellites weighing up to 2000 lb, with the associated rocket propulsion analyses, including preliminary performance evaluations of nuclear propulsion and ramjet propulsion. These naval studies all agreed that launching ton-class Earth satellites was feasible without any major advances in propulsion, being essentially a question of cost and detailed design.[8] During the next several years, many defense studies and project proposals followed these early reports, mainly in three areas: 1) overhead reconnaissance by satellites for intelligence gathering, 2) ballistic missile offense and defense, and 3) military man-in-space projects. Smaller studies continued in other aspects, such as communications and weather surveillance.

All these efforts to use rockets and satellites for military purposes are well documented in a wide variety of books, papers, and reports produced continually since the late 1950s. However, what is much less well documented, and certainly less widely understood even at this writing, is the military's response to the prospect of other nations (primarily Russia, in the beginning) using satellites for overhead reconnaissance against American forces. In retrospect, we can guess, but only guess, possible reasons for not giving this subject more exposure to date. One reason may be that space surveillance never involved extremely large budgets, at least in its early years, so that programs enabling protection or mitigation against satellite

[7]The Navy's Request for Study was based at least partly on an analysis completed in November [14]. At this writing, that 1945 study is difficult to access; however, it is discussed in context by Sloop [15], pp. 36–37. Additional background on the Navy's interest in developing a satellite is found in a RAND report [16], pp. 11–14. The Navy had become interested in a May 1945 report by Wernher von Braun, describing the German concepts for launching and using satellites, many of these going back to the ideas of Hermann Oberth in the 1920s and many of them his own ideas. On 3 October 1945, the Navy staff agreed on a formal proposal to develop a satellite, which led in quick succession to the Lancaster analysis in November and the Request for Study in December. Lancaster himself was a Naval Commander at the time.

[8]Augenstein [17], pp. 3–4. The Navy staff had already proposed a joint satellite project with the Air Force, but in 1946–1948, the senior leadership in both services still doubted the military utility of satellites. Augenstein points out that the Navy's immediate efforts to build and use satellites were abandoned in mid-1948. However, this happened only because of budget limitations; the Navy continued to research the subject. It is also interesting to note that satellites were important enough to the military that the next Annual Report (December 1948) of the Secretary of Defense expressly described research on "earth satellite vehicles." Beginning in late 1948, new Defense Department policies shifted most of the military satellite development work to the Air Force to bring better focus to these projects and get more usable results sooner.

reconnaissance were never directly competing against larger, more visible programs such as ICBMs or missile defense. Another possible reason is that space surveillance, whatever its cost, has been taken for granted by most persons simply because it is not a highly visible activity. Whatever the case, the military has become progressively more concerned about the surveillance *of* space as they have learned more about what kinds of intelligence that satellites can provide in a program of surveillance *from* space. These two distinct missions have always been linked in this manner, even though they are quite different activities pursued for quite different reasons. Surveillance *from* space is widely understood in principle, although the particulars are, by necessity, highly classified. In fact, even several generations of schoolchildren have taken the existence of "spy satellites" utterly for granted. Yet, very early in the Space Age, the American military also realized that, along with its emerging ability to collect intelligence via satellites, it had no viable alternative but to keep track of everything in orbit around the Earth, continuously and indefinitely. The next several sections discuss this realization. Moreover, as later chapters discuss, the problems that the military had to solve for the surveillance *of* space turned out to be technically difficult, unprecedented in the early years, and sometimes hard for key decision-makers to understand.

III. Every Service's Problem: Overhead Reconnaissance by Adversary Satellites

Even before Sputnik, planners in all branches of the American military had begun to realize the needs for satellite overflight warnings for all kinds of military forces and installations, whether land or sea based. The section after this one describes the Navy's particular problem of countering the reconnaissance threat against their mobile, sea-based forces. However, the problem for land-based forces was equally urgent, though not quite as technically difficult as the Navy's fleet-protection problem, as described in this section.

The first steps in any response to satellite reconnaissance, both then and now, are essentially the same as for any adversarial reconnaissance throughout history. The response had to be based on 1) knowing which reconnaissance satellites are in space and being able to distinguish them from nonthreatening satellites and 2) knowing where every one of these satellites is, at any time of interest, present, past, or future, and when certain ground locations are visible to them. Given such knowledge, military operations could be planned and carried out in ways to reduce the risk due to unfriendly satellite reconnaissance.

The practical answer to achieving points 1 and 2 was to maintain a current catalog of the orbital parameters for all so-called resident space objects (RSOs). Typically, both then and now, a variety of sensor systems at various locations around the world sent tracking data to a central facility, where all the data were used to estimate the current orbit of each satellite. For the Air Force, the first cataloging center was the interim NSSCC at Hanscom Field near Bedford, Massachusetts. The NSSCC began in 1957, shortly after Sputnik, as a developmental activity called Project Space Track. In the early 1960s, this center was moved to Ent Air Force Base in Colorado Springs, Colorado, and renamed. For the Navy, the cataloging center was at the Naval Space Surveillance (NAVSPASUR) headquarters at the Naval Weapons Laboratory (now Naval Surface Warfare Center) in Dahlgren, Virginia. That center had begun in 1958 as part of a project by Naval Research Laboratory to build a space surveillance system specifically to provide overflight warning for naval fleet units. The Air Force and Navy cataloging centers, and their relationship, are described in detail in later chapters.

Given the orbital information collated at the cataloging center, analysts could seek to identify which space objects were reconnaissance threats and to predict when these satellites would pass overhead within detection range of important ground locations. Such a capability would allow operational commanders at least some choices within the lower end of the spectrum of possible defensive actions against that reconnaissance, the spectrum being defined by the famous Five D's: Deceive/Deny/Degrade/Disrupt/Destroy. It was important for the military to have simple, economical countermeasures against unfriendly reconnaissance that did not involve aggressive or "irreversible" actions against the satellites in question. Such actions would be inherently likely to provoke or escalate strategic tensions, a poor prospect anytime, but especially so during the Cold War. Examples of the most desirable types of countermeasures are easy to list. Facilities could be camouflaged, disguised, or hidden. Troops and equipment could be exposed or moved when no reconnaissance satellites were overhead. Sensitive radio communications could be paused when satellites were in a position that would allow them to intercept the signals. Naturally, the success of this cataloging-and-countermeasure scheme depended critically on having complete surveillance tracking coverage of all RSOs, on being able to identify which space objects were capable of reconnaissance, and on being able to predict, on demand, positions of all RSOs with sufficient timeliness and accuracy.

The reader might reasonably ask, why would the military need to track *all* satellites in space? Why not track only those relatively few that constituted a reconnaissance threat? This question is a good one, and it is fundamental for understanding space surveillance. The answer is twofold. First, the military needed first-hand knowledge of any new satellite arriving on orbit, in time to assess and respond to the possible reconnaissance threat. This meant that, second, they needed a sufficiently accurate and up-to-date catalog of all currently detectable RSOs so that they could do the following:

1. Recognize the new satellite as actually new and not merely an artifact of failing to recognize an already-known cataloged satellite.

2. Add the new satellite to the catalog database.
3. Continue to correctly identify that new object whenever it was detected in the future so that it would not be confused with other detectable space objects.

From the very beginning, the military also needed specific information about the mission and technical capabilities of each satellite in space, especially new ones. In other words, the military needed to account for all possible reconnaissance threats as soon as they appeared, while also avoiding false reconnaissance alerts that would unnecessarily disrupt operations. Hence, the satellite cataloging operation inherently needed robust connections with the larger intelligence community where space mission assessments were usually done. This fact presages a development in strategic thinking that came decades later when military operations in the space domain began to be considered more systematically and on an equal footing with operations in the land, sea, and air domains. In essence, space surveillance, with its main job of maintaining the satellite catalog (plus a few related applications described in later chapters), began to be integrated in various ways with other intelligence activities to achieve a more comprehensive picture of the operating environment, a picture called SSA and later space domain awareness.

The Air Force developed software, SATRAN (Satellite Reconnaissance Advance Notice) which users could install on their own computer systems. Then they could receive satellite orbital information from the Air Force catalog center on a routine schedule and predict when their location would be visible to certain satellites soon, or, less proactively, when their location had already been visible to those satellites in the recent past. To that list of primary tactical users were added, at an early date, those who wanted overflight alerts in a wide variety of civilian agencies, such as U.S. government laboratories, intelligence agencies, and commercial facilities supporting the government. All these organizations wanted alerts about specific types of satellites, or even particular individual satellites. Not all of them were concerned about the reconnaissance threat per se; some simply wanted to monitor, observe, and study certain satellites of interest, whether adversarial or friendly. Of course, none of these other agencies were able to maintain a satellite catalog themselves, a task that called for a large, open-ended investment in tracking systems, computer and communication systems, and trained personnel. That investment, which only the federal government could afford, had already been made with the service departments, mainly the Air Force and Navy. There simply was no alternative for any end user, especially in the 1960s and for several decades after: they had to depend on the Air Force or the Navy catalog centers for space surveillance orbital data. Many persons are surprised to learn that even the nation's civilian space agency, NASA, which began operations on 1 October 1958 before the end of the IGY, also depended (and still depends) on the military space surveillance centers for satellite catalog data. Although NASA has always been equipped to track a few individual satellites for scientific purposes, only the military is equipped to provide a complete satellite catalog and the global SSA needed for planning and operating civil space missions. This situation began as a matter of deliberate policy with the Eisenhower administration during the IGY in its Cold War setting and continues today for essentially the same reasons. We return to this point in the next chapter.

Overflight alerting is merely the earliest example of a catalog-dependent application that became very widespread in a very short time. The result was that the catalog centers in the Air Force and Navy were continually called on to add more users to their customer lists, beyond their respective service-specific lists, and often on short notice. Just as awkwardly, there was (and, at this writing, still is) no bureaucratic mechanism by which the extra costs incurred by the catalog centers could be reimbursed from the external customers needing the support. In short, the catalog centers ended up providing extra services that they were not originally manned, trained, equipped, or funded to provide to a large list of users that they had not anticipated supporting, for a wide variety of quite legitimate, even urgent, reasons that were not always easy for the catalog analysts to understand in detail. Typically, these extra customers needed support that was not related to the basic Air Force or Navy reasons for maintaining a catalog in the first place. Needless to say, such a situation could not persist indefinitely. The original approach within the services was to provide external support based on the policy of best effort not to interfere with the primary mission. However, large unanticipated demand has a way of changing the economics of any situation, and that dictum certainly holds true for space surveillance. At this writing, a number of purely commercial efforts are being pursued to maintain catalogs for particular customers or classes of customers. Also at this writing, a unified U.S. government effort by the Department of Commerce (Office of Space Commerce) has been started to provide catalog-based services to nondefense users on a sustainable basis.

IV. THE NAVY'S PROBLEM: OVERFLIGHT WARNING FOR MOVING SHIPS

The need for warning against satellite reconnaissance was especially acute in the Navy. Essentially, overhead reconnaissance by satellites could remove one of the primary tactical advantages of naval forces, namely, the element of surprise. Throughout history, sea power had always been a key to national power, partly because a nation's adversaries seldom could get detailed knowledge of how that nation's naval forces were deployed until an actual engagement was imminent. The whole ocean was simply too big, and therefore too expensive, to monitor. However, after Sputnik, it no longer seemed unfeasible in principle for an adversary to maintain continuous surveillance of the entire global ocean to monitor and anticipate the actions of all our naval forces. The practical answer to this problem for the Navy was, as for the Air Force, to maintain a comprehensive catalog of the orbital parameters for all RSOs. Given that orbital information, analysts could seek to identify which of those space objects were reconnaissance threats and to predict when these satellites would pass overhead within their detection range of naval forces, just as was done for land-based forces. The complications for the Navy were that its fleet units were continually moving and often had limited communications. Assuming that these complications could be overcome, however, the same types of simple, economical, nonprovocative countermeasures used by land-based forces could be implemented at sea. Ships could be camouflaged or might change course temporarily while a reconnaissance satellite could see them, or submarines could simply stay submerged during periods when satellites might spot them on the surface. Perhaps certain sensitive communications or radar equipment could be muted when satellites might intercept the signals. Naturally, just as for the Air Force, the success of this naval cataloging-and-countermeasures scheme depended critically on having complete surveillance tracking coverage of all RSOs and on being able to predict positions on demand for all RSOs with sufficient timeliness and accuracy.

With its few hundred ships and submarines, and numerous fixed-base port and shore facilities, the Navy recognized immediately that the satellite overflight warning job was a large one, besides being unprecedented. Moreover, there were many constraints on how the job could be done. Because of the limitations of computers and communication systems in the late 1950s and early 1960s, the need for very quick notice of each new satellite appearing in orbit, almost by itself, drove the Navy to maintain its own catalog specifically for this purpose. In fact, the Navy initially chose to build an integrated and highly automated catalog-sensor system dedicated to the overflight warning mission, specifically to minimize latency in the reconnaissance warning process.

In the days before computers were routinely available to fleet operators, all the computation had to be done at the NAVSPASUR headquarters in Dahlgren, Virginia. The original approach called for each naval end user to send its set of positions and timespans of interest to the catalog center, where the most up-to-date satellite orbit information was used to compute the times and durations of potential reconnaissance coverage for each position and each satellite of interest. Each user received a tailored message containing a time-ordered listing of satellite overflights for each of his positions of interest. Oceangoing vessels would send a so-called PIM (position and intended movement, namely, time, position, heading, and speed) message to the catalog center and receive back a message similar in content to that for land-based users. Naturally, every time a ship wanted to change course or speed, it had to notify the catalog center so that its times of vulnerability to satellite reconnaissance could be recomputed. Then the vessel's commander had to wait for a reply or else choose to ignore the reconnaissance threat temporarily. This original scheme was cumbersome, slow, and prone to lapses. However, it was the best that could be done for the Navy's many mobile users until the 1980s, when small and highly capable personal computers quickly became ubiquitous throughout the service. In the mid-1980s, NAVSPASUR developed portable software, known as SATVUL (Satellite Vulnerability), which could be installed locally by each user. SATVUL allowed the user himself to input position or PIM information, even hypothetical data for planning purposes, without having to communicate that (possibly sensitive) information to anyone. Then, as had already been the case with the Air Force's SATRAN, the catalog center merely sent the necessary orbital information[9] from the most recent orbit estimate available, without needing to know any details of each user's operation. Not only was the workload at the catalog center drastically reduced, along with the global fleet communication load, but the whole alerting scheme was easier to use and more reliable. The number of naval SATVUL users increased rapidly as soon as the software was available. By the mid-1990s, approximately 800 users were receiving daily orbit updates from the naval catalog center.

V. THE NAVAL SPACE SURVEILLANCE SYSTEM OPERATIONAL REQUIREMENTS (1958–2013)

It is especially interesting that the Navy decided in mid-1958 to build a dedicated radar system with an associated computing and communications center to maintain its own catalog of satellites to alert fleet units about reconnaissance satellite overflights. We already mentioned the need to minimize latency of communications and computation in the overflight warning mission for the fleet. Yet, from our vantage point today, it might seem obvious that the Navy could simply have mounted a joint effort with the Air Force to support all defense end users, including fleet users. As it turned out, two separate catalogs were maintained for more than 50 years, albeit with common data standards, common procedures, and frequent working-level contact between the two service operational staffs. Many practicalities

[9]This information usually involved only eight data fields: the satellite identifier (catalog number and name), the timestamp of the orbit estimate, and six orbital element numerical values, besides several single-character software control flags. These data fit comfortably on a standard 80-column punched card (or electronic fixed-length record) in a human-readable ASCII format, making them easy to handle in the field. The SATVUL software included the same mathematical model of the orbit used by NAVSPASUR to estimate the orbit, and so the user at sea could predict satellite locations on demand almost as accurately as could the catalog center.

dictated this choice at the beginning. Different communication systems in each service provided orbital data to different end users at different times and, eventually, for somewhat different reasons as a variety of nondefense uses for catalog data emerged. Different chains of command pursued their respective missions with different priorities, different funding profiles, and different congressional budget appropriations. However, in the 1980s, the coordination between the Air Force and Navy space surveillance efforts began to be formalized under joint-service commanders. The working arrangement at that stage turned out to be for the Air Force cataloging facility in Colorado Springs to serve as the primary space surveillance center, with the NAVSPASUR headquarters in Dahlgren serving as the backup center, while each center continued to support its respective service-specific missions. The closer coordination needed in this arrangement was the first major step toward the formation of a separate Space Force as we know it today, whose mission includes space surveillance for all military purposes, in a single chain of command. We have more to say about this organizational evolution in later chapters.

To provide some background for appreciating this later development, and to provide some concrete information about how the military problems of space surveillance were understood at the beginning, we turn next to the Navy's Operational Requirement No. AD-01503: Satellite Detection and Reconnaissance Defense, dated 31 December 1959 [18]. This document also happened to be remarkably long-lived for a defense requirements document: the Navy never needed to update it during the entire operational life of the Naval Space Surveillance System that it mandated, approximately the 55-year period 1958–2013. Given our collective experience with the many military space defense programs that have come and gone since 1959, acquisition officials and system engineers might find such longevity of validated requirements nearly unbelievable. That longevity alone makes a closer look at this case worthwhile. Here we quote the most pertinent sections of the document verbatim, including the section headings as originally numbered, and record a few explanatory comments about them.

> **1. <u>OPERATIONAL CONCEPT</u>**
>
> **Reconnaissance information by satellites could provide the enemy with the necessary (now unavailable) information to press successful ballistic missile, air and submarine attacks against fleet forces. A continuing knowledge of the orbital position of all enemy satellites, when used in conjunction with shielding, camouflage, evasion, use of darkness and cloud cover, and other tactics will enable fleet units to operate with little increased vulnerability in the face of such enemy reconnaissance. This operational requirement specifies such a counter reconnaissance satellite system.**

In 1955, the NRL had proposed a scientific satellite program for geodesy [19], as part of the nation's efforts for the IGY. The tracking system for this effort became operational in 1957 as part of the Navy's Vanguard satellite program. For geodetic purposes, the satellites would have to be tracked frequently and accurately in their low Earth orbits, and for this purpose, NRL proposed a radio-interferometric system known as Minitrack (short for "Minimum-Weight Transmitter for Trackable Satellite," the name of an unpublished internal proposal at NRL).[10] This system was based on the satellite transmitting a signal continuously in a narrow frequency band around 108 MHz, the frequency recommended by the International Committee for the IGY. That continuous-wave transmission could be received by arrays of radio antennas deployed along precisely known baselines at each ground receiver site, so that the satellite direction from the site could be estimated with good accuracy by interferometry. Using such direction data, accurate orbits could be estimated by methods of orbital mechanics. In fact, Minitrack did work quite well in this application for the IGY [21]. Seven stations were deployed near the 75°-west-long. meridian, along the east coast of the United States and along the west coast of South America; four more were deployed around the world in San Diego, California; the West Indies; South Africa; and Australia. That tracking coverage was sufficient for all American satellites during the IGY. However, to some consternation in the United States and despite IGY recommendations, Sputnik 1 transmitted on frequencies of about 20 and 40 MHz, making it invisible to the Minitrack system. Suddenly, the requirement to detect satellites without depending on them to radiate signals on a convenient frequency could no longer be postponed. NRL's response was twofold. On one hand, they made some emergency modifications to one of the Minitrack stations to allow Sputnik 1 and later satellites in that series to be detected and tracked on the nonstandard frequencies. On the other hand, and more significantly for the present narrative, NRL also proposed to augment the Minitrack system with ground-based continuous-wave transmitters operating at 108 MHz so that the Minitrack stations could detect that signal reflected from the satellites. This so-called Dark Satellite Detection Program, also called the Active Minitrack Program, demonstrated the concept during early 1958, using a transmitter at Fort Monmouth, New Jersey, and the Minitrack developmental receiver site at Blossom Point, Maryland. On 29 June of that year, the Advanced Research Projects Agency (ARPA) assigned NRL the task of designing and deploying what was to become the Naval Space Surveillance System radar interferometer with transmitter and receiver

[10]Corliss [20], Chapter 1, pp. 6–33, gives much information about how Minitrack was the precursor for several critical tracking systems later used by NASA. Its origins at NRL are described in more detail later in our account.

stations deployed across the southern United States.[11] The first two NSSS stations were detecting satellites, and the NRL facility was cataloging orbits, within a month, by 29 July 1958. The NSSS and processing center were designated as the Navy Space Surveillance Facility on 19 April 1960 and formally established by the Secretary of the Navy as a naval command on 1 February 1961. The final station configuration planned at that time, three transmitters and four receivers operating in the same 108-MHz band as Minitrack had, was completed in June 1961. The result was that the operational concept and many of the operational requirements documented here were predicated on the solid foundation of what could already be done with the existing NRL system as of 31 December 1959.

2. OPERATIONAL REQUIREMENT

a. Threat:

1. The threat portion of this document has been removed to protect intelligence information from the broader distribution normally given to the Technical Development Plan.

b. Functions required to be performed:

1. Detect, track, and determine the orbits of all enemy satellites whether radiating or non-radiating. Friendly satellites and miscellaneous orbiting matter must be included to avoid confusion. New enemy satellites, including those making only one pass, must be included.

2. Determine that enemy satellites pose a reconnaissance threat and determine the degree of threat for each satellite.

3. Supply fleet units with the above information in catalog form and transmit amplifying information and changes to fleet units on an appropriate timely basis. Fleet units should be immediately apprised of potential compromise of their position or operations by new satellites or single pass satellites.

4. Develop camouflage systems and techniques such as paint, infra-red shields, decoys, ECM [electronic countermeasure] shields, jamming, etc.

[11]This report [22], the first in a series of NRL reports on the NSSS system during the period in which it was being built and tested, provides a complete summary of the intended six-station radar and the main features of the Dahlgren catalog center.

5. **Develop tactical evasion techniques such as reduced wake, utilization of darkness and cloud cover, electronic tactics, etc.**

6. **The development of active anti-satellite systems to actually destroy or damage enemy satellite is covered in Operational Requirement AD-07703 and is directly related to the functions of this operational requirement.**

Function (1) makes explicit that all detectable space objects must be tracked and cataloged to "avoid confusion," that is, to identify correctly all satellites being detected so that the reconnaissance threat could be correctly understood. The requirement for calculating a sufficiently accurate orbit estimate from data measured on only a single pass of the satellite through the tracking coverage of the system, implied in functions (1) and (3) and made more explicit in the performance specifications (4) and (5) that follow, was never fulfilled using the NSSS radar system alone. NRL worked on the problem intensively; however, the basic design of the system made this problem almost intractable. The transmitter and receiver stations were deployed along a great-circle arc across the southern part of the country, and the necessary sensitivity of the detection system was achieved by confining the transmitter and receiver beams very near to that great-circle plane. Hence, all detections were made in a very thin radar "fan" or "fence" that extended out into space. Essentially, only a single satellite position, though a reasonably accurate one, could be calculated by triangulation using the direction data from several different receivers on a single pass through the radar beam. This fence design also enabled the necessary altitude coverage with only modest transmitter power. However, every satellite passed through the thin fence beam too quickly for a comparably accurate velocity to be determined, so that the single-pass orbit estimates were inherently quite uncertain. By 1962, NRL concluded that even with then-projected system improvements (many of which never materialized), and even with the most favorable observing angles, the orbital period would be estimated within only about 5%, and the orbital inclination within only about 0.5 deg.[12] Consequently, the single-pass orbits from NAVSPASUR data alone had limited utility. At NRL, preliminary plans had been developed to deploy another radar fence in a north–south direction approximately perpendicular to the east–west fence, so that a usable single-pass orbit estimate could be computed from two accurately measured positions and the time of flight between them, a classic exercise in orbital mechanics. However, the Navy eventually abandoned this idea for cost reasons. Interestingly, the single-pass orbit requirement was eventually deemed to be met after tracking data from other space surveillance radar sensors were made available at the Dahlgren center. The other radar data could be combined with NSSS Fence data at Dahlgren to derive a sufficiently accurate first-pass orbit for the purpose, at the cost of some latency.

[12] Naval Research Lab. Rept. 5756 [23], pp. 13–14.

c. <u>Performance</u>:

1. The detection and tracking system is the primary component for this operational requirement. This system must continuously provide detection and tracking coverage from the lowest possible satellite altitude to the highest altitude that a reconnaissance satellite can be expected to pass through the detection system and still be at a low enough altitude when over fleet units at sea to conduct effective reconnaissance. For the 1960–1965 period this detection requirement extends from 75 to 2000 nautical miles altitude.

2. The smallest satellites for effective enemy reconnaissance must be detectable. Considering anti-radar coating techniques, one square foot effective radar cross section is considered appropriate for the altitudes given.

3. The system must be capable of detection and providing information on orbits of all inclinations.

4. The detection system must be capable of acquiring sufficient data from a single passage of a satellite to permit computation of an orbit.

5. Orbit computations must be made with sufficient accuracy from one pass of a satellite to enable a fleet unit to know where and when he is under satellite reconnaissance. Computation speed must be such that information on a single pass or a first pass of a satellite can be transmitted to the fleet unit in time to enable him to take evasive action against the first pass of the satellite after detection or change his course of action in time to prevent a successful enemy missile attack based on reconnaissance information from the first pass of the satellite.

6. A cataloguing and dissemination system for data on all enemy satellites is required to enable fleet units to predict when they may be under satellite reconnaissance. A rapid transmission system (preferably utilizing standard navy communication system) is required to supplement the catalog system with amplifying information and data on new satellites on a timely basis as precedingly (2).

7. Camouflage, shielding, countermeasure, and decoy systems should be sufficiently effective to enable fleet units to proceed and conduct required operations with impunity while under enemy satellite reconnaissance.

Specification (1), for the upper altitude of detection, was still out of reach for the NRL six-station radar design when it was fully deployed in 1959. However, upgrades were preplanned, and, over time, many reasons emerged to increase the altitude coverage of the radar. Fortunately, the NRL distributed interferometer design made it feasible (and attractive) to make incremental performance improvements without shutting down the whole system. For instance, increased power could be installed at one transmitter site without interfering with the operation of the other transmitters and receivers. Similarly, one receiver could be shut down for improvements while all the other stations remained operational. Eventually, the NSSS system would get a long series of incremental upgrades, both at the field stations and at Dahlgren, which would enable routine detection of rocket bodies and satellites at geosynchronous distance and sometimes beyond, well above 42,000-km altitude. Interestingly, no requirement of overall system availability was included, except that the word "continuously" in specification (1) obviously implies high availability. In fact, the distributed design of the field systems, and the redundancy of coverage among the individual sites, allowed major upgrades, maintenance, or repairs to be done while the system was operating and without significantly reducing the detection capability of the overall system. In today's terms, the availability of the NSSS radar exceeded 99%, reckoned over the entire 55-year life of the system.

Specification (2), the system sensitivity of 1-ft^2 radar cross section over the whole field of regard of the radar, was almost impossible to meet with the original system frequency of 108 MHz. The corresponding wavelength of about 9 ft meant that, in the same way that a thin pole standing in water near the seashore does not reflect much of an incoming water wave, no matter how tall the wave, and so a 1-ft^2 satellite would not reflect much of the incoming radar wave, no matter how powerful the transmitter. A few years later, the NSSS transmitter power was increased, and the center frequency of the entire system was doubled to about 217 MHz, so that the wavelength was reduced to about 4.5 ft. These improvements, and other signal processing improvements, moved the system acceptably close to meeting specification (2). Interestingly, no probability of detection is specified along with the 1-ft^2 radar cross section. Historically, for this system, the Navy treated the combination of range, radar cross section, and probability of detection as classified information, although range and cross section could be discussed together in an unclassified setting, as here. The probability, always included in detailed analyses, was often handled as an adjustable parameter or tradeoff variable.

Specification (3), that the NSSS radar be able to detect objects in orbits of all inclinations, involved some overriding nontechnical considerations. As noted earlier, the receiver and transmitter locations were on a great-circle arc across the southern United States, almost on a single east–west parallel of latitude, with the radar fence beam nearly containing the local vertical direction at each station. The plane of the great circle, and therefore the plane of the radar fence, was inclined at about 33 deg relative to the equatorial plane. Therefore, the NSSS radar, as deployed only in the continental United States, could detect low Earth satellites only in orbits above about 30 deg of inclination. A reasonable question would be, why were the stations deployed east–west? Granted that the "fence" form of the radar beam provided the most efficient use of radio energy for detection, why not orient the fence north–south, or at least on a more oblique great circle that cut across more parallels of latitude, thereby reaching satellite orbits of lower inclination? In fact, this operational requirements document did include a goal of having a north–south fence by 1961; refer to the last section of requirements quoted later (Sec. 9.b.), although that additional radar fence was never built. Consider also that most of the Minitrack receivers for the Vanguard satellite program had been deployed along a nearly north–south line from North America to South America, near long. 75° west, precisely to create a fence of radio-tracking coverage for the IGY satellites. Why not simply use the existing Minitrack stations? No single answer describes the situation completely. Partly, the Minitrack stations were dedicated for IGY work, while the NSSS stations had to be started before the IGY ended. Partly, several of the Minitrack stations were outside the continental United States, and their locations had been the result of delicate international negotiations involving the sharing of scientific data. For the Navy to take over those sites for operational fleet-support purposes, merely to track satellites in orbits of all inclinations, was not diplomatically feasible even if it were to prove affordable. Another partial answer is that the Eisenhower administration in 1958 was becoming increasingly worried about satellite reconnaissance of the continental United States for national security reasons, as we mention later. Deploying the NSSS stations across the southern United States alleviated some of that worry and kept initial system costs relatively low for the Navy. Moreover, especially in the early years, the main reconnaissance threat was from Russia. It turned out that the laws of orbital mechanics helped solve the inclination-coverage problem in that particular case: a satellite launched from any point on Earth will have an orbital inclination equal to or greater than the latitude of the launch site. Consequently, any satellite launched from the known ground sites within the Soviet Union would have an orbital inclination greater than about 50 deg, meaning that the satellite would inevitably pass through the east–west NSSS Fence. Eventually, as with the single-pass orbit requirement discussed earlier, the problem of covering all inclinations was solved by sharing space surveillance data from other radars.

d. Features of desired development work:

The development work must encompass the requirements of paragraph 2.b. [listed earlier].

1. **For the detection system the use of large antenna, more sensitive receivers, higher powered transmitters, and automatic data handling equipment must be investigated. Frequencies which will reduce the possibility of effective use of absorbent materials on enemy satellites and echo problems must be determined.**

2. **Means must be found to determine which enemy satellites are dead and which constitute a reconnaissance threat. The reconnaissance abilities and limitations of the enemy satellites should also be determined to the extent practicable.**

3. **The detectability of fleet units by satellites under various conditions must be determined as well as the effect of camouflage, shielding and other counter techniques. Camouflage, shielding, and other counter-measures must be developed and tested.**

4. **Operational tactics against reconnaissance satellites must be developed.**

5. **Tests with our own satellites appear necessary to simulate enemy satellites and determine enemy capabilities.**

The NSSS radar was intended for detection and tracking specifically for catalog maintenance, and the future upgrades listed in feature (1) were already being studied by the end of 1959. The other features listed here for future development were not part of the space surveillance mission per se. Rather, they were intended for other commands in the Navy that were involved with intelligence analysis and the development of tactics, techniques, and procedures for mitigating the reconnaissance threat.

e. Weight and space limitations:

There are no weight and space limitations for the land-based satellite detection fence. Weight and space limitations for the shipborne camouflage and shielding systems should be compatible with the ship and the importance of shielding and camouflage obtained.

f. Special environmental limitations:

The detection fence must operate in its environment continuously. The ship-based camouflage and shielding equipment must function wherever the ship can operate.

g. Security:

Maximum security should be provided to prevent furnishing intelligence to the enemy, and the system should have minimum susceptibility to counter-measures.

h. General information:

In all material developments, the Chief of Naval Operations considers timely availability and suitability of first importance. Considerations of cost, critical materials, and manpower are of almost equal importance. The performance figures given in this requirement are goals, except where specifically noted as minimums. During the course of planning for this development, it may be found that in meeting these goals a large and complex or costly paper will result; whereas it may be found possible to develop a simpler and therefore more readily available, reliable, and suitable equipment short of the ultimate specified, but which nevertheless will constitute a considerable advance over present available equipment. Determination and notification of such alternatives should be considered an essential part of the preparation of the Technical Development Plan. In submission of the Technical Development Plan, the developing agency shall inform the Chief of Naval Operations or Commandant of the Marine Corps, as appropriate, of the results, with respect to the factors enumerated in the preceding text, in order that consideration may be given to making an appropriate modification of this Operational Requirement.

3. SUPPORTING DATA AND RECOMMENDATIONS:

Navy Operational Requirement No. AD-07703 (Anti-Satellite Weapon System).
References listed in the preceding text.
Current data available from Naval Research Laboratory on operation of the Dark Satellite Detection Fence.
Current ARPA and USAF plans and data pertaining to "Interim National Space Surveillance Center" (ARPA).

4. PRESENT EQUIPMENT AFFECTED:

- **a. Navy Research Laboratory Space Surveillance System (ARPA Order No. 7-58).**
- **b. Bureau of Ordnance NORC computer.**
- **c. Ship painting schemes.**
- **d. Ship stack and exhaust systems.**
- **e. Naval communication system.**
- **f. Aircraft and ship smoke screen equipment.**
- **g. ARPA/USAF interim National Space Surveillance Center.**

5. MATERIAL UNDER DEVELOPMENT:

a. The Navy's Dark Satellite Detection Fence now located between Savannah, Georgia and San Diego, California consists of two detection complexes; one in the east and one in the west. The eastern complex consists of a transmitting station near Montgomery, Alabama, and receiving stations near Greenville, Mississippi and Savannah, Georgia. The western complex consists of a transmitting site near Phoenix, Arizona, and receiving sites near San Diego, California, and Engle, New Mexico. Between these two complexes there exists an appreciable gap. At present ARPA is expected to fund a transmitting installation near Wichita Falls, Texas. This installation will fil the gap effectively.

b. The present installation has detected all known satellites at heights below 400 nautical miles. A planned and funded improvement will approximately double this detection range.

c. Each receiving site has direct wire facilities to the Operations Center, U. S. Naval Weapons Laboratory, Dahlgren, Virginia. Analogue data from the stations is sent in real time to Dahlgren where the data are recorded, read and compared with predicted data. A planned, but unfunded, improvement involve changing the data transmission to a digital system which will scan and transmit the station data from Dahlgren automatically by means of the NORC computer. At present the data are reduced and put into the computer manually.

d. Three of the receiving sites utilized antenna having power gains of 250 while one station, Fort Stewart [in Georgia] uses antenna having gains of 1000. The higher antenna gain at Fort Stewart should increase its range capability by 50 percent. Funding for increasing all stations to equal capability of Fort Stewart is not yet provided.

e. **Further range capability increases can be obtained by increasing the transmitter power by a factor of 10. The cost of such an increase is small compared to other means available for increasing performance.**

f. **The improvement plans that have been made envision a range capability of 3000 miles on a 10 square foot target area.**

g. **Experiments underway indicate that from the viewpoint of accuracy and decrease of false echo returns, a higher transmitting frequency is indicated.**

h. **The NORC computer at Dahlgren is unique among digital computers in its great precision and high speed, both of which are essential to the processing of detection and tracking data for large numbers of satellites. The Naval Weapons Laboratory is programming the NORC computer for this work in its operations of the Space Surveillance Operations Center.**

6. TRAINING AND PERSONNEL CONSIDERATIONS:

Not yet determined.

7. RESEARCH CONSIDERATIONS:

a. **To a limited extent the near term phase of this operational requirement can be met with the existing basic and applied research and technical data. Technical information needs to be developed on the needs for the threat beyond 1961.**

b. **Further research is required in detection, orbit prediction, satellite analysis, and shielding and camouflage against reconnaissance satellites.**

c. **Research and development work in achieving this requirement is subject to the provisions of DoD Directive No. 5105.15 of 17 March 1959 and enclosures thereto which in part state that "the Director, ARPA will have exclusive authority in this field". Also applicable is a memorandum from the Director, ARPA to the Service Secretaries dated 14 September 1958 which authorized the military departments to enter into contracts for exploratory studies and feasibility investigations in APRA project areas within certain monetary limitations.**

8. INTERIM READINESS PERIOD:

A successful satellite detection fence with interim ability is now being operated by the Navy from Headquarters at Dahlgren, Virginia. Fleet units can presumably make use of darkness and cloud cover to avoid observation by ... [Some of the text in this section is not legible in the extant copies.] **... can take tactical measures ... infra-red radiation and visible wake.**

9. <u>DATE COMPLETION OF DEVELOPMENT REQUIRED</u>:

a. **The interim Navy Dark Satellite Detection Fence is now in operation and being evaluated.**

b. **Detection ability and geographic extension of this fence is presently required across the continental United States. A north–south extension of the fence across the United States is required by 1961 and a further extension to cover all possible reconnaissance satellite orbits including equatorial orbits is required by 1964. Interim shielding and camouflage systems against satellites are required by 1961 and more effective systems against sophisticated satellites are required by 1963.**

In commenting on the performance section, we already mentioned the issue of extending the NSSS radar coverage to include orbits of all inclinations. In this last section, the desirability of that extra coverage is reiterated. The east–west fence orientation that finally was built allowed efficient coverage of all satellites passing over the continental United States, and the north–south extension was finally not needed once the sharing of all space surveillance radar data was instituted. Nevertheless, a little-known option (never exercised) had been preserved by inclining the great-circle plane of the fence at 33 deg from the equator, with the apex of the circle near the main transmitter station in Texas. Fortuitously or not, that great circle crosses the island of Maui in Hawaii. Maui has been home to defense missile-tracking facilities for the Western Test Range since 1961, and by the late 1960s it had become a critically important site for optical space surveillance as well. At lat 20° north, telescope facilities there are well placed to observe satellites in orbits of all inclinations. The same would have been true of an NSSS receiver site built there, had one ever been built, although horizon limitations at the Maui receiver and the transmitter stations in the continental United States would have limited the fence coverage to satellite detections above about 700 miles altitude. At the very least, that location would have extended NSSS radar coverage to include many of the most common transfer orbits to and from the geosynchronous equatorial region. A Maui receiver also could have provided real-time cueing, independently of solar illumination conditions, day and night, for the narrow-field-of-view telescopes located nearby on the island.

VI. The Air Force's Other Problem: Potential for False Alerts in the Missile-Warning System

Strategic missiles, and the associated systems that provide warning against adversaries' missile attacks, were the legacy of the first strategic missile, the well-known V-2, used by Germany against cities in England, France, and Belgium during the last year of World War II. Although the V-2 came too late to affect the outcome of that war, the lesson about the military role of long-range missiles was taken to heart by all parties involved. In fact, the intense and mostly clandestine competition between Russia and the western Allies to impound the German rocket technology began well before the armistice in Europe and led directly to the decades-long Cold War afterward.[13] Within a few years after the armistice, the addition of nuclear bombs to the strategic portfolios of the Cold War nations created the main global military tensions that we still face today. In the United States, the tasks of managing American land-based ICBMs and providing early warning against attacks by foreign ICBMs were assigned to the Air Force. The first generation of that service's BMEWS was constructed in the period 1958–1961, in response to the threat of Russian ICBMs as that threat could be assessed at the time. To form a radar "fence" somewhat analogous to the NSSS Fence for detecting satellites, radar stations were deployed in Clear, Alaska; Thule, Greenland; and Fylingdales Moor in Yorkshire, United Kingdom.[14] Originally, BMEWS had been conceived independently of any satellite tracking mission. In fact, during the IGY, it was the rocket booster developments, rather than the satellite developments, that accentuated the perceived ICBM threat and accelerated the American missile-defense program.

Nevertheless, the existence of satellites presented a problem for the Air Force's missile-warning radar systems, which was different from the

[13]Neufeld [24] provides an indispensable account of historical developments involving German rocketry as World War II drew to a close and the Cold War emerged.

[14]Like the NSSS, the BMEWS stations used nearly planar fan-shaped beams for detection at long range, that being a naturally efficient way to use the radar energy. However, the detection fan beams at each station were tilted toward the Eastern Hemisphere and oriented relatively close to the horizon to allow the earliest possible detection of incoming missiles. Moreover, the missile-detection radars were of a wider-frequency-band monostatic design; that is, they operated independently of one another, each station having its own transmitter and receiver nearly colocated. In addition, each station could perform extended tracking on any object of interest that had been detected in the fan beam. In comparison, the NSSS stations were a narrowband multistatic design that allowed any receiver station to detect signals reflected by a satellite from any transmitter station, but no extended tracking of any object could be done on a single pass through the radar beams. The BMEWS coverage design also deliberately overlapped the beam patterns from the different station locations, just as was done for the NAVSPASUR stations, ensuring no system-wide coverage gaps for missile detection.

overflight warning problem. Had ICBMs been the only things that could be flying in near-Earth space, the missile-warning job in those early days would have been straightforward: any verified radar detection of an object in space implied an attack of some kind. However, a single satellite in orbit would fly through the radars' missile-warning coverage, not merely once, like a missile, but a few times per day, for weeks, months, or years into the future. Unfortunately, the radar's initial detection on each of these satellite passes was, by itself, hard to distinguish from a missile detection. The position of the new object was known reasonably well with only a brief radar track; however, that brief look usually did not provide enough information about the velocity of the new object to make confident predictions about its future path in space. The problem was very similar to the problem of calculating single-pass orbits using only the NSSS radar fence. Inevitably, some valuable missile-warning reaction time would be lost while further tracking and analysis would show whether the object's trajectory would 1) continue in orbit around the Earth, thereby constituting no immediate strategic threat, or 2) intersect the Earth's surface. In the latter case, one further had to distinguish whether the reentry trajectory implied an attack or was a harmless and predictable reentry of some RSO whose orbit was experiencing natural decay due to atmospheric drag.

Naturally, for strategic warning purposes, each of these distinctions had to be made with very high statistical confidence, even though it was based on a relatively small set of radar data having some irreducible level of random measurement error. The critical decision between "attack" and "no attack" for each new radar detection, which in practice could rapidly escalate into a "war/no war" decision, obviously had to be made as quickly as possible. Yet, achieving the necessary statistical confidence level for that decision required accurate and precise trajectory estimates, which, in turn, required extended tracking for each new detection. That extra tracking took extra time. Inexorable mathematics, such as the law of large numbers, prescribed that the more accurate the trajectory estimate needed, or the less accurate the tracking measurements available, the more data the radar needed to collect to allow sufficiently accurate and precise trajectory estimates, and hence the more time it would need for the follow-up tracking. Moreover, during the extra tracking time after a particular new detection, the radar would not be able to do as effective a job as it otherwise could do in simultaneously looking elsewhere for other new detections. The simple reason was (and, today, still is) that the radar could transmit only some limited total amount of energy into space in a given amount of time: whatever energy was being transmitted for extended tracking of one object could not also be transmitted elsewhere to search for other detections. All these tradeoffs caused by the frequent and unavoidable detection of satellites turned out to be bad for the Air Force's strategic missile-warning mission.

The main question was then, what could be done about these fundamental difficulties?

The problem had been anticipated to some extent in the Air Force's missile-defense planning before 1957, though it had become urgent with the launch of Sputnik. Fortunately, by the time the design and construction phase of the first BMEWS radars was in progress in late 1958, Project Space Track (as the interim NSSCC) at Hanscom Field and the NSSS processing center at Dahlgren were already providing continuously updated satellite catalogs and overflight warnings for their growing lists of end users. The answer to preventing false alerts in the missile-warning system due to satellite detections was conceptually simple and elegant, although somewhat complicated to implement: use those same catalogs to obtain prior information about when satellites were expected to transit through the missile-warning radar coverage. Whenever a BMEWS radar detection occurred, only a relatively quick check at the radar site was needed to determine whether that detection corresponded to an expected transit of a satellite. If the radar detection did match an expected satellite pass, then no missile alert had to be processed, and no extra tracking of the object was needed. Fortunately, in practice, because nuclear war has been avoided to date, so that have actual missile detections. Only in the (presumably) rare case that the radar detection did not match any expected passes of any cataloged satellite[15] would extra tracking of a newly detected object by BMEWS be necessary to decide between "possible attack" and "no attack."

To make this idea practical, the means had to be devised to retrieve and merge the catalog data continuously from NSSCC and NSSS, predict satellite transits through the BMEWS coverage for some period of time ahead, and make that schedule of transits available to the radars. The specialized computing and communication system that the Air Force built at Hanscom Field to implement these functions was known as the Space Detection and Tracking System (SPADATS). Already by early 1958 the Air Force catalog was being maintained by the interim NSSCC, and, by early 1959, the catalog data

[15]In fact, a fundamental feature of the cataloging operation is that this case of unexpected detections must be kept rare, at least on average, for the catalog to be able to fulfill its purpose. We return to this point several times in this history. Here we reiterate that when a new satellite is put into orbit, often the upper stage of the launch vehicle and several other smaller "launch debris" go into orbit along with the satellite payload itself. All these objects will be detected by the radars and will not match any space object already in the catalog. In the jargon of the radar community, the new objects are "uncorrelated targets" (UCTs), and extra time and resources must be expended to identify them and add them to the catalog database. Nowadays, the catalog maintenance operation is, by necessity, almost entirely automated. However, as we discuss in later chapters, in practice in the current surveillance system, if the number of UCTs on any day rises much above a few percent of the total tracking data being reported to the catalog data center, the operators and orbital analysts there may have to adopt special temporary procedures, including some manual interventions in the automated processes. If that occurs, the net effect is that the human crew ends up working much harder than usual for perhaps days at a time, simply to keep the catalog up to date. As we see later, certain historical events that did call for special "UCT processing" provide us not only with interesting lessons learned but also with unique insight into the general problem of maintaining the catalog.

being maintained at Dahlgren for the NSSS were also made available to enable the technical development of SPADATS by the Project Space Track research and development staff. The first version of SPADATS was ready for operation in 1960. Later, in 1961, because of reorganization by the Air Force, SPADATS and the NSSCC (by then no longer the "interim" by virtue of the reorganization) were moved to Ent Air Force Base in Colorado Springs, as NORAD took over the missile-warning mission in addition to air defense. The sense of newness and further development needed in the missile-warning mission area can be gauged by the fact that the Project Space Track staff continued to operate at Hanscom as the Project Space Track Research and Development Facility, although under the more prosaic name 496L System Program Office of the newly constituted Electronic Systems Division of Air Force Systems Command.[16] Initially, Project Space Track at Hanscom operated as a type of backup facility for the NSSCC in Colorado, maintaining a complete catalog besides doing a wide range of research and development for both the surveillance mission and the missile-warning mission. NSSS catalog data continued to arrive at both the Hanscom facility and the NSSCC facility, so that even at this early date, a quite robust multiservice cataloging operation was fielded quite rapidly, supporting both space surveillance and missile warning as integrated enterprises. Conveniently, and at every stage of the SPADATS development and implementation, satellite tracking data from the missile-warning radars could also be reported back to the cataloging centers, in effect using the BMEWS radars as space surveillance sensors, to further improve the estimates of those orbits. Then the times and positions of future transits of those satellites through the missile-warning coverage could be computed, on demand, with slightly better accuracy on average over time. Almost the entire data-processing load for missile warning amounted (and, today, still amounts) to maintaining the catalog of satellites to prevent false alarms.

As we describe in the following chapters, the requirements for maintaining the catalog with the necessary reliability (timeliness, accuracy, precision, and completeness) for overflight warning, preventing false alarms in the missile-warning system, and supporting new applications as time went on, did put large demands on radars, computer systems, and communication systems. The same requirements also implied the need for new mathematical modeling of orbital motion and extensive new software development. Moreover, all this capability for the surveillance of space had to be developed within a few years, almost from scratch because nothing like this had ever been needed before by any government.

[16]Almost every documentary source seems obliged to admit that the two different names for Space Track were used interchangeably without any confusion for some years after the reorganization.

VII. NOTE ON NATIONAL SECURITY CONSIDERATIONS

Besides the tactical problems facing each branch of the military in the late 1950s, the Eisenhower administration grew increasingly urgent about detecting, predicting, and characterizing satellite overflights of American territory, especially the continental United States, for national security reasons. At issue were two main concerns: 1) the prospect of continuous satellite reconnaissance of all American defense activities, including industrial activities, and 2) the possibility of adversaries putting nuclear bombs, or even "conventional" warheads, into low Earth orbit. Such orbital bombs could be disguised as less-threatening space vehicles that could be de-orbited at any moment to complement, or substitute for, ground-based ballistic missiles. In the late 1950s, the latter possibility was not well defined, and the actual risk to the nation from orbital bombs could not be assessed reliably. Hence, a comprehensive counterstrategy could not be formulated. All these national-level concerns, perceptions, and beliefs did condition how the space surveillance mission developed in the United States. For one example, mentioned earlier in this chapter, it was not only the Navy's immediate tactical needs but also the service's awareness, however formal or informal, of national-level strategic concerns, which influenced the operational requirements and engineering design for the NSSS radar fence.

The Eisenhower administration admitted to these national security issues but avoided public discussion of them. What Eisenhower did do explicitly was 1) try to educate the U.S. government, and by extension the public, about the nature of the space environment and the means needed to get there and operate there, and 2) articulate the reasons for a comprehensive space program that emphasized civil uses of space, manned exploration, and scientific research, rather than national security, although without excluding the possibility that a satellite might have both civilian and military uses [25]. By the end of Eisenhower's administration in 1960, the main military use of satellites was reconnaissance. At that time, it would have been hard to guess that the increasing technical capabilities of reconnaissance satellites, and the military implications of these capabilities on both sides, would not create worse Cold War tensions. However, what happened was that sophisticated overhead reconnaissance with satellites by both sides, that is, surveillance *from* space, and the ability of both America and Russia to monitor the positions of each other's satellites through the surveillance *of* space, became major factors in moving both countries toward a formal nuclear détente beginning in about 1969, reducing the risk of global nuclear war for the following several decades.[17]

[17]Pindjak [26] provides some background for this outcome. To date, the subject of the intended versus the actual ramifications of Eisenhower's first-generation space policy has unfortunately not been studied in the detail needed to understand all the national-level motivations for developing both surveillance *of* space (our subject) and surveillance *from* space during his administration. We tend to take both surveillance missions for granted and to regard them as equally necessary for national defense. However, the global military situation in Eisenhower's time was very different from that in our time. This is why it can be hard for us now to understand important instances of perception and reaction between adversaries, even in well-documented cases, and even before we go on to consider the more difficult questions of actual causes and actual effects, for any aspect of strategic engagement near the beginning of the Space Age.

REFERENCES

1 Van Allen, J. A., "Genesis of the International Geophysical Year," *Eos, Transactions American Geophysical Union*, Vol. 64, No. 50, 13 Dec. 1983, p. 977. https://doi.org/10.1029/EO064i050p00977-01 [retrieved 9 Feb. 2024].

2 Van Allen, J. A., "Genesis of the International Geophysical Year," *Johns Hopkins Applied Physics Laboratory Technical Digest*, Vol. 6, No. 4, Oct.–Dec. 1985, pp. 367–368. https://secwww.jhuapl.edu/techdigest/Home/Archives [retrieved 9 Feb. 2024].

3 Van Allen, J. A., "James Van Allen Oral History," Ohio State Univ. Special Collections Registry, recorded 18 Nov. 1997 (final ed. of transcript, 13 July 2001). https://library.osu.edu/collections/spec.pa.56.0074.

4 Van Allen, J. A., *Origins of Magnetospheric Physics*, Smithsonian Inst. Press, Washington, D.C., 1983. Republished, with new preface by Van Allen, by Univ. of Iowa Press, Iowa City, IA, 2004.

5 Ludwig, G. H., "The First Explorer Satellites," 2004. Lecture presented on the occasion of James Van Allen's 90th birthday celebration at Univ. of Iowa, 9 Oct. 2004. Transcript available online at https://space.physics.uiowa.edu/plasma-wave/van90/[retrieved 11 March 2024].

6 Hagerty, J. C., Press Secretary to the President, White House, 29 July 1955, statement regarding the intention of the United States to launch a scientific satellite during the International Geophysical Year. This terse two-paragraph statement is available, as a scanned copy of the original, from the Dwight D. Eisenhower Presidential Library, https://www.eisenhowerlibrary.gov [retrieved 13 Dec. 2023].

7 Hughes, K., "Celebrating 65 Years of the Army in Space: The Launch of EXPLORER I," U.S. Army Aviation and Missile Command History Office, courtesy story, Redstone Arsenal, AL, 24 Jan. 2023. Posted online by Defense Visual Information Distribution Service, https://www.dvidshub.net/news/437144/celebrating-65-years-army-space-launch-explorer [retrieved 30 Jan. 2024].

8 "Reaction to the Soviet Satellite: A Preliminary Evaluation," 16 Oct. 1957. This confidential internal memorandum was declassified on 29 Aug. 1979. It is available as a scanned copy of an original mimeograph, from the Dwight D. Eisenhower Presidential Library, https://www.eisenhowerlibrary.gov [retrieved 13 Dec. 2023]. A confidential final version of this intelligence assessment was issued the following day, 17 Oct. 1957, by the U.S. Information Agency under the title "World Opinion and the Soviet Satellite: A Preliminary Evaluation." This U.S. Information Agency document was declassified on 26 Nov. 1993 and later cited and discussed in a NASA Special Publ. by David Portree [9]. It is available online through the Dwight D. Eisenhower Presidential Library, https://www.eisenhowerlibrary.gov [retrieved 31 Jan. 2024], and at *NASA's Origins and the Dawn of the Space Age*, https://www.nasa.gov/history/monograph10/, menu selection Documents, Document 1 [retrieved 31 Jan. 2024].

9 Portree, D. S. F., *NASA's Origins and the Dawn of the Space Age*, NASA SP-4510, Monographs in Aerospace History, No. 10, Sept. 1998. The complete work is available at https://www.nasa.gov/history/monograph10/[retrieved 31 Jan. 2024], including appendices giving an event timeline and scanned copies of important source documents that are not available as part of the print edition.

10 Lipp, J. E., "The Time Factor in the Satellite Program," Douglas Aircraft Co., Santa Monica, CA, Project RAND Rept. RAD-51, 18 Oct. 1946. This paper was included in the 1947 collection by Lipp [11]. His quotation about the preeminence of the first nation to launch a satellite, cited in this chapter, is found on p. 8.

11 Lipp, J. E. (ed.), "Reference Papers Relating to a Satellite Study," Douglas Aircraft Co., Santa Monica, CA, Project RAND Rept. RA-15032, 1 Feb. 1947. The last paper in this collection, pp. 41–49, is by Lipp's paper of 18 Oct. 1946 [10].

12 Campbell, V., "How RAND Invented the Post-War World," RAND Corp., Invention and Technology, RP-1396, Summer 2004, pp. 50–59. https://www.rand.org/content/dam/rand/pubs/reprints/2009/RAND_RP1396.pdf.

13 Neufeld, J., *The Development of Ballistic Missiles in the United States Air Force 1945–1960*, Office of Air Force History, U.S. Air Force, Washington, D.C., 1990. Available online through the Defense Technical Information Center, Accession No. ADA439957. https://apps.dtic.mil/sti/tr/pdf/ADA439957.pdf [retrieved 22 Feb. 2024].

14 Lancaster, O. E., Moore, J. R., Olmstead, P. N., and Taylor, M. M., "Investigation on the Possibility of Establishing a Space Ship in an Orbit Above the Surface of the Earth," Bureau of Aeronautics, Dept. of the Navy, Washington, D.C., Rept. ADR R-48, Nov. 1945,

15 Sloop, J. L., "Liquid Hydrogen as a Propulsion Fuel," NASA History Series, NASA SP-4044, 1978. https://ntrs.nasa.gov/api/citations/19790008823/downloads/19790008823.pdf [retrieved 3 Sept. 2024].

16 Davies, M. E., and Harris, W. R., "The Role of RAND Research in the Origins of Space Reconnaissance and Related Applications for Arms Control Verification, Communications, Meteorology, Mapping, Lunar and Planetary Exploration," Research Note N-1532, July 1987. Sponsored by the RAND Corporation in Commemoration of Project RAND 1946–1986 and the RAND Corporation, 1948–1988.

17 Augenstein, B. W., "Evolution of the U.S. Military Space Program 1945–1960: Some Key Events in Study, Planning and Program Development," RAND Corporation Paper P-6814, Sept. 1982. https://www.rand.org/content/dam/rand/pubs/papers/2008/P6814.pdf [retrieved 28 Feb. 2024].

18 Operational Requirement No. AD-01503 (Satellite Detection and Reconnaissance Defense). Dept. of the Navy, Office of the Chief of Naval Operations, Washington, D.C., Charles B. Martell (by direction), 31 Dec. 1959.

19 Naval Research Lab. Memorandum Rept. 466, "A Scientific Satellite Program," 13 April 1955. Prepared by Rocket Development Branch, Atmosphere and Astrophysics Div., Naval Research Lab., Washington, D.C. Scanned copy available online in the Defense Technical Information Center Archive as Accession No. ADA605755, https://archive.org/details/DTIC_ADA605755.

20 Corliss, W. R., "Histories of the Space Tracking and Data Acquisition Network (STADAN), the Manned Space Flight Network (MSFN), and the NASA Communications Network (NASCOM)," NASA CR-140390, June 1974. Scanned copy available online at https://ntrs.nasa.gov/api/citations/19750002909/downloads/19750002909.pdf.

21 Mengel, J. T., and Herget, P., "Tracking Satellites by Radio," *Scientific American*, Vol. 198, No. 1, pp. 23–29, Jan. 1958. Describes the Naval Research Lab.'s Minitrack system in detail for a general readership, including several photos of the ground facilities at Blossom Point, MD.

22 Naval Research Lab. Memorandum Rept. 896, "Space Surveillance System," Technical Summary Rept. No. 1: ARPA Order No. 7-58, 31 Dec. 1958. Prepared by Applications Research Div., Naval Research Lab., Washington, D.C. Scanned copy available online at Defense Technical Information Center, Accession No. ADA459637, https://apps.dtic.mil/sti/pdfs/ADA459673.pdf.

23 Naval Research Lab. Rept. 5756, "Space Surveillance System," Technical Summary Rept. No. 5, 1 March 1962. Prepared by Applications Research Division, Naval Research Lab., Washington, D.C. Scanned copy available online at Defense Technical Information Center as Accession No. AD0328442, https://apps.dtic.mil/sti/citations/AD0328442.

24 Neufeld, M. J., *The Rocket and the Reich: Peenemünde and the Coming of the Ballistic Missile Era*, Free Press, New York, 1995.

25 President's Scientific Advisory Committee, "An Introduction to Outer Space: An Explanatory Statement," U.S. Government Printing Office, Washington, D.C., 26 March

1958. This 15-page report includes an opening statement, signed by President Eisenhower, describing his purpose in releasing the report to the public. Copies of the report are available online at a wide variety of websites, including, for example, http://www.plosin.com/beatbegins/pdf/IntroductionToOuterSpace.pdf [retrieved 7 Dec. 2022] and https://lunar.colorado.edu/jaburns/publicfiles/IntroOuterSpace.pdf [retrieved 7 Dec. 2022].

26 Pindjak, P., "The Eisenhower Administration's Road to Space Militarization," graduate thesis, University of Pittsburgh Graduate School of Public and International Affairs, 14 April 2011, https://d-scholarship.pitt.edu/7698/1/Thesis_PindjakPeter2011Updated.pdf [retrieved 28 Feb. 2024].

Chapter 2

The Key People and Organizations

When Sputnik was launched, there were no experts in space surveillance. Moreover, no existing organization, either military or civilian, was structured, manned, trained, equipped, or funded to deal with this mission. In fact, in the early years, only a few persons outside the defense establishment understood the need for detecting, tracking, and characterizing every man-made object in space, not merely the research satellites for the IGY and later scientific programs. In this chapter, we examine how some key individuals in the Air Force, Navy, and other governmental agencies learned some complicated technical lessons very quickly and devised surprisingly effective space surveillance capabilities for the United States in the first few years after Sputnik. Yet, although the human expertise in this mission area developed rather fast, the organizations took much longer to adapt to the emerging defense needs. That organizational evolution, in terms of budget, numbers of people involved, size and complexity of facilities, and increasing demand for computation and communication, reflected the growing importance of space surveillance in American military thinking. Comprehensive SSA, enabled by space surveillance, graduated from being an essential component of military tactics to being an essential component of military strategy and even of national policy. Consequently, the physical tracking systems, computational centers, and operational crews were constantly pressed to support this expansive thinking. Although highly capable individuals have been involved from the very beginning of space surveillance to the present day, the early history owes more to brilliant individuals than does the later, more complicated history closer to our time. As the years went on after Sputnik, the advances needed in space surveillance came to depend less on the talent of individuals and more on the talent of teams and organizations collectively, simply because ever more work had to be done to make those advances. Such a trend is a natural consequence of the success of the space surveillance effort, but it also means that, in this historical survey, we mention individuals more in the early history and less in the later history.

Another theme that we mentioned in the previous chapter, and which will be evident in this chapter, is that two different military services were involved in space surveillance. Although space surveillance operational units eventually became joint-service activities drawing staff from all the branches of service, it still fell mostly to the Air Force and Navy to man, train, and equip these activities. For reasons that we mentioned in the previous chapter, and which we examine more closely here, these two services pursued almost independent developments in space surveillance, with different requirements, different schedules, and different funding profiles. Necessarily, the structures of the organizations were also different. However, both services turned to their respective laboratories for the initial efforts, and that is where we begin.

I. EARLY SPACE SURVEILLANCE FOR THE AIR FORCE

By 1957, in preparation for the IGY, many of the Air Force's experts in subjects related to satellite tracking, such as electronics, geophysics, computers, communications, astronomy, and mathematics, had already been assembled at the Geophysics Research Directorate of Air Force Cambridge Research Laboratories at Hanscom Field (later Hanscom Air Force Base) near Bedford, Massachusetts. AFCRL had been organized in 1945, immediately after World War II, by merging the wartime research efforts of the Massachusetts Institute of Technology Radiation Laboratory and the Harvard Radio Research Laboratory.[18] In fact, what some consider the first American attempt at military space surveillance per se began with radio systems at AFCRL only days after the launch of Sputnik 1. The head of the directorate, Milton Greenberg, led an attempt by his staff to detect and track the satellite using several radio interferometers and experimental Doppler radar.[19] Judging from the paucity of reports and records of that experiment, we can conclude that their results must have demonstrated mainly the need for a much more comprehensive effort to track satellites. A two-day conference at the headquarters of Air Research and Development Command,

[18] Altshuler [1] gives a firsthand account of work life at that new Air Force laboratory, although that author worked in applied physics, mainly radio propagation, rather than in space surveillance. The laboratory facilities, originally in Cambridge and later at Hanscom, were also known collectively as Air Force Cambridge Research Center (AFCRC). For clarity in this short account of the early years, we have used the plural name.

[19] Sturdevant [2], p. 8. As we discuss at more length later in this chapter, the tracking of satellites by optical means was, in almost every way, more advanced than tracking by radio and radar, in the earliest days. A practical, though limited, means of detecting and tracking satellites never observed before had already been devised for a network of optical sensors even before Sputnik 1 was launched. However, the opportunities for making ground-based optical observations were (and today still are) highly limited by local conditions of weather and sun elevation angle at each observing station. Radio/radar tracking inherently does not have either of these limitations.

18–19 November 1957, resulted in consolidation at AFCRL of all Air Force activities for space surveillance. Hence, the first formal space surveillance effort, Project Space Track, began on 30 November 1957. The work was organized and led by two expatriate German astronomers, Gerhard Robert Miczaika and Eberhard W. Wahl. Miczaika, a Harvard professor who had agreed to take on this interesting extra duty, was already a prominent astronomer at that time.[20] Wahl's doctorate had been in meteorology, but he also worked in astronomy at AFCRL and became the technical director for Project Space Track. Initially, the project staff were assigned offices and floor space that were out of the way[21] in the basement of the directorate's Building 1535 at Hanscom, because no one knew exactly what facilities would ultimately be needed. Space Track's general objectives[22] were to 1) maintain surveillance of satellites and 2) provide observational data and orbit predictions for scientific study of satellite vehicles. To achieve these goals, the project had to inventory and assess the existing sensors that were able to track satellites, develop ways to collate the observational data from multiple tracking sites, coordinate the overall tracking operation, recommend how to expand the sensor coverage worldwide, and identify priorities for new tracking systems and methods that should be developed.

Miczaika and Wahl had been chartered to examine and develop all types of tracking systems, and they realized from the beginning that space surveillance for military purposes would have to include both optical- and radio-based methods. However, when they started, optical tracking methods were more advanced and better understood by the larger technical community than were radio methods; nowadays we would say that optical methods had a higher "technology readiness level." Moreover, new and innovative work on optical tracking systems had already been inspired by the IGY research programs, as we describe shortly. For the Air Force, most of the early work on radio and radar systems suitable for tracking satellites would occur a few years later, as part of ballistic missile defense projects, independently of the IGY research

[20]Miczaika's 1961 book *Tools of the Astronomer*, is, at this writing, still available from Harvard University Press as the first book in its series *Harvard Books on Astronomy*. Altshuler [1], p. 55, records that Miczaika had come from Prussia and that Wahl had come from Berlin.

[21]Mansfield [3] mentions that the offices were "squeezed behind Shipping and Receiving," although the project made rapid progress despite these circumstances.

[22]Miczaika and Curtis [4] . This report is probably the earliest detailed description of the scope of work for Project Space Track, written barely six months after the project's inception. That the specific details in this report could be compiled as definitively as they were in that short amount of time indicates the intense effort involved in achieving an operational capability for space surveillance as quickly as possible. In June 1958, radio/radar tracking techniques had already been developed by the Naval Research Laboratory, as we describe in more detail later in this chapter, but actual sensor performance data for those new sensor types were still very preliminary. Hence, Miczaika's 1958 report describes only optical tracking sensors as they were then being readied to support the IGY satellite programs. However, the report includes detailed discussions of optical observation types, optical sensor accuracies and precisions, and substantive recommendations for the highest-priority sensor developments needed for space surveillance, including those for radio/radar sensors.

programs.[23] Thus, Project Space Track found itself in the happy position of being able to take immediate advantage of work on satellite tracking sensors, which was already in progress elsewhere for other reasons besides military space surveillance. It would eventually turn out that Project Space Track, or, more precisely, its operational element, would support both missile-defense and space surveillance. In an interesting counterpoint, the Navy ended up developing a radar system for space surveillance as an extension of the first American effort to launch a satellite during the IGY, Project Vanguard. As we discuss later in this chapter, the Navy was not pursuing space surveillance per se as part of Vanguard; they intended to use radio beacons on the Vanguard satellites to enable tracking of those satellites by receivers on the ground. However, before the first Vanguard satellite was launched, they were confronted by the launches of the Sputnik series, which led them quickly to evolve a transmitter-receiver-type radar system from their existing Vanguard receiver-only tracking system. That Naval space surveillance radar would also eventually support both missile-defense and space surveillance, alongside the Air Force systems.

Meanwhile, the natural starting place for Project Space Track was the scientific community's preparations for tracking research satellites during the IGY. Conveniently, two such activities had been organized during the previous two years nearby in Cambridge, Massachusetts, by Fred Lawrence Whipple at Smithsonian Astrophysical Observatory (SAO). These were 1) the development of a new system, the Baker–Nunn cameras, for optical tracking of IGY satellites, and 2) Project Moonwatch. Whipple was already one of the foremost astronomers in the world, whose distinguished career had begun in 1931 at Harvard College Observatory. Moreover, he was one of the relatively few astronomers who had developed a personal and professional interest in artificial Earth satellites. For example, he had consulted with Project RAND[24] on its famous study [5], "Preliminary Design of an Experimental World-Circling Spaceship," published in May 1946, the same report that we mentioned in Chapter 1. Later that year, Whipple [6] published a paper in a

[23]Although it is true that missile-defense radars would, in the long run, provide a large fraction of all space surveillance radar observations, it is important to realize that all kinds of radars were (and still are) pressed into service as needed for various aspects of the space surveillance mission. An entire category of so-called contributing sensors emerged to support the cataloging and characterization of satellites, as the on-orbit population grew. We discuss a few of these sensors later in Chapter 6 on the Space Surveillance Network. Contributing sensors, both optical and radio/radar types, were often one-of-a-kind research instruments or systems built for other missions entirely, and they often represented the pinnacle of technology available at the time they were needed. Nevertheless, the total number of observations reported by them was usually only a few percentage of the total reported by missile-defense radars. One important example of a contributing sensor was the Millstone Hill steerable dish-type radar operated by MIT Lincoln Laboratory near Westford, Massachusetts. The Millstone Hill Radar was primarily a research instrument, but it came online in time to detect Sputnik 1 in orbit in 1957. It has an illustrious history of its own, and, at this writing, continues to provide unique support to various aspects of the space surveillance mission.

[24]Project RAND, the postwar research organization originally hosted at Douglas Aircraft Company, was reorganized as an independent nonprofit, RAND Corporation, in May 1948. It retains that status at this writing.

supplement to that RAND report, entitled "Possible Hazards to a Satellite Vehicle from Meteorites," drawing on his years of astronomical research in meteoritics. In this paper, he introduced the concept of using multiple layers of thin metal sheets to serve as a kind of bumper, dissipating the energy of small meteors that might happen to strike a satellite. He later patented the concept. Today the device, known as the "Whipple shield," has assumed more importance than he likely could have guessed, in the mitigation of damage to satellites from man-made debris too small to track in Earth orbit.

In 1955, in preparation for the IGY, the Smithsonian Institution in Washington, D.C., had decided to move its SAO headquarters from Washington to Cambridge, adjacent to the grounds of the Harvard College Observatory,[25] to allow closer collaboration between its scientists and those at Harvard. An important part of the arrangement was that Whipple, by that time the chairman of the Harvard Astronomy Department, would serve as director of SAO while retaining his professorship in the Astronomy Department. The idea was that these two research organizations would remain distinct but have a senior staff member in common to coordinate, informally, if possible, the respective programs and shared use of instruments and facilities.[26] It is easy to imagine that only someone of Whipple's professional stature would have been able to make such an arrangement work, as, in fact, he did until his retirement 18 years later.

Well before they moved to Cambridge in July 1955, the staff at SAO had proposed to the U.S. National Committee for the IGY, based in Washington, D.C., to track all satellites launched during the IGY, using optical sensors and computational methods familiar to all astronomers but suitably modified for tracking Earth satellites. Whipple welcomed the idea, believing that optical tracking of satellites was certainly feasible and could be made practical. He based his optimism on his long experience with photographing meteor trails from multiple stations simultaneously to estimate their trajectories. Yet, there were many new considerations. IGY planning had identified a multitude of uses for scientific satellites, especially for ionospheric research, measuring the radiation environment in space and determining accurate geodetic parameter values for the size, shape, and mass distribution of the Earth. In fact, virtually every space-related scientific application depended on knowing the position of the satellite at any time of interest, and more accurate satellite positions translated directly into better geophysical science. Hence, it is no surprise that the scientists at SAO had become interested in the tracking problem. In a

[25]Until 1959, the SAO staff had their offices in Kittredge Hall, the same building that now houses Harvard University Press. In that year, construction began on a permanent facility for the SAO, and the staff moved into temporary quarters in the neighborhood.

[26]The same basic arrangement exists at this writing. In 1973, the year that Whipple retired from these dual jobs, the collaboration between the Harvard observatory and the SAO was formalized as the present Harvard Smithsonian Center for Astrophysics, with both organizations preserving their distinctive staff, programs, and specialties.

general sense, optical tracking was a natural subject for astronomers, who, for several centuries, had been tracking objects in the solar system by telescope and using celestial mechanics to understand and predict the motions of those objects. However, the transition from solar-system astronomy to satellite tracking was not a simple matter. Astronomical telescopes typically had very small fields of view and were designed to follow the slow apparent motion of objects in the sky, typically about 0.005 degrees per second or less and mainly due to Earth rotation, which greatly simplified the instrument designs needed for this purpose. In 1955, it was not immediately clear what instrumentation would be needed to detect, track and record Earth satellites moving across the sky hundreds of times faster, conceivably up to 4.0 or 5.0 deg/s, and in any direction, not merely in the east-to-west diurnal apparent motion. Most analysts at the time guessed that although IGY satellites would move generally west-to-east in the sky, because launches would be generally toward the east so that Earth rotation could help the boost into orbit, a wide range of north–south or south–north apparent motions could also be expected from satellites exploring the high northern and southern latitudes of the Earth's space environment. Moreover, the well-developed mathematical orbit models long used for calculating orbital motion in the solar system had a much different character from that required in models of Earth satellite motion. The truth was, Earth satellite orbit models basically did not exist before the IGY preparations and had to be developed in time to support the scientific programs.

Whipple knew that his SAO team would take on all these problems, with tight schedule constraints that were variable beyond their control and with uncertain costs, if they succeeded in getting the tracking assignment. Fortunately, his approach to holding down two full-time jobs was to delegate major tasks to the best talent that he could find. Sometime in November 1955, Whipple confided to Josef Allen Hynek, a physicist and astronomer at Ohio State University, that SAO might get the IGY assignment for satellite tracking and offered that Hynek could become the associate director of SAO in charge of that project. In fact, before the end of December 1955, only a few months after they had moved to Cambridge, the SAO was assigned the task of tracking all American satellites launched during the IGY, with six months of grant money available on 1 January 1956.[27] Further funding would be contingent on SAO producing viable plans for a comprehensive tracking program and the

[27]The assignment came from the U.S. National Committee of the IGY in a somewhat bureaucratically burdened way involving the Smithsonian Institution, the National Academy of Sciences, and the National Science Foundation, according to Hayes [7], Chapter 1, p. 8. That author, in his book and several of his earlier reports to the Smithsonian Institution headquarters in Washington, D.C. [8], provides one of the most detailed firsthand accounts of the early development of space surveillance by optical sensors, beginning with the IGY up to the mid-1960s. A related book by W. Patrick McCray, *Keep Watching the Skies! The Story of Operation Moonwatch & the Dawn of the Space Age*, Princeton University Press, 2008, was aimed at a more popular audience. It emphasizes the personnel and the human aspects of the SAO and Moonwatch efforts. Together, these two books provide a comprehensive picture of the early optical tracking of satellites.

design of suitable instruments and facilities to carry it out. Hynek promptly took a leave of absence from Ohio State and reported in January to Cambridge as Associate Director, a post he held until August 1960.

As an aside, it is worth noting here that one of the first lessons learned in space surveillance would be that although SAO's official responsibility may have been limited to tracking American satellites, someone really did have to keep track of all the satellites anyway, regardless of their national origin, simply to avoid confusing satellites as their total number in orbit increased. The SAO staff, and later the staff assembled for Project Space Track, did understand this problem and were able to demonstrate the necessity and practicality of solving it. It is also interesting to note that in the satellite tracking literature, during preparations for the IGY, there is only scant mention of this need for comprehensive space surveillance. However, by the end of the IGY, at least a few leading scientists were discussing that need in detail.[28] Even if it was only a natural progression of thinking under the circumstances, it is fortunate that this emerging understanding in the American scientific community came at almost the same time that the military staff were beginning to realize the true scope of the space surveillance job that they faced.

Within a month, probably before the end of January 1956, Whipple and Hynek had agreed on the main principles of optical tracking to support the IGY:

1. A new Schmidt-type camera, using the same basic optical layout as Schmidt telescopes, would be designed specifically for the IGY purposes. The camera would need a very wide field of view. It would also need high angular measurement accuracy even for faint, fast-moving objects such as most satellites were expected to be. Achieving these two somewhat conflicting goals for the camera would turn out to be the main source of both innovation and schedule delays for the SAO program.
2. On demand, the camera would be able to record the streaks of satellite images on film, using a high-speed shutter controlled by an accurate electronic crystal clock to introduce precisely timed interruptions (dark gaps) in the recorded streaks. Those gaps in the streak's image would allow accurate determination of apparent angular rates simultaneously with the apparent angular position.
3. The electronic crystal clock would need to be highly accurate for long periods of time between calibrations against a master clock.

[28]For one example, Karl Henize, in his often-cited review paper published in 1960 [9], devoted an entire section to the need for space surveillance, beyond the needs of satellite research programs. His comments are still germane today; refer to Henize [9], pp. 120–121. At the time he wrote that paper, Henize, whom we will mention later, had, for more than three very active years, been the senior astronomer at SAO in charge of installing and operating the first dozen Baker–Nunn camera stations around the world. Those stations supported essentially all satellite applications during the IGY, besides Project Space Track, and so Henize was in an especially good position to understand the ramifications of tracking and estimating orbits for every space object that his cameras could detect.

4. A network of 12 camera stations would be built, with locations in various countries around the world.
5. A communication and computation center would be established at SAO in Cambridge, to (a) receive and process camera observations at a central point, (b) estimate definitive values of all the orbit-related parameters for each satellite and monitor changes in those orbits over time, and (c) produce predictions of the times and directions of future satellite passes visible at each station, so that the camera stations could plan and collect further observations of each satellite.

Whipple asked James G. Baker, a consultant for the Perkin-Elmer Corporation in Norwalk, Connecticut, to do the optical design of the new satellite camera. Baker had previously designed a Schmidt-type camera for Whipple's research on meteors, and Perkin-Elmer had manufactured that instrument. Moreover, Whipple had already discussed the problem in general terms with Baker during the previous summer. Hence, the way ahead for the camera design seemed obvious. As soon as Whipple issued the formal invitation to him in February 1956, Baker began work on the optics for the new camera, which Perkin-Elmer stood ready to produce. Meanwhile, Whipple and Hynek discussed the mechanical and mount design of the camera with Joseph Nunn in Pasadena, California, whom both had known from his previous work in telescope design. Their main question for Nunn was, should the camera be steered to follow 1) the satellite, letting the stars appear as streaks on the film, or 2) the stars, letting the satellite appear as a streak on the film. His answer was that the mount mechanism should point in both modes, but alternately, allowing the recording on a film strip of a series of both angle measurements and angle rate measurements on every pass of a satellite over a station. This concept had not been used in astronomy before, and Nunn's mechanical design was considered as innovative at that time as the camera's optical design. It was manufactured by the small telescope-engineering firm Boller and Chivens Company locally in Pasadena,[29] who also assembled the complete mounted camera system.

In its final design,[30] the Baker–Nunn camera had a field of view of about 5 deg by 30 deg centered on the camera boresight. This field was quite large compared to nearly all astronomical instruments, besides being unusual in its oblong aspect. In regular operation, the longer dimension of the field of view was to be oriented along the apparent path of the satellite in the sky because the greatest errors in the predicted positions from the computation center were expected to be along that path. The still-rather-broad width of the field of view

[29] In a move perhaps not too surprising in retrospect, the Boller and Chivens Company was acquired by Perkin-Elmer Corporation in 1965. That corporation continued to grow and for several decades was a main supplier of specialized optics for the U.S. government, including the main mirror for the Hubble Space Telescope.

[30] Hayes [7], Chapter 1.

was intended to accommodate the remaining prediction errors. The aperture of the camera was 20 in., small compared to the most famous astronomical telescopes, and the focal length was short, also about 20 in. (500 mm), to allow the wide field of view. Moreover, the short focal length allowed an image of the faintest satellite anticipated for the design to be recorded in, at most, several seconds, rather than requiring the minutes- or hours-long exposures used in most astrophotography. The mount was gimballed so that the camera boresight could follow the stars in the background, letting the satellite image streak on the film, or it could follow the satellite itself, based on the predicted path of the satellite in the sky according to the SAO computation center. In the latter mode, the stars streaked on the film, but the satellite image was (if the predicted pointing of the camera were perfect) concentrated as a single dot so that fainter satellites could be detected. By taking exposures in each of these two pointing modes in close succession, the apparent angular position of the satellite and the apparent angular direction of the streak, relative to the stars, could be determined accurately by viewing the two photographs in rapid alternation in a blink microscope. The blink microscope was a familiar piece of astronomical equipment that had long been used to detect objects such as asteroids moving slowly relative to the background of fixed stars. Famously, in 1930, Clyde Tombaugh at Lowell Observatory in Arizona had discovered the planet Pluto using a blink microscope in just this way. Now essentially the same method would be used to enable routine detection and observations of Earth satellites.

How accurate did optical observations of satellites need to be, and how many observations would be needed?[31] These questions had been debated at least since SAO first considered offering a proposal to track all IGY satellites. The answers were critical to both the optical design and the mechanical design of the tracking systems, besides the requirements for accurate computation of the orbits. Some consensus finally emerged in time for Baker and Nunn to adjust their designs accordingly, and for Whipple and Hynek to examine the tradeoffs involving total budget and number and locations of stations. For merely monitoring the orbits, making predictions for tracking stations to be able to reacquire satellites regularly, and for avoiding confusion among multiple objects in orbit, a modest measurement accuracy of about 1 arc-minute was thought to be acceptable. This figure presented no special design difficulties for the Baker–Nunn cameras, although it did assume that observations would be distributed over most of the orbital path, that is, from stations widely distributed around the Earth. In comparison, the requirement

[31] Brandenberger [10]. The author of this paper was from the Department of Geodetic Sciences at Ohio State University. Although he was associated more with the Air Force Baker–Nunn stations than with the SAO IGY stations, his paper gives a nearly complete early discussion of how requirements for optical observations were analyzed in that period and how the Baker–Nunn observations were recorded and processed.

for IGY scientific applications of satellites in geodesy turned out to be quite stringent; most analysts wanted a tracking measurement accuracy of about 2 arc-seconds, with as many observations as possible and well distributed around the globe. This accuracy was close to the limit of what was possible at the time. For the Baker–Nunn cameras, the necessary optical design for that level of accuracy led to complications with the recording film and its handling, in addition to questions about attainable manufacturing tolerances for the optical elements. Added to that were questions about uncontrollable distortions of the satellite's image due to viewing it through the turbulent atmosphere and about the amount of orbit computation that would be needed to make practical use of such high-precision data. Of course, it was well known that, in purely statistical terms, having more observations could partly make up for having less precise observations, but quantifying this tradeoff was controversial at the time (as it still is today). Everyone understood that the IGY budget would allow only some limited number of observations, in any case, and that those observations would not be ideally distributed around the Earth. Moreover, this desired angular measurement accuracy also dictated that the measurement of time, of both the absolute time and time intervals, had to be accurate to 0.001 s or better. The basic reason was that satellites moved across the sky several hundred times faster than typical astronomical objects: a near-Earth satellite passing directly overhead at a tracking station could move as much as several arc-seconds of angle in a single millisecond of time. There simply was no choice but to record the times of each observation with this unprecedented accuracy and then somehow make such clock accuracy routinely available at remote field stations without high-speed communications to calibrate the clocks against a master time reference. The latter problem was especially difficult to solve. It took an extra year after the initial deployment of the cameras to make the station timing systems uniformly and reliably accurate. As things turned out, the Baker–Nunn cameras went into service without high confidence that such extreme angular measurement accuracy could be delivered routinely. Nevertheless, the SAO tracking data proved good enough overall for the IGY to produce several fundamental improvements in knowledge of the Earth's size, shape, and mass distribution, improvements that could not have been made in any other way at the time.

Incidentally, the 1-ms timing requirement for IGY tracking was left in place for military space surveillance for the next few decades. The main reason is that many modern surveillance tracking systems evolved directly from the IGY tracking systems, as we describe in more detail later. The 1-ms standard is still embedded in some operational space surveillance software at this writing. Meanwhile, the demands of navigation and scientific research have led to clocks, both terrestrial and space-borne, which are literally a million times better, having nanosecond or better accuracy. Only recently, with the possibility of large numbers of optical surveillance instruments becoming available with 1-arc-second angle accuracy, and the need for even better accuracy for a few newer tracking applications, has the operational space surveillance community begun to reconsider its timing requirement.

A particularly knotty problem that Whipple and Hynek had to solve was the so-called acquisition problem for newly launched satellites. When a new satellite was launched, how would the camera stations find it? In other words, how would the stations know when and where to point the cameras and start recording, since no orbital predictions would be available at that initial stage of the tracking? The problem was more difficult than it might seem. The whole camera assembly with its mount weighed several tons and could not be reoriented to another part of the sky quickly, meaning that, by design, advance planning and pointing were necessary. Some satellites were expected to be too dim for the human eye to detect unaided, and so optical aid was mandatory to spot them, but what kind of optical aid? The hemispherical night sky at any ground station contains nearly 22,000 square degrees, and satellite passes could appear in any part of that expanse. Whipple and Hynek finally decided to enlist as many volunteers as they could from amateur astronomy groups internationally to spot satellites, an auxiliary effort that they called Project Moonwatch. Ever since he had become the SAO director, Whipple had been turning over in his mind ideas about how amateurs might somehow contribute to satellite observing programs, but now he and Hynek worked out a few specific details. The volunteers would receive notice of new satellite launches from SAO and scan the skies using their own equipment, mostly binoculars and small tabletop telescopes, although SAO eventually loaned a very few army-surplus spotting scopes to several groups. The volunteers would report immediately back to SAO any sighting information, including estimated directions relative to known stars and the times of the observations. Typically, the times would be obtained with handheld stopwatches previously synchronized manually with a reference clock. Given multiple reports from different groups at known locations, it was possible for SAO to compute a crude preliminary estimate of the orbit. Whipple and Hynek came to believe that if the amateurs could, on average, supply directions accurate to within 1 deg and observation times accurate to within 1 s, such a preliminary orbit estimate would be accurate enough to cue the Baker–Nunn stations to collect the first accurate tracking data. Then progressively better orbit estimates would follow, so that the stations could continue to reacquire the satellite reliably in the future. Today, most persons hearing about this scheme for the first time are surprised to hear how well it worked, or that it even worked at all. Whipple and Hynek had to convince plenty of skeptics at the time, too, including members of the American IGY panels who reviewed their proposal. A few years later, Whipple recalled,[32]

[32]Green and Lomask [11], Chapter 9, p. 149.

Some of my colleagues were convinced that too few amateurs would volunteer, and that those who did would not always perform satisfactorily. Time, I'm happy to say, has proved these fears to be groundless. The amateurs joined up in droves all around the globe. They did a splendid job for Project Vanguard, and they have been doing an increasingly more effective one for the American space effort ever since.

In April 1956, the Working Group on Tracking and Computation, part of the American IGY's Technical Panel on Earth Satellite Programs, approved the SAO tracking plan, including Project Moonwatch. The SAO, having already in February 1956 appointed Armand N. Spitz to lead the Moonwatch effort, had him announce the amateur program publicly at a news conference in September 1956. To say that the idea caught on would be a drastic understatement. By the time that Sputnik 1 was launched, 80 Moonwatch teams in the United States and 84 additional teams in foreign countries had been organized and briefed on how to do the job. These teams were, in fact, the only trained observing teams that were ready to track satellites at that time.[33] The Navy's Minitrack radio interferometer stations (discussed later) were ready, but they detected radio signals only on the American IGY satellite frequency, not the frequencies used by the Sputnik series. By the time the first American satellites were being launched in early 1958,[34] 230 Moonwatch teams had been enlisted, including 121 in the continental United States, 1 in Canada, 13 in South and Central America, 77 in Japan, 5 in Australia, 5 in various Pacific islands, 3 in Asia, and 5 in Africa. By the time the third and final Vanguard satellite was launched in September 1959, nearly 8000 amateurs had been organized into 250 teams in 27 countries worldwide, including 126 teams in the United States. In fact, Moonwatch continued to be popular in amateur astronomy circles for many years after the IGY ended, so that SAO continued the program formally until 1975, at which time more than 100 teams were still active.

In the same month that Moonwatch was announced, September 1956, Karl Gordon Henize[35] joined SAO as the astronomer (becoming, a year later, the senior astronomer) in charge of Baker–Nunn stations. He stayed in that position until the fall of 1959, when he took an academic post in astronomy at Northwestern University. His SAO job was a large one, both technically and administratively, because the stations were to be located worldwide. Hynek, Henize, and others had to visit more than 20 countries to organize

[33]Hynek's memorandum on the anniversary of Sputnik 1 [12]. The complete text of that memorandum is in Appendix A. See also Corliss [22], Chapter 2, pp. 35–36.

[34]Hayes [7], Chapter 3. Additional background on Project Moonwatch is found in Hayes [8], Annual Report for the Year Ended 30 June 1961.

[35]Soon after his arrival at SAO, Henize published a paper [13] for general readership in January 1957, in which he told some of the Baker–Nunn story. He became better known a decade later when he was selected as a NASA astronaut in August 1967. At this writing, Henize is still the only astronaut to have been associated with space surveillance in a major way.

collaborations, identify sites for the stations, and arrange for the construction of facilities and the installation of equipment and communications.

The original schedule called for all the Baker–Nunn stations to be ready to track satellites by the start of the IGY, July 1957, an ambitious 18 months after formal project start. Almost inevitably, given such an innovative telescope design for such a challenging task, the deployment of the Baker–Nunns was delayed, although the cameras did eventually perform according to specifications. Difficulties cropped up in their design and fabrication, and logistic and construction problems recurred at some of the remote sites. None of the cameras or stations was ready was ready when Sputnik 1 was launched on 4 October 1957. Embarrassingly, the first complete camera had just been disassembled for some minor rework after its initial testing in Pasadena, and Henize had to reassemble that camera in time to photograph Sputnik 1 while it was still in orbit. Perhaps needless to say, several tense program review meetings followed with the IGY oversight panels; however, the option to cancel the project was finally laid to rest. The first Baker–Nunn station began operating at Organ Pass, New Mexico, in November 1957. Still, despite terrific pressure from all interested parties, it would be June 1958 before the remaining stations in the planned list of 12 were operating. Table 2.1 summarizes the SAO Baker–Nunn sites at that time.

Soon after the Air Force organized Project Space Track in November 1957, that project joined the optical tracking effort, taking advantage of the SAO work already done, and promptly ordered three additional Baker–Nunn cameras from the companies already contracted by SAO. Two more were added to the Space Track inventory by 1961, in time to support regular operations at a new space surveillance center that was established in Colorado Springs, Colorado, in July of that year, which we discuss later. Those five

TABLE 2.1 THE FIRST SAO BAKER–NUNN STATIONS (JUNE 1958)

Country	Location
United States	Organ Pass, New Mexico
United States	Jupiter, Florida
United States	Haleakala, Maui, Hawaii
Australia	Woomera
South Africa	Olifantsfontein
Spain	Cadiz
Japan	Mitaka, Tokyo
India	Naini Tal
Peru	Arequipa
Iran	Shiraz
Netherlands West Indies	Curaçao
Argentina	Villa Dolores

TABLE 2.2 U.S. AIR FORCE BAKER-NUNN STATIONS (JULY 1961)

Country	Location
United States	Edwards Air Force Base, California
Chile	Santiago
Norway	Harestua, Oslo
United States	Massachusetts*
Canada	Royal Canadian Air Force Base Cold Lake, Alberta*

*Added in a second phase after the first three stations.

instruments were installed and operated by the Air Force at the sites listed in Table 2.2. Although the two organizations operated their respective stations independently, and the predicted satellite passes for each station came from their respective computation and communication centers, all observations and orbital data were shared routinely between SAO and Project Space Track.

During 1958, as part of their task to coordinate the overall tracking effort at Project Space Track, Miczaika and Wahl continued to compile a list of groups and facilities that were able to track satellites, and to form alliances and collaborations with them. This was a large effort: by 1960, more than 150 sensor sites, including astronomical optical observatories, radio telescopes, radars, and Moonwatch stations, were contributing observations of satellites to the Space Track processing center at Hanscom Field.[36] Besides the Air Force Baker-Nunn sites listed earlier and the Naval Space Surveillance System discussed later, some especially notable connections were established with then-new radar sites at Millstone Hill near Westford, Massachusetts, operated by Lincoln Laboratory, and Jodrell Bank Observatory in Cheshire, United Kingdom, operated by the University of Manchester. The Millstone Hill Radar had been developed in the mid-1950s to support the BMEWS missile-defense program. The radar electronics and the antenna were ready in the fall of 1957, just in time to detect Sputnik 1 and measure the range to the satellite on 5 October 1957, the day after the launch, while the orbit of Sputnik was still not well determined. This feat was impressive because the radar construction would not be complete until later in 1958: the 84-ft-diam radar dish antenna had to be manually steered to allow the detection, meaning that no actual tracking could be done. However, construction was finished in time to detect and track Sputnik 2, launched on 3 November 1957, while it was still in orbit.[37] Although Millstone provided the first American radar detection of

[36]Fitzpatrick and Findley [14], p. 6. This number is much higher than the number of stations contributing space surveillance observations in later years, as we discuss in later chapters. It reflects the facts that in the early days, there were no high-volume tracking sensors, communications were limited and sometimes intermittent, and the whole satellite cataloging enterprise was still experimental. It is all the more remarkable that Miczaika and Wahl and their staff were able to compensate for these limitations, bringing their effort to what we now think of as "operational status" by late 1959, as we mention later.

[37]Sridharan and Pensa [15], Chapter 1. See also Stone and Banner [16], p. 224.

Sputnik 1, the first radar tracking had been done a few hours earlier by the newest radio telescope at Jodrell Bank Observatory. The largest fully steerable radio telescope in the world at the time,[38] the 250-ft-diam dish antenna known since 1987 as the Lovell Telescope after its founder Bernard Lovell, who was knighted for his role in leading the project, came online in mid-1957. It tracked Sputnik 1 on 4 October 1957, and a week later was able to track the Sputnik upper stage [17]. The latter success was considered especially important at the time because it indicated that unaided radars could detect and track ballistic missiles, the very same problem that the Millstone Hill Radar had been built to investigate for the BMEWS program. Jodrell Bank was quite important in tracking satellites in the early years of space surveillance and, at this writing, continues to track and monitor space probes at interplanetary distances, besides its basic mission of radio astronomy. Likewise, since the beginning of space surveillance, the Millstone Hill Radar has been, and continues to be, crucial in several aspects of space surveillance and catalog maintenance.[39]

The year 1958 also brought a top-level change in how the space surveillance mission was organized, the full implications of which took some years to develop. Eisenhower established NASA on 29 July 1958, the third anniversary of his announcement that the United States would launch a satellite during the IGY. In large part, this was a continuation of his policy to establish the "freedom of space." Specifically, he wanted to distinguish and clarify the roles of the American military space programs and civil space programs. (We have more to say on this point later in this chapter.) The new agency would be dedicated specifically to civil space programs. Work at NASA began formally with the start of the government's next fiscal year on 1 October 1958, and the agency's charter included sponsorship of the SAO when the IGY ended in December of that year. In practice, this meant that the Satellite Tracking Program at SAO would continue and be funded through NASA's congressional appropriation rather than through the National Science Foundation. The actual effects of this change for space surveillance were minor for several years after because observations and orbital data continued to be shared between SAO and Project Space Track at Hanscom. Eventually, however, the military space surveillance effort would grow enough in its scope and capability that it would not be dependent on the SAO's optical tracking of satellites, allowing the SAO to focus more on its research programs. For example, SAO became part of the joint Harvard-Smithsonian Center for Astrophysics in 1973 and ended its amateur-based Moonwatch program in 1975. That evolution would continue until, at this writing, SAO continues close collaboration with NASA for scientific purposes and is funded through a variety of nondefense sources. In short, Eisenhower's clarification of military and civilian roles in American

[38] At this writing, the Lovell Telescope is still the third-largest fully steerable radio telescope in the world, behind the 100-m-diam instruments at Green Bank, West Virginia, and Effelsberg, Germany.

[39] Grometstein [18], Chapter 10, p. 174.

space programs meant that the military would inherit the entire space surveillance mission along with ballistic missile defense, while NASA would inherit most space-related scientific research and the manned and unmanned space exploration programs.

By December 1958, Space Track had improved its routine tracking and computer processing activities to the point that the project began functioning as an "interim" NSSCC, producing regular reports about satellites and their orbits. Of course, the goal of establishing such a center had been the main driving force behind the work of the project all along, and everyone involved was eager to see it happen. A year later, in December 1959, the goal was fully realized. Project Space Track moved into a new building at Hanscom, where the NSSCC was declared operational on 9 February 1960, with a total staff of about 70 persons. Harold O. Curtis, from the nearby Lincoln Laboratory and a colleague of Miczaika, was the first director of the NSSCC, and Wahl continued as technical director. Although round-the-clock operations placed large demands on the staff, the center maintained a significant level of research and development work.[40] Perhaps not too surprisingly, although the name of the program had changed officially to "NSSCC," the more familiar name "Space Track," or sometimes even "Project Space Track," continued to be used in correspondence and some reports long after 1960.

Project Space Track, despite remarkable progress in its first few years, suffered from one of the perennial shortcomings of nearly all in-house government, and, for some reason, particularly defense-related, technical projects. Although highly capable people were creating and implementing entirely new methods for detecting, tracking, and predicting satellite motion to fulfill a new and as-yet ill-defined mission, nobody was documenting the work in a form usable by anyone outside that immediate small group. The situation was different at SAO. There, in the more academic atmosphere fostered by the IGY and their university setting, many incentives and opportunities existed for staff members to present and publish their new methods and results in journals, at academic meetings, and at scientific conferences. Most of the SAO papers and publications from that era can still be found in the scientific literature today. In comparison, only a few systematic accounts of any aspect of the technical developments at Project Space Track are known to exist, and these are mostly internal reports that were not published widely. The main exception to this rule came in 1960 when, apparently because of the widening interest of Air Force people in the new

[40] Wahl [19]. No one person in the Air Force cadre at Hanscom had a more direct influence than Wahl, the author of this paper, did on how military space surveillance developed in its first several years. He stayed with the Space Track operation at Hanscom until 1963, when he decided to return to meteorology and took an academic position at the University of Wisconsin at Madison, in the Department of Atmospheric and Oceanic Sciences. He worked in that department until his retirement in 1984. Besides producing a large body of research at Madison, he was noted for frequently teaching the freshman survey course in weather and climate.

space surveillance mission and the consequent emerging awareness of the lack of documentation for it, two staff members were dispatched from the Special Studies Section, Directorate of Aerospace, Air Proving Ground Center, Eglin Air Force Base, Florida, for an extended stay at Hanscom Field, specifically to document the tracking operation at the NSSCC, under a task funded as part of Project Space Track. Philip M. Fitzpatrick, a physicist, and his colleague George B. Findley were welcomed at the NSSCC by both the senior and working staff there, and they later remarked in their report about the generous help that they received from many NSSCC persons for their task. One of the younger NSSCC staff, Hans Beat Wackernagel, received particular mention by Fitzpatrick and Findley in their report [14]. In 1958, Wackernagel had been recruited by the Air Force right out of graduate school at the University of Basel in Switzerland, almost as soon as he had earned his doctorate in astronomy, physics, and mathematics. His first job for the Air Force was at Project Space Track at Hanscom, which at that date was still located in the basement of the Geophysics Directorate in Building 1535. Although Wahl had already formulated the basic method of computing orbital predictions first used by Space Track, Wackernagel developed many of the other computational methods and computer routines used by Space Track and later the NSSCC. He was also in charge of testing and calibrating the Air Force's Baker-Nunn cameras. Consequently, soon after Fitzpatrick and Findley arrived, no doubt wondering how to get started on their task, they must have been pleased when they realized that they "had found their man" in Wackernagel. Although the latter had written short technical reports on individual satellites and on other limited topics for Project Space Track by that time,[41] Wackernagel took time to discuss with Fitzpatrick and Findley, at length, every aspect of the mathematical formulations, computer processes, and operational procedures for the tracking operation and catalog maintenance process and helped them get that information into writing.[42]

---

[41] Such as Wackernagel's report [20], for one example, containing a total of eight pages.

[42] How a Swiss citizen, born and raised in Basel, could, upon graduating from his hometown university, have become instrumental in the Air Force's space surveillance mission within that same year, is a question that leads to many other questions. What we can assert at this writing is that Miczaika and Wahl must have known exactly where to look for the expertise they needed. Wackernagel's extensive knowledge and willingness to document complicated systems was valuable in later years as well. In fact, their value persists until now. In 1961, the NSSCC operation was moved to Colorado Springs, leaving the Project Space Track research and development staff to continue its work at Hanscom. Wackernagel chose to go with the operational unit and worked in many aspects of space surveillance there until his retirement in 1990 [3]. In the latter part of his career, he collaborated with C. G. Hilton on writing the documentation that became the first edition of *NORAD Technical Publication TP-SCC-008 Mathematical Foundation for SPADOC Computation Center Astrodynamic Theory*, first published in 1982 at North American Aerospace Defense Command Headquarters for use by persons working with the Space Defense Operations Center. This document is still known, both in its original version and in many later updates, as a definitive reference by everyone involved with the space surveillance mission in Colorado Springs. It is really a much-expanded version of the kind of document, on essentially the same subject, that Fitzpatrick and Findley [14] produced with Wackernagel's help in 1960.

Late in 1960, General Curtis LeMay, then Vice Chief of Staff of the Air Force, directed that a military cadre be trained to begin space surveillance operations as part of Air Defense Command in Colorado Springs. In that year, missile warning and space surveillance were being added to the command's mission. Twelve military personnel began training at the NSSCC on 7 November 1960, under the tutelage of Wackernagel and several others. The graduates were assigned to a new squadron, the 1st Aerospace Surveillance and Control Squadron, on 6 March 1961. Because Air Defense Command was the American part of the bi-national U.S.–Canadian air defense effort, the space surveillance squadron became operational as part of NORAD on 1 July 1961 at Ent Air Force Base in Colorado Springs, occupying the first two floors of Building P4 Annex, adjacent to the NORAD command center. Known as the Space Defense Center, the facility and its staff functioned mainly as part of the Space Data and Tracking System (SPADATS) for missile warning, which we mentioned in Chapter 1, although the full range of space surveillance applications was added during the following year. Satellite tracking operations continued at the old NSSCC at Hanscom, with that center serving as a backup center for the NORAD squadron. Satellite observations from the sensor sites were sent directly to both centers, and to complete the flow of space surveillance data, the primary center in Colorado Springs would send orbital data bulletins to the sensors to aid their data collections, unless the backup center at Hanscom was being called on to do that job.

The operations center at the old NSSCC at Hanscom supported research and development by the Space Track staff, just as it had always done for Project Space Track. By this time, Hanscom had become the home for another new Air Force organization, Electronic Systems Division of the recently organized Air Force Systems Command (succeeding Air Research and Development Command). Within that division was an office, the 496L Systems Program Office,[43] dedicated to overseeing what was, at the time, called more formally the Space Track Research and Development Facility. The 496L SPO arranged for contract support to this facility and to the NSSCC, devised long-range development plans for both the NSSCC and the 1st Aerospace Surveillance and Control Squadron, and coordinated the ongoing equipment upgrades and software development at both centers.

In April 1966, the Space Defense Center moved from Ent Air Force Base into NORAD's underground station at Cheyenne Mountain on the southern outskirts of Colorado Springs. Here the assembled resources of space surveillance, missile warning, and air defense could be coordinated to form the basis of America's attack warning and assessment in the nuclear age, in a unique subterranean compound that had been designed to survive an ICBM

[43]The cryptic nomenclature "496L" referred to a paragraph number in a government budget document. It was merely a convenient shorthand that allowed easy designation of this one among the many other Systems Program Offices that began to appear in that era.

strike little more than a mile away. The solemnity of the combined mission was not lost on anyone involved and could only have been heightened in the singularly atmospheric setting of the immense man-made 5-acre cavern a full quarter-mile inside “the Mountain,” as the Cheyenne Mountain Complex was known by everyone connected with the mission, with its long tunnels, two-story blast doors ready to close automatically on command at any moment, and more than a dozen buildings mounted on huge springs. Of course, even in such awesome circumstances, there are always a few souls whose best relief includes comic relief. Around the time of the move into the Mountain, or perhaps a few months later, Lawrence Cuthbert, a civil servant from Project Space Track at Hanscom, who, like Wackernagel and others, had joined the operational staff of the 1st Aerospace Surveillance and Control Squadron in Colorado Springs, wrote a short essay satirizing the government’s bureaucracy and inefficiency in handling the space surveillance mission. His essay ended with the following words:

> *And now with Spacetrack operations moving into Cheyenne Mountain, evolution of the System has swung full cycle. It began in the basement and ended in a cave.*[44]

Naturally, the evolution of space surveillance did not end where Cuthbert’s essay left off. However, the Air Force’s command-and-control center for the mission remained in Cheyenne Mountain for the next four decades. In 2007, the Air Force space surveillance center moved from the NORAD facility in Cheyenne Mountain to Vandenberg Air Force Base in California, the result of several interrelated decision processes that we discuss in more detail in a later chapter.

II. EARLY SPACE SURVEILLANCE FOR THE NAVY

The Air Force, along with the SAO, had been explicit from the beginning about developing space surveillance per se, not merely satellite tracking for the IGY, “beginning” usually being reckoned as 1955 for SAO and 1957 for

[44]Cuthbert [21]. Cuthbert had been one of the persons cited by name by Fitzpatrick and Findley [14] for “important contributions” to their report on NSSCC operations at Hanscom. His inimitable humor is still treasured by many of those who have been involved with the space surveillance mission over the years. It is hardly necessary to record that his essay (with photos) was never published officially, despite what his title page says. However, it was photocopied and passed on from hand to hand, until many later *n*th-generational copies were barely legible. Someone reprinted the essay in 1982 expressly in commemoration of the 25 years elapsed since Sputnik 1 and the start of Project Space Track. At this writing, repeated photocopies and scans of even that reprint are getting faint and hard to read. Cuthbert himself was important to the evolution of space surveillance, as described in a later chapter. He made important early developments in what we now call uncorrelated target (UCT) processing, that is, identifying which satellites could (or could not) be associated with observations that, for a variety of possible reasons, do not match any orbit in the satellite catalog. Rather often, such observations indicate that a new satellite has appeared and needs to be added to the catalog. In any case, UCT processing is the most technically difficult part of maintaining a catalog of satellites.

the Air Force. Inherently, the Baker-Nunn systems would detect everything in their field of view which was bright enough to be recorded on the camera film, and so it was not possible to be selective about which satellites had to be accounted for. The fact that the upper stage of the launch vehicle would also remain in orbit, gradually drifting away from the satellite payload, meant that it, too, would often be detected and would have to be tracked. This had to be done to avoid confusion and correctly identify each object every time it was detected. In fact, the inert upper stages of launch vehicles were always much larger, brighter, and easier to detect than the active satellites being used for research, making it difficult to avoid tracking them. Moreover, tracking for proper identification became increasingly important and, as described in later chapters, disproportionately more difficult, as the number of objects in orbit increased. In comparison, the Navy's plan to put a satellite into orbit, and to track it, was tied to the national IGY scientific objectives and executed mainly through its service laboratory, the NRL. Their plan had essentially nothing to do with space surveillance. NRL's design for the radio beacon tracking system for Project Vanguard selectively tracked only those relatively few satellites transmitting on a particular frequency and was blind to any other space object. That design turned out to work extremely well for the IGY programs that it supported, and so well that it was used for other projects by NASA for some years afterward.[45]

What is quite astonishing from our point of view today is that this radio beacon tracking system turned out to be easily converted, without any prior intention or planning to accommodate such an evolution, into the crucially important NSSS, a continent-wide radar interferometer. That system in its final form would consistently track almost 85% of the entire orbiting population and do it continuously for the next half century. NSSS did not need any tracking acquisition aid from any other sensors; rather, it provided acquisition cues for many other Air Force sensors of all types. It began operating in mid-1958, and reached full capability by mid-1965, when there were still only relatively few space objects to monitor. By the end of its service life in 2013, the on-orbit satellite population had grown to more than 16,000 objects, and the Dahlgren headquarters facility was automatically recording and processing about 200,000 NSSS observations per day, besides at least that many more being shared from the Air Force space surveillance sensors. Yet, the direct operating cost of the NSSS radar did not increase as its tracking load increased, because the radar, by design, emitted a fixed amount of unmodulated, continuous-wave energy into space. This feature allowed the direct cost per observation to decrease in almost inverse proportion to the number of observations. Moreover, to cope with the large and rapidly growing number of space objects, the NSSS never needed any basic design change and

[45] Corliss [22], Chapter 2.

never added any stations for extended geographical coverage. The redundancy of coverage among the stations was improved, the altitude coverage was extended several times, and several modifications to the transmitters and antenna arrays for better performance were made over the years. Basically, all that was needed to sustain the system, besides electrical power and routine maintenance, was to upgrade the computation and communication systems periodically to keep up with the growing amount of tracking data and to replace obsolete electronics at the ground stations. Achieving such a comprehensive outcome from such a constrained beginning is a remarkable combination of serendipity and creative engineering that calls for some background explanation.

The seeds of Naval space surveillance were sown as early as March 1955, when the Eisenhower administration quietly began practical preparations for launching a U.S. satellite during the IGY. When that U.S. intention was announced publicly on 29 July 1955, the DoD had already been assigned overall responsibility for the launch program, primarily because the only rocket boosters available then were missiles developed by the military. A technical committee chaired by Homer J. Stewart of the Jet Propulsion Laboratory (at that time, the Jet Propulsion Laboratory was sponsored by the Army), with representatives from the Army, Navy, and Air Force, was convened to consider, in realistic terms, what could be done for launching a satellite within the IGY schedule. In fact, there were only two teams that were within reach of the goal. Although the Air Force had proposed to use a version of the future Atlas ballistic missile, that was a high-risk option at that date because the booster had not had any test flights at all. The idea was shelved immediately.[46] One team, with a much more viable proposal called Project Orbiter, was headed by Wernher von Braun of the Army Ballistic Missile Agency at the Redstone Arsenal near Huntsville, Alabama. His group had been developing several different ballistic missiles that, von Braun realized, could be assembled rather straightforwardly as booster stages of a vehicle to put a satellite into orbit. The other team with a promising proposal was headed by Milton W. ("Milt") Rosen in the Rocket-Sonde Branch of the Atmosphere and Astrophysics Division at NRL in Washington, D.C. Both Rosen and von Braun had previously been involved with the cadre of German rocket engineers (among them von Braun himself) who had been brought to the United States immediately after World War II, along with a collection of parts and materials for about 100 German V-2 rockets. For a few years, the captured V-2s had been assembled, reassembled, reverse engineered, variously modified and launched in test flights at White Sands Missile Range in southern New Mexico. Rosen was interested in using rockets for high-altitude atmospheric research and

[46]In 1962 and 1963, a version of the Atlas missile was used to launch the latter four of the first six American astronauts into orbit for Project Mercury.

other space science and had developed an improved version of the V-2, the Viking sounding rocket,[47] for that purpose. In contrast, von Braun had continued work on military rockets for the Army, at the time mainly on the Redstone family of ballistic missiles,[48] the first large ballistic missiles built in the United States, rather than on scientific research per se. Given this background and the orientation of each team, we can well understand how von Braun estimated that Project Orbiter could put a satellite into orbit within several months, while Rosen, in the Navy's proposal for Project Vanguard,[49] estimated that about two years would be needed to produce and test an upgraded Viking launch vehicle.

Even today, the Stewart Committee's decision between these two proposals is a subject of discussion and some speculation among historians. On 3 August 1955, the committee voted five to two, with one member absent, to recommend to the DoD that Project Vanguard become the American satellite program for the IGY. This decision came despite strong protests from the Project Orbiter team, from a host of other interested parties, and even from Stewart himself. On 9 September, Eisenhower and the Defense Department formally accepted the recommendation, and work started at NRL on 6 October. The overriding consideration for Eisenhower seems to have been to establish the precedent of "freedom of space."[50] This meant emphasizing the peaceful, scientific uses of outer space and deemphasizing, though without ignoring, the potential military uses. This policy was articulated in general terms several years later, not in a formal policy statement but in a 16-page pamphlet written for the general public by the President's Scientific Advisory Committee [25], with a very short introduction signed by Eisenhower. The fact that the Viking booster was being developed explicitly for scientific research for the IGY, while Project Orbiter inherently retained its military ballistic missile aspect, certainly favored Vanguard in this kind of consideration. Moreover, there were other factors in Vanguard's favor. Scientific research with the satellite was a late consideration for von Braun. Since well before World War II, he and other German visionaries had dreamed of putting a satellite into orbit; the specific reasons for doing it in any particular time period were a separate consideration, and perhaps not the main reasons for the rocket engineers themselves. In fact, in the postwar years in the United States, von Braun had been willing to put a

[47]Rosen recounts this effort in detail in his book *The Viking Rocket Story* (Harper and Brothers, New York, 1955), which was published well before the IGY started.

[48]The Redstone booster, in its Jupiter-C variant configuration, had been used for suborbital reentry tests as early as 1955. The Jupiter-C/Juno variant would later launch America's first satellite, Explorer 1, two months ahead of the Navy's Vanguard satellite. In 1961, the Mercury-Redstone variant would launch the first two American astronauts, Shepard followed by Grissom, on their respective suborbital missions for Project Mercury.

[49]Naval Research Laboratory Memorandum Rept. 466 [23]. The Viking-upgrade rocket development schedule is proposed on pp. 4–5.

[50]As discussed at length in Neufeld [24].

small, lightweight, completely inert satellite into orbit,[51] if that ploy would avoid the extra delay of waiting for a scientific payload to be designed and built. We noted near the beginning of Chapter 1 that von Braun had contacted James Van Allen in 1954 about becoming his project scientist and designing a scientific payload for a satellite. Van Allen did not join von Braun's group, although he did design an experiment that eventually flew on the first American satellite, Explorer 1. In any case, those scientific considerations came inconveniently late in the IGY planning period. By comparison, Rosen had spent a decade thinking about research with rockets and satellites, and the Vanguard proposal had much carefully conceived scientific content. Perhaps as importantly, Project Vanguard also had a detailed plan for tracking the satellite payload using a small radio beacon transmitter onboard and a set of high-gain radio receivers on the ground. Accurate tracking was, in fact, crucial for the scientific objectives of the Vanguard program. The Project Orbiter proposal hardly mentioned tracking and assumed that other IGY programs would supply all the observational evidence needed to prove that a satellite, inert or active, was in orbit. In short, even without considering future space policy, the Stewart Committee had a reasonable basis for recommending Project Vanguard for the IGY.

In retrospect, it seems virtually certain that Project Orbiter could have launched a satellite before Sputnik, had it been chosen initially instead of Project Vanguard. In fact, what eventually happened was that Project Vanguard ran into developmental delays with the Viking booster, which, by mid-1957, forced its projected first satellite launch into 1958. The launch of Sputnik 1 was all the excuse that a plurality of government and defense officials, including the new Secretary of Defense Neil McElroy,[52] needed to urge the president himself to give the von Braun team their chance. Eisenhower was quite reluctant at first to approve such a radical change of plans, but, finally, on 7 November, after Sputnik 2 had been launched four days earlier, he gave the go-ahead to the Secretary of Defense. The order to proceed with Project Orbiter reached the Army Ballistic Missile Agency on 8 November, after some additional reluctance in the Army chain of command had been overcome,[53]

---

[51] In 1994, an actual use for very small, inert satellites was demonstrated. The Massachusetts Institute of Technology Lincoln Laboratory conceived and executed a tracking experiment called ODERACS (Orbital Debris Radar Calibration Spheres). Six small, completely inert metallic spheres, having diameters of 15, 10, and 5 cm, were released into low Earth orbit from the space shuttle (STS-60). A second series of six spheres of the same sizes, along with metallic rods (dipoles) of the same lengths, were released from another shuttle the following year. Their purpose was to assess how well certain special ground-based telescopes and radars could track extremely small objects such as orbital debris. At the time, very few regular space surveillance sensors, either optical or radar, had any significant likelihood of detecting objects that small, although the issue of untrackable, yet potentially lethal, man-made orbital debris was becoming more important for a variety of reasons.

[52] Hayes [7], Chapter 2, in discussing Project Vanguard, and Ludwig [26], in discussing the rush to launch Explorer 1.

[53] Stuhlinger and Ordway [27], Chapter 6, p. 122. A fuller account, with much more information about Eisenhower's reasoning and intentions, is given by Neufeld [28], Chapter 13, pp. 311ff.

and von Braun's group feverishly set to work. As if to highlight just how flummoxed the American response was at the time, in the same post-Sputnik flurry, a Vanguard booster with a real payload was unwisely rushed to the launch pad at Cape Canaveral to try to answer the Soviet feat. It exploded on the pad in a very public failure[54] (as Rosen had feared) on 6 December 1957. Meanwhile, von Braun's team continued their hectic, round-the-clock preparations, and the Army launched Explorer 1 successfully, less than two months later, on 31 January 1958.[55] Of course, from our retrospective vantage point, we can also see that the Sputniks, whenever they happened to be launched, would have immediately opened the entire problem of space surveillance for the Americans. The subsequent development of space surveillance tracking probably would have been about the same, regardless of which nation turned out to be first into orbit.

The Vanguard radio beacon tracking method was designed by Roger Lee Easton in the Rocket-Sonde Branch of the Atmosphere and Astrophysics Division, in collaboration with Rosen, the Viking program manager in that branch, and John Thomas ("Jack") Mengel, head of the Tracking and Guidance Branch, in the same division at NRL. These three men represented the intellectual core of Project Vanguard, with Rosen becoming the Technical Director of the project. The backstory of why they chose a radio-tracking method for Vanguard, instead of relying on the substantial development of optical tracking methods going on simultaneously at SAO, reveals important

[54]Remarkably, the satellite payload on that infamous Test Vehicle 3 (TV-3) launch failure, America's first attempt to launch a satellite, was recovered by the NRL engineer, Martin Votaw, who had built the tracking beacon transmitter. Votaw noticed that the tracking signal was still present despite the booster explosion. After the fire had been extinguished, Votaw found the 6-in.-diam satellite lying in some bushes among other wreckage near the launch pad, damaged beyond reuse, but mostly intact mechanically and still transmitting its radio beacon signal. He carried it in a brown cardboard box to Roger Easton, the NRL engineer who had designed the Project Vanguard tracking system (Minitrack). In a quandary about what to do with it, Easton hand-carried the damaged satellite in the same box on an overnight commercial flight home to Washington, D.C., left it on his kitchen table, and went to bed. Later, Easton gave the recovered satellite to John Hagen, Director of Project Vanguard. That recovered payload is essentially identical with the Vanguard 1 satellite that was launched successfully several months later. It can be seen by the public today because Hagen eventually donated it, bent antennas and all, to the Smithsonian's National Air and Space Museum in Washington, D.C. The story is told in highly engaging style by Roger Easton's son in Easton and Frazier [29], Chapter 1, pp. 7–8.

[55]Accounts of the launch of Explorer 1 routinely neglect to mention that the lack of attention to tracking in the original Project Orbiter proposal had been remedied by the time the satellite was actually launched. Explorer 1 carried two radio beacon transmitters using the 108-MHz IGY standard so that it could be tracked by the radio receiver system (Minitrack) developed by Naval Research Laboratory and discussed later in this chapter. The Minitrack ground receivers had already been designed and built by Rosen's group for Vanguard, before Sputnik 1 was launched. Hence, the difficulties implicit in Project Orbiter's original proposal for tracking never materialized. We should mention that Explorer 1 was also tracked by a Doppler-based radio-beacon system developed by the Army at the Jet Propulsion Laboratory and Aberdeen Proving Ground in Maryland, a system known as Microlock, which happened to be somewhat less accurate than Minitrack at that time. Later, after NASA had taken over operation of the Jet Propulsion Laboratory from the Army, the Microlock development was canceled in favor of Minitrack as their primary method of tracking payloads in Earth orbit. However, Doppler-based tracking became quite important in the following years as NASA began to pursue interplanetary satellite missions, and so the Microlock experience and legacy were not lost. See Corliss [22], Chapter 2, pp. 33–35.

aspects of the later development of space surveillance, even though space surveillance was not an issue at the time for NRL.

Both Mengel and Rosen were fully aware of SAO's optical tracking project, but both had doubts about whether optical sensors would be adequate for IGY applications, especially for Vanguard. At Rosen's request, another NRL colleague, Richard Tousey, head of the Instruments Branch, scrutinized the analysis done by Whipple's team. Tousey convinced the NRL group that the Baker-Nunn cameras could indeed detect small satellites, including the Vanguard series, but if, and only if, they could acquire the satellites in the first place. A small, informal NRL consensus put the probability of the Baker–Nunns acquiring the Vanguards (or any other satellite never observed before) soon after launch, without any other aid, at about one in a million.[56] Moreover, SAO's approach of using Moonwatch teams of amateur astronomers to acquire the satellites for optical tracking seemed implausible, or unreliable at best, to the NRL analysts.

Mengel's branch had other concerns about optical tracking besides the difficulties of acquisition. Optical telescopes in general, and the Baker–Nunns in particular, were useless for tracking faint space objects in the daytime sky.[57] Moreover, they could not detect a satellite if it were not illuminated by the Sun, that is, if it were in the shadow of the Earth. For low Earth orbits, almost half of the orbital path could be in the shadow, depending on the orientation of the orbit in space, although the fraction of the orbit in sunlight increased as the mean altitude of the orbit increased. Hence, the opportunities for optical tracking at any one station were limited to those periods of time when the station was not in sunlight but the satellite was. Consequently, for most low Earth orbits, optical tracking (then and, except for a few specialized instruments, even now) turned out to be possible only for about two hours before local sunrise at a station and about two hours after local sunset. These constraints were known generally as "terminator conditions," meaning that low satellites could be tracked optically only near the terminator line separating the daytime and nighttime hemispheres of the Earth. Perhaps more obviously, the success of optical tracking also depended critically on the lack of cloud cover when the satellite was otherwise visible to the station. Naturally, these lighting and weather constraints also applied to the acquisition phase of optical tracking, if that phase itself depended on optical methods, as Project Moonwatch did.

---

[56]Green and Lomask [11], Chapter 9, pp. 145–147, give a fuller account of these deliberations.

[57]It would be another 50 years before researchers at Air Force Research Laboratory observatories at Kirtland Air Force Base, New Mexico, and Maui, Hawaii, could demonstrate how to track some of the brighter Earth satellites in the daytime with existing large-aperture telescopes, and it took longer yet before such tracking could be done routinely. Tracking a satellite optically when it was in the Earth's shadow became possible several years before that by using new cameras and detectors sensitive to infrared wavelengths, but again, making it a routine capability required much extra time.

These problems led to other problems. For one example, if all these limiting constraints for optical tracking had to apply to the Vanguard program, then the choice of launch times would also have to be highly limited to deploy the satellite when the optical sensors would have a reasonably good chance of acquiring it during the first few passes over tracking stations after launch. Probably, only stations near the terminator line would be able to see it. Given, then, a possible delay measured in days before the satellite could even be confirmed as actually in orbit and given the large number of poorly controllable factors involved in launching rockets at all in those days, tight launch time constraints would be unworkable.

At a conference presentation in March 1956, Mengel [30] described the difficulty of the tracking problem for Project Vanguard, a difficulty that had occupied the team since early 1955:

> *The final realization of man's efforts to place a satellite in orbit about the earth will immediately pose a new series of problems: how to determine the precise orbit that it is following; and how to measure what is happening within the satellite from the vantage point of a ground station. The immensity of the first of these programs, how to prove that the satellite is in fact orbiting—the acquisition phase—can be realized by . . . an analogy. . . . Let a jet plane pass overhead at 60,000 feet at the speed of sound, let the pilot eject a golf ball, and now let the plane vanish. The apparent size and speed of this golf ball will closely approximate the* [apparent] *size and speed of a satellite 3 feet in diameter, at a height of 300 miles. . . . The acquisition problem is to locate the object under these conditions, and the tracking problem is to measure its angular position and angular rate with sufficient accuracy to alert non-acquiring tracking stations, and those trying to follow the satellite by optical means, as to the time and position of expected passage of the object.*

Hence, Rosen had plenty of rationale for asking Roger Easton to design a radio-tracking system for the Vanguard satellites, because radio techniques inherently had no lighting or weather constraints. Some radio systems allowed the possibility of very wide fields of view, which would neatly solve the acquisition problem. Moreover, Rosen suspected that the design job might not have to be done from scratch. For several years previously, he had been involved with developing a new guidance system for the Viking rocket. The new system used a ground-based radio interferometer to track a transmitter on board the rocket.[58] Easton found that, indeed, that design could be rescaled to

[58] Corliss [22], Chapter 2, p. 12. In the same place, Corliss, who personally interviewed Rosen in 1966 about this subject, explains that in the early 1950s, the Navy was interested in developing a ballistic missile from Rosen's Viking rocket design, but that they needed a new guidance system for the prototyping. Such a guidance system was designed and built by the Convair Division of General Dynamics Corporation. The contractors' approach was to use a radio interferometer consisting of several ground receiver antenna arrays to track a very small, low-power transmitter carried on the rocket, to estimate the position, and then to

use the IGY-recommended standard of 108 MHz in the VHF band between 30 and 300 MHz. Naturally, he had to be involved with designing both the radio transmitter that was to fly on the satellite and the radio receivers at the ground stations, so that Easton became a key member of the Vanguard team and helped write the NRL proposal for Project Vanguard, which was accepted in August 1955. The tracking system was known as Minitrack,[59] a name invented informally by Mengel as engineering shorthand for "use of an oscillator of minimum size and weight . . . to illuminate pairs of antennas at a ground station which measures the angular positions of the satellites using phase-comparison techniques."[60] In other words, Minitrack was a radio interferometer system. The direction-finding technique of comparing the phases of an incoming wave train measured at different points in a field of receiver antennas is identical in fundamental principle with what goes on unconsciously in the human brain to allow persons to tell the direction of a source of sound using two ears. Getting a wide field of view for the interferometer antenna array was not difficult. Even the early Minitrack receiver prototypes could detect signals over fields of view of 100 deg or more, and later versions could detect signals anywhere above about 5 deg of elevation angle. By recording the radio beacon phase data continuously, Minitrack could provide a sequence of accurate directions from the station to the moving satellite, during the entire time the satellite was in the field of view.

---

transmit that position to the rocket where steering commands would be generated automatically to keep the rocket close to a preestablished trajectory. The tracking system was named Azusa by its main design engineer Robert C. Weaver, who lived in southern California near the town of that name. Although the Navy's project finally did not produce a ballistic missile, the Azusa system continued to be improved and eventually found extensive use by the Air Force's Atlantic Missile Range at Cape Canaveral, and at many other places. By an unlikely coincidence, Weaver and his colleagues at Convair had long been in regular contact with a group in the Sound Division at NRL, including H. L. Saxton who was analyzing the use of sound interferometry to do direction-finding underwater for objects reflecting waves from a sonar transmitter. According to Rosen, many ideas were exchanged between these two groups in the critical period of the development of these two otherwise unrelated systems in different parts of the country. Later, in a journal paper giving the first open-source description of the Naval Space Surveillance System, Easton and Fleming (1960) [31] would mention the Azusa tracking system and the work at Convair Division but would credit Saxton at NRL for developing the interferometric techniques used in the Minitrack receivers and soon after in the Naval Space Surveillance System.

[59]In later years, Easton always insisted that he had not invented Minitrack, although he was often credited with doing so. He maintained that the idea and some hardware had already been invented and that all he did was to redesign a version of it for Vanguard. In any case, he had to admit to inventing two major systems that were inspired by the Minitrack experience, namely, the Naval Space Surveillance System (NSSS) and a series of navigation satellites leading to the Global Positioning System (GPS). Easton had joined NRL right out of undergraduate school in late 1943. He first got involved in space-related work in 1952 in the Rocket-Sonde Branch, helping with postwar rocket experiments at White Sands Missile Range, followed by Project Vanguard at NRL headquarters in Washington, D.C. From 1958 until his retirement in 1980, he was the head of the Space Applications Branch at NRL. His illustrious career reached a notable high point on 13 February 2006, when President Bush presented him with the National Medal of Technology. A few years later, in 2010, at age 89, Easton was inducted into the National Inventor's Hall of Fame.

[60]Mengel [30]. See also Mengel and Herget [32], pp. 23–29, and Corliss [22], Chapter 1, pp. 6–33, for complete discussions of the Minitrack concept in its original context.

As Mengel implied in his 1956 presentation [30], Minitrack turned out to provide tracking acquisition aid for many optical sensors if the satellite was transmitting a signal close to 108 MHz. In fact, the Minitrack receiver stations were available before the Baker-Nunn cameras were ready to be deployed. However, Whipple decided not to colocate Baker-Nunn cameras at most of the Minitrack stations, even though doing so would have eased some logistic problems for the tracking programs at both SAO and NRL and might have allowed more efficient coordination between the two sensor networks. For example, the Minitrack raw signal data could be rapidly processed at the NRL Vanguard computing center, and the resulting direction data, at least in principle, could be transmitted back to the station to provide pointing cues for a colocated Baker-Nunn. Whipple's difficulty was that most of the Minitrack stations had frequent cloud cover at certain times of the year. As it turned out, only the station in the arid region near Woomera, Australia, ended up with both radio and optical sensors during the IGY, and, even there, tight coordination between the local radio and optical sensors was not implemented.

Exactly where to put the Minitrack receiver stations for best support of the IGY programs was a complex question that challenged Mengel and his colleagues. The project budget limited the number of stations for combined tracking and telemetry at 108 MHz to about 10, perhaps 12. Blossom Point, Maryland, about 40 miles from NRL in Washington, D.C., became the prototype station where the electronics and antennas were developed and tested. Stations on the East and West Coasts of the United States were a natural choice, and the stations in South Africa and Australia, which were needed for more global coverage, were expedited by existing agreements with those governments. However, what Mengel also wanted was a "picket line" (his own words) extending well into both the Northern and Southern Hemispheres along some line of longitude to detect and track satellite passes reliably on each orbital revolution. That goal inevitably involved new negotiations with several national governments. The most affordable idea was to extend the line suggested by the two East Coast stations in the United States, through the Caribbean and down the west coast of South America, roughly along the longitude 75° west line. The U.S. Army Corps of Engineers offered Mengel a list of possible locations that they deemed workable for his purpose. At that stage of the deliberations, a remarkable process ensued, which nowadays we might expect to take, at the very least, a few years to complete, but which at the time was handled in less than a year by the Department of State and the DoD working together. Perhaps the efficiency was due simply to the enhanced national and international cooperation surrounding the IGY. Whatever the reasons, the site surveys and foreign negotiations were completed without any serious delays. Except for three stations, the Army Corps of Engineers handled the site preparations, transport, and construction, aided by every branch of the military as opportunities and needs for assistance arose. Meanwhile, NRL took direct responsibility for the three stations at Blossom Point, San Diego, and, with contractor help from Radio Corporation of America, Antigua. The plans for all the prime Minitrack sites shown in Table 2.3, including communications and logistics, were finalized by January 1957. By May, an ongoing analysis at NRL had shown that one of the proposed sites, located in Panama (Río Hata) and not shown in Table 2.3, could be omitted without compromising the overall tracking coverage, so that plans for that station were scrapped. The Minitrack network was completed by 1 October 1957, except for some final installations at the Woomera station. In fact, when Sputnik 1 was launched on 4 October 1957, all the Minitrack stations except Woomera were in final checkout of their communications and electronic calibration procedures. Mengel summarized in a paper for the public about two months later, not only the completed Minitrack network, but also the following specification of his picket line[61]

> *All the Vanguard satellites will be fired from Cape Canaveral, Fla. They are to be shot into an east–west course near the equatorial region, and the angle at which they can be fired is limited to within 35 degrees of the Equator because of the presence of the Bahama Islands offshore. The picket line of Minitrack stations across their expected path is strung out in a north–south line up the west coast of South America and the east coast of North America . . . Seven stations form this picket line, and there are four others scattered over the world, in San Diego, the British West Indies, South Africa and Australia. With this arrangement we have a 90 per cent chance of intercepting every pass of a satellite which is higher than 300 miles.*

TABLE 2.3 MINITRACK STATIONS (OCTOBER 1957)[a]

Country	Location
United States	Blossom Point, Maryland
United States	Fort Stewart, Georgia
United States	San Diego (Navy Electronics Laboratory), California
British West Indies	Antigua (Coolidge Field)
Cuba	Havana (Batista Field)
Ecuador	Quito (Páramo de Cotopaxi/Mt. Cotopaxi)
Peru	Lima (Pampa de Ancón)
Chile	Antofagasta (Salar del Carmen)
Chile	Santiago (Peldehune Military Reservation)
South Africa	Johannesburg (Esselen Park)
Australia	Woomera, South Australia[b]

[a]Adapted from the narrative in Green and Lomask [11], Chapter 9; the map in Mengel and Herget [32]; and private communication with Robin Smith of NRL [33].
[b] Woomera station was fully functional in November 1957.

[61]Mengel and Herget [32], pp. 27–28.

Mengel's description explains also why the Minitrack receiver beams were designed to be wide in the north–south direction and narrow in the east–west direction. Although, in principle, the receivers could have been designed to cover the whole sky at each station, the extended north–south beamwidth at all the stations, not just at the picket-line stations, allowed higher probability of detecting a satellite moving in a generally west-to-east direction, given that only a limited number of stations were built. The same consideration would arise later in 1958 when the Navy Space Surveillance System would be designed using the Minitrack experience as a guide. That later system would end up using broader east–west coverage and narrower north–south coverage at the receivers, for reasons that we discuss later.

It is interesting to ponder what the Minitrack designer Roger Easton and his colleagues, including Mengel, eventually came to think of SAO's strategy of having amateur astronomers help their IGY optical tracking program via Project Moonwatch. The Moonwatch groups had used a type of picket line to improve their chances of seeing newly launched satellites, although not involving an alignment of different stations. Rather, the observers at a station would work together to cover a long strip of the night sky, with each member assigned to watch one small patch of sky within the strip. Although the NRL staff remained highly skeptical of Moonwatch as a viable approach to optical tracking acquisition in the long term, as noted earlier, Easton himself seems to have concluded that the idea of proliferating tracking sites in a manner like Moonwatch had done must have some merit after all. While he was engaged in designing and building the Minitrack network, he also designed a simplified receiver and antenna system. The prime Minitrack stations listed in Table 2.3 were designed both to track the 108-MHz beacon and to receive telemetry from the satellite about conditions both in the local space environment and on board the satellite itself. However, the tracking function alone was much simpler to implement than the dual functions of the prime stations. Easton called his simplified tracking system Minitrack Mark II, and he believed that it could be built affordably by amateur radio clubs.[62] He even offered a graduated series of three different simplified designs that would fit the ambitions, capabilities, and budgets of different amateur radio groups. For a short time, in tacit acknowledgement of SAO's successes with Moonwatch, NRL officially sponsored a program known as Project Moonbeam that was intended to contribute amateur Minitrack Mark II observations to the NRL Vanguard computing center. However, the effort was short-lived; only a few amateurs ever undertook to build Moonbeam stations. Several of Easton's

[62]Easton [34] is the sole author of this report, although he acknowledges, by name, contributions from 18 different staff members of NRL, including Mengel and Martin Votaw, and from eight more individuals outside NRL. It may be that no one was interested in being coauthor with Easton on this report.

Mark II systems would find use as government-operated auxiliary stations for the prime Minitrack network for the Vanguard program, although they would never be part of the Naval Space Surveillance System that was about to be born from the Minitrack concept. That distinction would belong to the San Diego and Fort Stewart prime Minitrack stations listed in Table 2.3.

1. MINITRACK TRIES TO COPE WITH SPUTNIK

Mengel was candid even in his public statements about the "disorder" (his own word) introduced into IGY plans and into the Navy's plans for tracking IGY satellites, by the launches of the Sputnik series. A sudden and very urgent national-level priority for NRL became to determine what they could do to detect and track Sputnik 1 before its batteries gave out, probably, it was thought, in two or three weeks (three, as it turned out). In the first place, the Minitrack receivers already deployed around the world were blind to the 40- and 20-MHz signals from the Sputnik series. Fortunately, the detection problem could be remedied by simply shipping new receiver electronics packages to the stations and connecting that equipment in parallel with, or in place of, the existing equipment. The Army did that within a few days. However, in the second place, the entire array of antennas for any radio interferometer must be designed for a specific frequency band, in the case of Minitrack, a narrow 10-KHz band centered on 108 MHz. Although the existing antenna layout would allow the stations to detect and record the Sputnik signals during the satellite passes, the configurations of the antennas and their spacings from each other were wrong for determining the direction to the satellite. Until the antenna field could be augmented with additional antennas designed and properly spaced for 40 MHz and 20 MHz, no tracking could be done, and the orbit could not be estimated from Minitrack data. This antenna-augmentation task was not a simple one: the longest prime Minitrack antenna spacings were up to 500 ft, and spacings of more than twice that distance would be needed for the lower frequencies to get the same angular measurement accuracy. Moreover, the antennas had to be accurately aligned with each other and leveled relative to each other across the entire array to within a small fraction of the wavelength. The locations and spacings among all the antennas also had to be accurate to within essentially the same small fraction of a wavelength. The installation and testing jobs had to be done without disturbing the 108-MHz antenna array or its coaxial cable connections and electronics. In a few days of intense work, NRL was able to design and provide new antennas tailored for the stations at Blossom Point, San Diego, and Lima, and the Army Corps of Engineers installed them. Those three stations were able to track Sputnik 1 for several days before its beacon transmitter went silent on 26 October. They also began tracking Sputnik 2 as soon as it was launched on 3 November.[63]

The disorder to which Mengel referred included more than just the initial disruptions of IGY planning. An obvious question raised by NRL's Sputnik tracking experience was as follows: How would they handle the case of a satellite that was not transmitting any kind of signal at all? Naturally, SAO's and Project Space Track's Baker-Nunn cameras could detect and, presumably, track a satellite whether it was transmitting or not. However, most of the NRL staff believed that, besides the optical acquisition problem, there were, and probably always would be, too few optical trackers placed around the world to be able to observe any particular satellite frequently, ideally on every orbital revolution, given the constraints of weather and terminator lighting conditions at each station. Minitrack had been designed to do that every-revolution tracking with high probability, though only for satellites that transmitted a radio signal on a specific frequency, per IGY standards. Yet, even for IGY applications, it was easy to imagine researchers devising a scheme for shutting off a beacon transmitter for periods of time to conserve the battery power until the satellite needed to be tracked accurately at some later time. It happened that Vanguard 1 was the first satellite to use solar cells to maintain its battery charge over a long period of time, and the Vanguard series took advantage of that same innovation to avoid the limitations of single-use batteries. Many more satellites would use solar cells in years to come after the IGY. However, many other satellites could be expected to use lower-cost, single-use batteries for short-term applications, just as the Sputnik series had. Moreover, the military concerns about satellite reconnaissance, false alarms in the missile-warning systems, and correctly identifying every satellite every time it was detected remained essentially the same, even for satellites that did not radiate any signals. Hence, although the question was outside the concerns of the IGY programs, it persisted in many minds at NRL.

2. Calibration Leads to a New Invention: The Naval Space Surveillance System

Meanwhile, Easton had been working on methods to calibrate the Minitrack receiver antenna arrays.[64] In engineering practice, this subject can get quite complicated, even though the whole point of it is merely to make sure that the reported tracking data were always consistent with accurately known external reference data sources. If that consistency could be assured over time, the tracking data could be used reliably for estimating orbits. To be sure, this subject would not divert us at all in our story, except that Easton's calibration

[63] Mengel and Herget [32], p. 28.

[64] Most of the following information is gleaned from the NRL report by Easton [32], pp. 25–29, and the later journal paper by Easton and Fleming [31], pp. 664–665.

work uncovered the clue for converting the Minitrack design into a true space surveillance system that did not depend on signals transmitted by satellites.

The specific calibration problem for Minitrack that Easton faced was to make sure that the direction angles to satellites, as calculated by comparing signal phases measured on different antennas at a station, could be correctly converted to the usual azimuth and elevation angles (or other conventional direction angles) used to compute the satellite orbits. Naturally, in the ideal situation, no calibration problem existed: the antenna arrays as built would be perfectly aligned with the local geographic directions, the antenna elements would be aligned perfectly with respect to each other, the whole array would be exactly planar and level with respect to the local horizontal plane, and none of the electrical connections or components anywhere in the system would introduce phase errors. In practice, every actual array would have small misalignments and electrical imperfections of these types, with some of the effects subject to small but complicated variations according to air temperature and time of day at the site, and some of them remaining more or less constant. All these systematic errors would have to be discovered, monitored over time, and accounted for whenever the reported tracking data were used for orbit calculations.

Of course, Easton already had a good understanding of the calibration scheme used by SAO for their Baker-Nunn optical trackers. In the optical case, the pointing-calibration problem was to make sure that the telescope mount, its rotational axes, and the internal alignment of the optical elements could produce measured directions that always compared consistently with the reference directions provided by the stars in the night sky. Often, the apparent position of the satellite as recorded on photographic film could be measured directly against the well-known positions of stars recorded on the same film, in which case mechanical pointing-calibration was barely an issue at all. The night sky provided a wealth, even a surfeit, of precisely known optical pointing-calibration reference points, provided only that the star images on the film could be identified correctly and referred to a star catalog. The difficulty in Minitrack calibration was that there are very few natural radio sources in the sky that 1) have accurately known directions and 2) could be detected by the Minitrack system. The Sun happens to be a fairly strong radio source at 108 MHz. However, the physics of the Sun, especially its differential rotation and highly active fluid surface, meant that the center of radio emission was known a priori only within the half-degree apparent diameter of the solar disk and could not be used as an accurate, stable reference source over time. A few radio stars were known, such as Cassiopeia A and Cygnus A, which might have been used, but the detectable ones were too few to consider for routine calibration. Easton had tried attaching a small flashing light bulb to a small, low-power continuous-wave beacon transmitter designed for use on IGY satellites and having this apparatus towed at altitude behind an airplane at night. As the beacon passed over the receiver array, the Minitrack-derived direction to the transmitter would be reckoned as a function of time, and a camera in the center of the array would photograph the flashing bulb repeatedly as it moved against the star background. Every photograph compared with the measured Minitrack direction at that time provided a direction-calibration check.[65] Needless to say, this cumbersome method could not be used as often as it was really needed, even though it was, by default, the main receiver calibration method for all the Minitrack stations until the end of 1957. The advantage of the method was that, under the best conditions, it could deliver calibrations at the few-arc-second level.

Easton's improved calibration solution was to use the Moon as a reference direction, illuminated by a high-powered radar transmitter on Earth. By this means, most of the Minitrack stations in the entire Moon-facing hemisphere of the Earth would be able to receive the reflected signal at essentially the same time. This fact greatly simplified the logistics and scheduling of the calibration function at the stations. Although the Moon was, like the Sun, about half a degree wide in the sky at any station, it was a solid body with an inactive surface, which always presented the same face toward the Earth, reducing the variability of the center of reflection. Like the Sun's apparent position, the Moon's apparent position (center of its apparent disk) at any time at any station was accurately known far in advance. To test his idea, Easton arranged a joint effort between the Blossom Point Minitrack station and the Army Signal Electronics Laboratory (SEL) 250 miles away in Fort Monmouth, New Jersey. A radar with a 50-ft dish antenna, rated at 50 KW and transmitting at 151 MHz, was available at the Army's facility, while Blossom Point adapted a developmental Minitrack receiver system to the 151-MHz frequency. Tests began in May 1957. Although the large, half-degree apparent diameter of the Moon was a limiting factor in the accuracy of the calibration reference direction, simply reckoning the reflection at the center of the apparent disk turned out to be good enough to keep the Minitrack accuracy within the 1-arc-

---

[65] Hynek's memo to his SAO optical tracking staff on the first anniversary of Sputnik 1's launch, recorded in Appendix A, describes essentially this scheme (without the transmitter) of a towed light bulb of one-tenth candlepower being used to train the Moonwatch crews to find dim satellites moving among the stars. He remarked that once Sputnik was up, the airplane was not needed: the crews could practice with the real thing! A more subtle problem with the towed direction-reference beacon for Minitrack was that the airplane could not always get high enough above the array to simulate a realistic satellite pass. Rather, the transmitter was often in the so-called near field of the array antenna pattern, which introduced into the measured data a variety of effects that were hard to compensate. Easton found that the aircraft would have to be at least 15,000 ft above the Minitrack array to reduce the near-field effects enough to neglect. Later, when the Navy Space Surveillance System was built, based on the Minitrack design, the receiver antennas had to be much longer than the Minitrack antennas to have the higher gain necessary for that application. However, this meant also that the near field of a receiver array now extended much farther up. In turn, the required altitude of the towed calibration transmitter and flashing light bulb became impractically high, 50,000 ft or more. A completely different approach to calibration had to be adopted for the Naval Space Surveillance System, as we describe later.

minute specification desired by the IGY programs, at least on average.[66] By December 1957, the Moon-reflection technique had been made reliable enough that NRL and the Army set up an ongoing program for the SEL radar to transmit on the Minitrack frequency of 108 MHz. The calibration immediately became part of the regular operation of all the Minitrack stations.

Sometime in early 1958, Easton and the Army engineers decided to turn the radar dish toward Sputnik 2 as it transited over the east coast of the United States. The satellite's batteries had been depleted rather soon after its launch the previous November; however, the satellite would not reenter the atmosphere until 14 April 1958, making it a prime example of a radio-silent, payload-sized space object. The Blossom Point test receiver picked up the reflected radar signal and tracked the satellite direction rather easily. It must have been a memorable moment for Easton. Multistatic radar, in which the transmitter and receiver are widely separated, had been conceived long before the famous radar developments during the 1930s and World War II, and the concept was already well understood. The key realization in Easton's case was that the job of designing a system to track "dark" satellites, that is, those not transmitting any signal, was already more than half done. All that was needed for the existing Minitrack receiver design to be able to detect and track them was a suitable source of radio energy illuminating the satellite while it was visible to the receiver. Imitating the Minitrack beacons, the transmitters could be the simple, unmodulated, continuous-wave type, which were relatively economical to design, build, operate, and maintain. They were, in fact, easier to design than the receivers had been. Easton's colleague Martin Votaw would later recall thinking that this emerging space surveillance concept was just "Minitrack with the transmitter on the ground rather than in the satellite."[67] The reflected signal arriving at the receiver would always be somewhat weak, mainly because the total path transmitter-to-satellite-to-receiver was long, and satellites were both small and imperfect radio reflectors. However, this effect was easy to offset by making the receiver antennas longer, giving them higher gain. A detailed proposal endorsed by NRL followed within weeks and was sent to the newly established ARPA of the DoD.[68]

---

[66] Easton [34] had stated in September 1957, without publishing details, that the method could deliver calibrations at the 20-arc-second level. However, it is difficult, even in retrospect, to know exactly how accurate the Moon-referenced pointing-calibration was for Minitrack. It was acknowledged from its beginning to be less accurate than the towed-beacon method, but it proved decisively more practical and affordable. The Moon-referenced calibration was not needed for long, in any case, as discussed later. By the time the Navy Space Surveillance System was in the first phase of its operation at the end of July 1958, the preferred calibration reference objects had become a few well-tracked Earth satellites whose orbits were known especially accurately. This approach to calibrating space surveillance sensors for pointing and ranging is still used today, as noted later in this chapter.

[67] From an interview with Votaw in 2009, reported by Easton and Frazier [29], Chapter 3, p. 47.

[68] ARPA had been established as a response to the defense implications of Sputnik 1, by Secretary of Defense McElroy on 7 February 1958. In 1972, the agency was renamed Defense Advanced Research Projects Agency (DARPA). According to the DARPA website, McElroy specifically listed three priorities for ARPA: space technology, ballistic missile defense, and solid rocket propellants. From that beginning,

On 29 June 1958, the NSSS was born, known initially and variously as the Active Minitrack System and the Dark Satellite Detection System. Following the NRL's proposal exactly, ARPA issued on this date Order No. 7-58 (one of their earliest ones), with funding, for NRL to build two tracking complexes, one in the eastern United States and one in the western United States. Each complex would consist of one transmitter station between two receiver stations. All the stations would be located on a common great-circle arc across the southern United States, with large distances, usually several hundred miles, between adjacent stations. This was really Mengel's picket-line idea, now adapted to a generally east–west line of stations intended to provide tracking coverage of all reconnaissance satellites passing over the continental United States, and, specifically, tracking coverage of all satellites that could be launched from inside the Soviet Union. The beam patterns from the transmitters and receivers would be confined very close to the plane of the great circle, forming a thin vertical "fan" pattern rather than covering the whole sky above each station, to maximize the altitude-reach, given a total transmitter power available and the required east–west coverage of the surveillance system. On this approach, a satellite would transit through the "Fence" (as the system's coverage was called) quickly, usually in several seconds or less for LEO satellites. However, that was enough time for each of the receiver stations that detected it to get an accurate measure of direction. If it was necessary to estimate the position of the satellite at the observation times, then the lines of sight from two or more stations, which, by design, necessarily lay in or very near the great-circle plane and were almost simultaneous in time, could be combined in a straightforward triangulation method.[69]

The first two sites included a receiver station at Fort Stewart, Georgia, one of the passive Minitrack sites for the Vanguard program, and a transmitter site at Jordan Lake, Alabama. Both sites came online on 29 July 1958, barely a month after the ARPA order was issued. The existing antennas at Fort Stewart were physically rearranged from north–south to east–west coverage to suit the new Fence design, and a completely new antenna array had to be built for the transmitter site at Jordan Lake. Easton remarked in his 1960 journal paper [31], conveying the urgency felt at the time, that the only way to meet that accelerated schedule was to move the Army's 50-KW radar transmitter electronics from Fort Monmouth to the Jordan Lake site to feed the new antennas there. In any case, with that equipment in place, the system began tracking satellites immediately. The stream of tracking data in the form of

---

the scope and diversity of ARPA's mission expanded rapidly. Today DARPA manages a large and constantly evolving portfolio of projects covering the whole range of defense concerns.

[69]This is the reason why, in writing, Easton always described this surveillance system as a "radal" system, short for "radio direction and location," rather than as a "radar" system, short for "radio direction and ranging", in which distance is measured directly. Strictly speaking, he was correct, although today that more specialized terminology has fallen out of use, and "radar" is used as a general term.

phase measurements from the individual receiver antennas was sent in continuous analog form over standard telephone lines to the developmental computing center at NRL. The same scheme was eventually used to report analog data from all the receivers.[70]

That processing facility at NRL had been established for the Vanguard program and still served as a backup for the main operational Vanguard processing center operated by IBM contractors in downtown Washington, D.C. Being a research facility, the NRL center operated on a regular 40-hour work week, and so data from the NSSS receiver sites were not being taken or processed continuously. However, already in early August 1958, data from the Fort Stewart receiver were being used to build and maintain an NRL catalog of satellites visible to the new system in a manner essentially the same as that used in the fully operational NSSS implemented over the next three years. Soon the NRL center was being called the Space Surveillance Control Center by laboratory staff, and the regular operations involving incoming Fence data and catalog updates had to be segregated from developmental activities for Vanguard still being pursued in a separate partition of the processing system.

The Eastern Complex, as it was called, was completed with the activation of a new receiver site at Silver Lake, Mississippi, at the end of November. In the west, the San Diego station, modified from its Minitrack configuration, was completed in mid-December 1958. The Gila River transmitter station and the Elephant Butte receiver station were completed at the end of January 1959, allowing the Western Complex to come online in early February 1959. At that point, the Eastern and Western Complexes and the processing center at NRL began operating 24 hours per day. With more data to process, NRL needed more computing power. Fortuitously, and as discussed in more detail later, by the time the Western Complex was nearing completion, the base commander at the Naval Proving Ground in Dahlgren, Virginia, learned of NRL's space surveillance efforts and offered them the use of a very powerful computer that had been installed at Dahlgren for ballistics research. NRL gladly accepted the offer and immediately began planning for the primary computation center for the NSSS to be at the Proving Ground, with the NRL center serving as a backup facility.

NRL continued to operate the system under its ARPA order and to maintain a catalog of satellites detectable by the system, until, in October 1960, the DoD transferred responsibility for operating the NSSS from ARPA to the Navy's Bureau of Naval Weapons. However, NRL retained responsibility for further development and improvement of the system. The bureau, in turn, followed through on NRL's intention to set up the primary computation facility on the Virginia base by establishing a combined operations/

[70]By about 1990, as part of a system upgrade, the analog data system was replaced with a digital system that sent discrete phase data, down-sampled in signal-processing stages from 75 KHz at the antenna feeds to 55 Hz (later 72 Hz) for transmission, over the same standard telephone lines.

computation center for the NSSS at the former Naval Proving Ground in Dahlgren.[71] Shortly after, on 1 February 1961, the Navy established the Naval Space Surveillance System (later, Center) as an operational command with headquarters at the NSSS computation and communication center. The new organization was charged to meet specific naval fleet and fleet-marine-force tactical requirements as well as to support broader requirements for space surveillance, including missile warning (SPADATS) in particular, under the operational control of NORAD in Colorado Springs. Connoisseurs of Naval abbreviations will especially appreciate the name by which the Navy's first command dedicated to military space activities was routinely styled in official records and correspondence: NAVSPASUR.

The NRL proposal to ARPA had included a list of upgrades to the basic system just described, and these projects proceeded while NAVSPASUR operated the system from Dahlgren. This fact illustrates an important feature of the NSSS, which was mentioned briefly in Chapter 1. Because the transmitters and receivers were separated, but with overlapping coverage between the stations, and any receiver could detect signals originating from either transmitter, the whole system did not have to be shut down when one station needed special work: space surveillance could continue while major construction or maintenance was underway at any particular site. Naturally, the biggest impact to operations occurred when a transmitter had to be shut down for some reason, but, even then, both complexes did not have to be shut down at the same time. As NRL's upgrades progressed over the next several years, the net effect was to increase the redundancy of the coverage between receiver stations, making the final system quite robust to outages, including communication outages, at any one site.

In 1960–1961, the coverage gap over Texas between the Eastern and Western Complexes got first attention from NRL.[72] A third transmitter, having more than 10 times the radiated power of either of the existing transmitters at Jordan Lake and Gila River, about 560 KW, was installed at the Lake Kickapoo site. That new transmitter began operating in June 1961. Over the next several years, two new receivers were built near Hawkinsville, Georgia, and Red River, Arkansas. The latter filled a coverage gap that, until then, still existed

[71] The base was later renamed the Naval Weapons Laboratory, and much later the Naval Surface Weapons Center. At this writing, the base is known as the Naval Surface Warfare Center–Dahlgren Division.

[72] Much of the following information came to one of the authors in a private communication from Carroll C. Hayden [35], dated 1 February 2011, besides in many discussions over the years, beginning in late 1988 and early 1989. The 2011 date coincided with the 50th anniversary of NAVSPASUR as a Navy command. Hayden had joined the NSSS activity in Dahlgren in 1960, while he was still enrolled in engineering school at the University of Virginia, working mainly with the Fence operation. He eventually became the chief engineer for the Fence, a role that continued until his retirement from government service in 1999. After that, he continued to work in private industry, supporting the Fence operation as a contractor, until the field stations were shut down by the Air Force at the end of their service life in 2013. Hayden's 2011 summary was a material help in understanding this history, tying together many disparate facts found scattered throughout more official NRL documents.

for LEO satellites passing through the Fence between the Silver Lake and Elephant Butte stations. The Hawkinsville station was a new receiver design, with even longer antennas than the other receivers. The intention was to take better advantage of the higher-power transmitter at Lake Kickapoo; depending on the size of the objects, the higher receiver gain would allow satellites to be detected as high as 15,000 miles or more, what we would call today medium-Earth-orbit (MEO) altitudes. Shortly after, the Elephant Butte station was converted to the same "high-altitude" receiver design, so that even the highest detectable satellites could have their positions determined by triangulation using these two stations.

By mid-1965, the frequency of the NSSS was doubled to 216.980 MHz. This was done primarily to reduce the effect of ionospheric refraction on the received signals, thereby improving the overall accuracy of the tracking data. It was a big step, because it required reworking the antenna arrays at every station, transmitters included, although the receiver gains remained about the same. Also, the radiated power at Lake Kickapoo was increased to 768 KW, making it certainly one of the most powerful, if not the most powerful, continuous-wave transmitter in the world at the time. The boost meant that the NSSS could detect most satellite payloads and essentially all rocket bodies at 22,000 miles altitude or higher, depending on the object size, if they passed through the Kickapoo transmitter beam. Although that altitude matches what we now call geosynchronous equatorial orbit (GEO) altitude, the transmitter beam was inclined at 33.57 deg to the equator. No satellites in the actual GEO were illuminated unless they happened to drift across the intersection of the inclined beam and the GEO region, which did happen, but only occasionally. The main use of the NSSS at such altitudes was to detect satellites and rocket bodies in highly elliptical transfer orbits to and from the GEO region. Satellites in those orbits were, and today still are, for numerous reasons, notoriously difficult to track. The Kickapoo beam also turned out to be powerful enough that all the NSSS receivers routinely detected the Moon when it passed through the Fence coverage. The Moon moved quite slowly in the sky compared to satellites, so that the continuous-wave signal effectively saturated, for many minutes at a time, all the receiver channels picking up the Moon's Doppler-shifted reflection. Special provision had to be made to predict the Moon's transit through the Fence coverage and to ignore those saturated channels. One additional modification completed by mid-1965 was to offset the frequencies of the two lower-power transmitters by about 10 KHz, one above and one below what was now the system center frequency of the Lake Kickapoo transmitter. This was needed because, now with three transmitters and six receivers located on the great circle, located in such a way that any receiver could detect signals originating from illumination provided by any transmitter, some ambiguities could arise in processing the Doppler-shifted receiver data, regarding which transmitter had contributed the illumination.

TABLE 2.4 NAVY SPACE SURVEILLANCE SYSTEM STATIONS[a] (T = TRANSMITTER, R = RECEIVER)

Station Location	Mode	Date Operational; Comments
Fort Stewart, Georgia (near Savannah)	R	29 July 1958; station later moved to nearby Tattnal County, Georgia
Hawkinsville, Georgia (near Macon)	R	Mid-1965; high-altitude station
Jordan Lake, Alabama (near Wetumka)	T	29 July 1959; 108 MHz, 50 KW Mid-1965; 216.990 MHz, 50 KW
Silver Lake, Mississippi (near Hollandale)	R	30 November 1958
Red River, Arkansas (near Lewisville, Arkansas, and Texarkana, Texas)	R	Mid-1965
Lake Kickapoo, Texas (near Archer City)	T	June 1961; 108 MHZ, 560 KW Mid-1965; 216.980 MHz, 768 KW
Elephant Butte, New Mexico (southeast of Elephant Butte Reservoir)	R	26 January 1959; later converted to high-altitude station in 1965
Gila River, Arizona (near Maricopa)	T	26 January 1959; 108 MHz, 50 KW Mid-1965; 216.970 MHz, 50 KW
San Diego, California (Brown Field near Chula Vista)	R	15 December 1958

[a]Besides information from the report by Easton [34] and the journal paper by Easton and Fleming [31], some information for this table has been gleaned from NRL reports [36, 37]. Later dates shown in this table are approximate.

Offsetting the frequencies of the two transmitters eliminated the ambiguities while keeping essentially all signals within the 30-KHz detection bandwidth centered on 216.980 MHz. Table 2.4 shows the list of NAVSPASUR field stations at the end of 1965.

It is interesting to note that although the calibration problem had led Easton to the development of the NSSS in the first place, operating that new system led back to another calibration problem to be solved. In fact, calibration was more important and more difficult for the NSSS than it had been for Minitrack, and this fact was appreciated quite early in the system's development. The basic reason was that the surveillance signals at the receivers were fainter even than the direct Minitrack-type signals originating on the IGY satellites.[73] This meant that the surveillance signals were harder to distinguish from radio noise originating from natural and artificial sources in the environment. Not all these noise sources were predictable or even well understood at the time: ionized

[73]In an interview reported by Sturdevant [38], p. 53, Easton explained that the individual Minitrack antennas were about 50 ft long, and that to get the higher gain needed for the surveillance receivers, antennas about 100 times longer had been needed. Although the point in question is made correctly in the interview, and 5200-ft antennas were considered in the early designs, the actual receiver antenna lengths in the final version of the Navy Space Surveillance System turned out to be a little less than 40 times as long as the Minitrack antennas. In any case, besides the requisite higher gain, the other net effect of the longer antennas was that the near field of the surveillance receiver arrays was far too high for an aircraft-towed calibration beacon.

meteor trails in the atmosphere, aircraft, electrical storms, satellites broadcasting on 108 MHz, radio stars, solar radio emissions, refraction of the signals' paths in the ionosphere, direct feedthrough from the NSSS transmitters under some atmospheric and surface conditions, and other effects. Each of the noise sources had to be identified, quantified, and filtered out or mitigated somehow, and many of them had larger amplitude (in everyday terms, were "brighter" or "louder") than the surveillance signal itself. The first step was to achieve more accurate, or at least more consistent, calibration and identification of the desired signal at an NSSS receiver than had been available from the Moon-referenced calibration of Minitrack, such that the remaining signal could be confidently identified as noise. The answer was to use as calibration objects a few satellites whose orbits were known more accurately than others in the catalog so that the calculated satellite positions at any time of interest would be more accurate than for other satellites. The first choice of a calibration satellite for the NSSS was none other than Vanguard 1, which, thanks to its solar cells maintaining its battery charge, was still transmitting (until May 1964, as it turned out) and could be tracked accurately enough by Minitrack, besides the much more accurate Baker-Nunn cameras. Vanguard 1's orbital inclination of 34.25 deg was slightly greater than the 33.57-deg inclination of the NSSS great circle, meaning that this satellite crossed the Fence frequently and always at a shallow, grazing angle, so that it remained in the transmitter beams for many seconds at a time. The result was an abundant NSSS data set for Vanguard 1, which allowed Easton's team to do the necessary noise analysis and mitigation. It is worth noting that, still today, the calibration for accuracy and precision of American space surveillance sensors is usually done by periodically tracking a few selected satellites that have unusually accurate orbits known by other, independent means. If the data set to be evaluated is statistically independent of the data set used to calculate the reference orbit, the method can produce reliable results.

3. SPACE SURVEILLANCE AND THE INVENTION OF GPS NAVIGATION

Some proposed NSSS upgrades never materialized, partly due to competing priorities for funding and partly due to evolving operational circumstances. One of these upgrades was the idea of a second Fence operating in tandem with the existing one to allow first-pass orbit determination with the NSSS alone, a concept that was mentioned briefly in Chapter 1. One ambitious proposal described a possible second fence stretching diagonally across the continental United States from Miami, Florida, to Fargo, South Dakota.[74] A

[74]NRL report [39], p. 12, describes this southeast--northwest second-fence idea briefly. Two other reports mention the parallel second-fence idea in connection with different projects, [40] and [41]. NRL had begun work on a southern complex in 1960, but that project was eventually cancelled before it was operational.

more modest proposal called for a "southern complex" comprising a transmitter flanked by two receivers, thereby forming a shorter second fence parallel to, but slightly south of, the first one, near Raymondville, Texas. In either design, the motivation for a second fence close by the first was that, with two separate triangulated satellite positions available on a single pass over North America, an approximate orbit could be calculated by standard methods of orbital mechanics within several minutes of the first detection. Without the extra fence available, the second position would have to be obtained on some future pass through the existing Fence, at least one orbital revolution later. As noted in Chapter 1, this initial-orbit-determination problem was ultimately solved by sharing tracking data from all space surveillance sensors in both the Air Force and Navy, so that an initial estimate of a new orbit could usually be obtained during the first orbital revolution, that is, within a few tens of minutes after first detection, for a new satellite after its launch. Nevertheless, until the second-fence idea was shelved several years later for operational and budget reasons, NRL staff, mainly Easton and his colleagues, devoted some creative effort to the NSSS-only single-pass orbit problem. In 1960, they began developmental work on the southern complex.

Their interest in a southern complex was motivated by two different technical issues. One was the problem of synchronizing the clocks at the NSSS receiver sites and keeping them calibrated to true time, now that they were distributed along an arc nearly 3000 miles long from Savannah to San Diego. The somewhat stressing time accuracy requirement of 1 ms or better, for both absolute time and time intervals, for IGY applications had been accepted by NRL as a matter of course, even though Minitrack, and now, by extension, NSSS, were not expected to meet this requirement initially. In those years, even the famous WWV radio-broadcast timing signal from the National Bureau of Standards (now the National Institute of Standards and Technology [NIST]), begun in 1945, was not able to meet that requirement because of one-way radio propagation delays: light travels 186 miles in 1 ms, and so even in the continental United States, a clock set by WWV would always be at least a few milliseconds behind reference time, unless the path delay could be estimated somehow. Minitrack had used WWV with techniques to try to account for propagation delays, but the problem was never completely solved for that system. For the NSSS, the only alternative to WWV was to transport (by ground vehicle, because of the station locations) a cesium-based reference atomic clock to each station in turn. The job had to be kept up continuously, because the station clocks would start to drift randomly from the calibrated time as soon as they were set. Moreover, even the portable cesium reference clock was known to drift measurably compared to the national atomic time standard at the U.S. Naval Observatory (USNO) in Washington, D.C., in the time it took to drive between stations.

A second technical issue for the NRL staff, mainly Easton and his colleagues, was the notable advancements in timekeeping technology happening at the time. The first cesium-based atomic clock had been demonstrated in 1955, and was, under ideal conditions, accurate to well under a nanosecond for very long periods of time. That fundamental accuracy was to improve by four orders of magnitude by the time a more advanced cesium clock became the standard in 1967.[75] Although such extreme accuracy could not be expected for mobile clocks, accuracies of about 1 μs for long periods of time might be expected. Since light travels only about 0.186 miles in 1 μs, clocks with that accuracy could, in principle, be used to measure the bi-static, transmitter-to-satellite-to-receiver range to within about 300 m, whatever the value of the range. This capability would require the transmitter to be modulated somehow, and the clocks at transmitter and receiver to be synchronized to within a microsecond, so that accurate times of both the transmission of a signal modulation and its reception would be known. With the bi-static range known to this level of accuracy, the initial orbit determination from NSSS-only data would also be much more accurate. The NRL staff decided to test these ideas with equipment installed in south Texas, serving as an experimental southern complex. Although this work would never evolve into an operational space surveillance capability, it would lead to another serendipitous realization for Easton, a navigation system with timing and ranging (later becoming the acronym NAVSTAR), essentially what we know today as GPS (Global Positioning System).

Sometime in September 1964, Easton was pondering the problem of synchronizing station clocks when he hit on the idea of putting an atomic reference clock in a satellite that could broadcast accurate time to several stations at once.[76] The orbiting clock would also be easier to calibrate frequently against the USNO's master timekeeping facility and could broadcast accurate time signals to many NSSS stations at once, as often as every orbital revolution. Later, in October, while he was fulfilling his Naval Reserve duty, Easton realized that an atomic clock on a satellite can also be made into a navigation system for anyone with an equally accurate clock and the ability to receive the satellite's time signals. The key was the range

---

[75] This is according to the paper posted by NIST at https://www.nist.gov/atomic-clocks/brief-history-atomic-time [retrieved 21 November 2024]. The long-term accuracy of that first cesium clock was about 1 s in 300 years.

[76] The story is told in Easton and Frazier [29], Chapter 3, pp. 49–50, with extensive background information. A persistent embellishment of the story, never actually mentioned by Roger Easton in the hearing of any of the present authors, and not well attested otherwise, nevertheless has much appeal. Allegedly, Easton himself was driving a cesium clock from one station to another across Texas in a "woody" station wagon one hot day in September, thinking there had to be a better way, when he first conceived of putting an atomic clock in a satellite. What we can say with certainty is that Easton was absorbed enough in the project that he would not have needed such an experience to move him toward the satellite-clock solution.

calculation enabled by the accurate clocks at both satellite and receiver. Of course, the complete story is beyond the scope of this discussion. Many controversies, and many persons from many organizations, were involved before the operational Block 1 NAVSTAR GPS satellites began to be launched in 1978. In any case, the final design of that system had been preceded by years of testing by NRL, first with the Timation (Time Navigation) series of satellites, followed by the Navigation Technology Satellite series, before the Air Force was assigned the lead role for launching, operating, and maintaining the GPS constellation.

4. WHY DAHLGREN?

We mentioned earlier that the Navy chose Dahlgren, Virginia, as the location for the NAVSPASUR headquarters because an especially powerful electronic computer was there. The machine was the Naval Ordnance Research Calculator (NORC). In those days, it was easier to bring people to the computer than to bring the computer to the people. The Navy's Bureau of Ordnance had taken delivery of the NORC from IBM Corporation at a ceremony on 2 December 1954 at the Thomas J. Watson Scientific Computing Laboratory, then located at Columbia University in New York City. By March 1955, the computer had been disassembled and packed for shipment to Dahlgren on a convoy of 17 trucks. At Dahlgren, a crew of 35, mostly local Navy computer professionals and a few IBM staff from New York, gathered to reassemble the machine. By July 1955, the NORC was in regular operation, calculating mainly artillery-shell trajectories for the Computation and Exterior Ballistics Laboratory of the Naval Proving Ground–Dahlgren.

For several years after, the NORC was the only one of its kind and was widely rated as the most powerful computer in the world. IBM referred to the NORC as a "supercomputer," even in its own literature,[77] although that term has taken on more specific meaning in our day than it had in 1955. Well before it began operating at Dahlgren, it was already well known in scientific circles, thanks to IBM's presentations around the country, but there was not universal comprehension about why the Navy needed such a machine. For one example,[78] about a month before the NORC was even finished in New York, the famous Dr. Edward Teller, the "father of the hydrogen bomb" and director of Lawrence Livermore National Laboratory, had written to the Bureau of Ordnance to explain that the real need for the NORC was at his laboratory in California, that ballistics research at

[77] According to IBM's pamphlet "NORC Booklet," 1954, Introduction, as posted at https://web.archive.org/web/20050321140201/http://www-03.ibm.com/ibm/history/exhibits/norc/norc_intro.html [retrieved 10 September 2024].

[78] The story is told in Niemann [42], pp. 27–28, and in Rife and Carlisle [43], Chapter 4, pp. 101–102.

Dahlgren was not as important to the national defense as the thermonuclear research at Livermore, that surely those ballistics calculations could be done on some other computer, and that he wanted the NORC shipped to Livermore. Interestingly, Teller's letter did not intimidate the bureau staff as much as it made them realize just what a capability they had with the NORC. They informed him that the computer would not be shipped to California but that the Livermore staff could submit computer programs (by mail) to be run on the third shift of the 24-hour-a-day operation at Dahlgren. The Livermore staff did just that, until they got more computers of their own by mid-1956.

By late 1958, the new commander of the Dahlgren base, Navy Captain M. H. Simons, Jr., had learned of the space surveillance efforts by NRL. He proposed to the laboratory that they should build a processing center in Dahlgren and use the NORC for reducing the Fence raw phase data and calculating orbits. NRL promptly accepted the offer and in May 1959 began setting up their primary computational center for the NSSS in Dahlgren. Subsequently, the Navy established the NAVSPASUR command in Dahlgren because the main computation center for the NSSS was already there. As it turned out, the NORC was used for space surveillance for less than two years. By mid-1961, NAVSPASUR was using a pair of new IBM 7090 computers. The design of the 7090 had been greatly informed by the IBM/Navy experience with the NORC, as mentioned again in Chapter 4. More than two decades later, in the 1980s, the Navy would establish a more senior command, Naval Space Command, in Dahlgren, adjacent to the NAVSPASUR headquarters, because the latter was already there. After another two decades, in the early 2000s, the Navy would transfer their entire space surveillance mission, including the NSSS operation, to the Air Force. The Air Force would continue to operate the NSSS from Dahlgren for another decade, until the system was decommissioned in 2013. The Air Force, and today the Space Force, still occupies and depends on some of the computing and communication facilities created for NAVSPASUR and Naval Space Command in Dahlgren. Hence, it is interesting to ponder that the Navy's entire 45-year legacy in space surveillance in Dahlgren, and everything that followed from that, came about because the best computer in the world for the job in the late 1950s was already in Dahlgren, although no one realized it at the beginning.

5. Minitrack and Naval Space Surveillance System Data Called for New Processing Techniques

In the decade leading up to the IGY, the term "astrodynamics" had come into use to distinguish the new analyses of artificial satellite motion from the older analyses used in classical celestial mechanics and dynamical astronomy.[79] Fundamental aspects of calculating orbits and modeling the orbital motion had to be revisited, adapting traditional astronomical practices to the new mission. For merely one example, because of the limitations of electronic computing systems in those days, the use of scaled numerical values for units of distance and time became essential in formulating and implementing sufficiently accurate space surveillance calculations for Earth satellites. Astronomers had for more than two centuries used such scaling for celestial mechanics calculations, even when the calculations were done by hand, to standardize the results. However, in the mid-20th century, a new issue arose. The earliest electronic computer systems typically had short register lengths and sometimes used fixed-point arithmetic. In that case, the large distances and high speeds involved in satellite motion could easily lead to numerical truncations (overflow or underflow), eventually causing large errors to accumulate in a long chain of calculations. The remedy was to choose suitably scaled basic units of distance and time for the input data to keep all calculated numerical values between the minimum and maximum values allowed on the computer. Nowadays, when multiple-precision number representations and floating-point arithmetic are the standard in scientific computation, the scaling of units is a much less critical issue. Nevertheless, analysts then and now have realized that for any computation, it is always a good idea to normalize one's variables in a suitable way. In fact, space surveillance was successful in its earliest days partly because of this simple expedient of choosing basic units thoughtfully for the extended calculations that were required. All this was understood at Project Space Track, of course, and they had adopted scaled units for their orbit calculations, though without any special nomenclature for the practice. At the NAVSPASUR processing center in Dahlgren, the scaled time unit for Earth satellite motion was given the special name "herg," in honor of the astronomer Paul Herget of the University of Cincinnati and Cincinnati Observatory. Herget consulted with NRL beginning in October 1955, on Project Vanguard, the Minitrack

[79]According to Junkins [44], Samuel Herrick (1911–1974) of the University of California at Los Angeles (UCLA) originated the term "astrodynamics." Herrick is known for his now-classic two-volume textbook having that title. Except for one year at MIT in 1961–1962, Herrick taught astronomy at UCLA for 37 years, in a career filled with unexpected and fascinating contributions right at the dawn of the space age. Following are a few examples: In 1945, he cofounded the Institute of Navigation. In 1946, he taught the world's first university course in astronautics, titled "Astrodynamics and Rocket Navigation." In 1951, he consulted with Hollywood producers on the film *The Day the Earth Stood Still*, one scene of which shows elaborate (and correct) celestial mechanics formulas written on a chalkboard, undoubtedly due to Herrick himself. Later, he consulted with NASA on the first American manned space program, Project Mercury. In 1962, he accepted a dual appointment at UCLA as professor of astronomy and engineering, an unusual combination of disciplines that well characterized the early days of the space program. Although Herrick never worked in space surveillance himself, he devised what has become the most common method of determining an initial estimate of a satellite orbit from a single track of radar data, the Herrick–Gibbs method, which bears his name and is used thousands of times per day in current space surveillance operations. Many of his students worked in the field or collaborated on space surveillance research projects.

system design and operation, and the developmental phase of the NSSS. He later spent time periodically with the NAVSPASUR operational and software staff at Dahlgren and became highly respected both there and at NRL.

As an astronomer, Herget specialized in orbit estimation for minor planets, such as asteroids, comets, and meteors. He founded the Minor Planet Center in 1947 and oversaw its operation at Cincinnati Observatory until his retirement in 1978. The basic job of that center was to maintain a catalog of the orbits of all known asteroids and comets.[80] This astronomical activity had, and still has, many processing and procedural similarities to the military space surveillance cataloging effort. The main difference is that the astronomical job can be done at a far more leisurely tempo than can the strategic and tactical military surveillance job: most asteroids complete one orbit around the Sun in several years, while almost all man-made satellites complete one orbit around the Earth in a day or less. At least 85% of all satellites, including practically the entire near-Earth population, complete an orbital revolution in less than two hours. From its beginning, therefore, space surveillance necessarily required a sophisticated level of automatic computation, just because of the short response times involved. Fortunately, Herget was one of the earliest advocates of automated computation in astronomy and celestial mechanics and was widely regarded as an expert in that emerging specialty.[81] Especially for the latter reason, he became deeply involved in estimating orbits for artificial Earth satellites, beginning with the first one, Sputnik 1 in 1957, continuing with Explorer 1 and Vanguard 1 in 1958, and following with many others in later years. Starting in 1959, he also consulted on orbit estimation for the Project Mercury manned spaceflight program.[82]

Recall that the Minitrack system was tailored for tracking IGY satellites.[83] Even in that somewhat narrow application, however, making practical use of that new data type for satellite tracking required ingenuity in both the mathematical formulation and the automation of the orbit estimation process, which Herget was able to provide. In the early days of space surveillance, only radar systems, including the NAVSPASUR Fence, could supply large numbers of observations of space objects in near-real time. Hence, the most pressing

[80] Upon Herget's retirement, the Minor Planet Center was transferred to the Smithsonian Astrophysical Observatory in Cambridge, Massachusetts, where the catalog of orbits for asteroids and comets is maintained today. Refer to http://www.minorplanetcenter.net/iau/mpc.html [retrieved 16 July 2024].

[81] Niemann [42], p. 28, records that Herget had been a frequent visitor at IBM's Watson Laboratory in New York City while the NORC was being checked out. Herget waited patiently to get his turn on the NORC before it was shipped to Dahlgren. His rationale was solid: the NORC was so much faster than the machines he normally used for the minor-planet catalog, that traveling and waiting for the NORC was still better in the long run than using his older machines. After the NORC was operational at Dahlgren, Herget became a regular visitor there also, for the same reason.

[82] Some of this story is told in the publication by Green and Lomask [11], Chapter 9. Additional background information is found in the paper by Osterbrock and Seidelmann [45].

[83] Mengel and Herget [32], pp. 23–29.

computational and estimation problems to be solved in those early days necessarily involved the new radio and radar data types. Herget's methods for those data types, and even some of his software, were still in use at NASA and other U.S. government agencies as late as 1970, that is, several computer-generations after the IGY.

Partly because of his connection and expertise with Minitrack and the NSSS, Herget was invited to spend a few weeks each summer at Dahlgren in the late 1950s and early 1960s to get computer time on the NORC. In fact, it was Herget's favorite computer, as he put it, "the greatest love of my life, after my wife and daughter."[84] The NORC was used around the clock, with the regular daylight hours reserved mainly for ballistics research and Polaris missile development by the Naval Weapons Laboratory. However, after regular business hours, usually in the second shift, the NORC was used to calculate Earth satellite orbits and to update the NAVSPASUR satellite catalog database using mathematical formulations and software adapted from Herget's original methods. During Herget's residencies at Dahlgren, the NORC was also used to estimate the orbits of asteroids for the Minor Planet Center. Asteroid 1625, officially named "The NORC," is the first astronomical object named after a computer, thanks to Herget.

6. THE NAVSPASUR LEGACY IN DAHLGREN

The logistics of moving the NSSS processing center from NRL to Dahlgren took about a year, despite the pressure to get more computing capacity. The new center was allocated space in Building 111, the NORC facility, at the Naval Proving Ground in May 1959, around the time the base was renamed as the Naval Weapons Laboratory.[85] At last, in June 1960, the name Space Surveillance Operations Center (SSOC) could be applied to the new facility, with the NRL computational facility serving as the backup. Several months later, in October 1960, the SSOC moved to the somewhat more accessible and spacious Building 218, which faced the parade field of the base. The reason for the move was to share access to the new IBM 7090 computer that had been installed for NWL, and which was used in the same manner as the NORC had been. When the NAVSPASUR command was formally established on 1 February 1961, a second IBM 7090 was installed, exclusively for the space surveillance mission. Various computer upgrades followed over the next three decades, but NAVSPASUR remained in Building 218 during that time.

In 1983, a new senior command, Naval Space Command (NAVSPACECOM), was established as another tenant command at the same

[84] Osterbrock and Seidelmann [45], p. 67.

[85] Some of this information about the early days of the space surveillance mission in Dahlgren is gleaned from Swift [46], pp. 2–3.

base, which was by then known as the Naval Surface Warfare Center. In fact, in its first year, the small initial NAVSPACECOM staff, led by astronaut Rear Admiral Richard Truly, shared office space with NAVSPASUR in Building 218. However, in 1984, a new headquarters building adjacent to NAVSPASUR was completed and dedicated for NAVSPACECOM, known as the Shepard–Glenn Building.[86] In 1993, NAVSPASUR became part of NAVSPACECOM in a reorganization that meant the disestablishment of the separate space surveillance command. The idea was to consolidate naval space activities under one admiral who, presumably, could better coordinate them with other military space activities. Thus, the space surveillance mission gained some status in the Navy, becoming at least as well known as other formerly independent commands, such as the Naval Satellite Operations Center (NAVSOC) at Point Mugu, California, also brought into the admiral's responsibility.

That consolidation with NAVSPACECOM in 1993 coincided with a redesign and update of the computer systems used to support space surveillance. We have more to say about this redesign in Chapter 4 on Processing Systems; however, the intention was to replace the aging mainframe system in Building 218 with a more modern distributed-processing system in a new building. The building itself had been designed specifically to support all the NAVSPACECOM headquarters activities, with space surveillance being the most computation intensive, although other computer applications such as communications were also growing rapidly. In 1995, the space surveillance operation and staff moved into Building 1700 across the base, which was by that time known as the Naval Surface Warfare Center–Dahlgren Division.

In 2002, NAVSPACECOM was reorganized as the Naval Network and Space Operations Command (NNSOC), to better align with the Navy's emerging strategic view of network-based and information-based warfare. Space activities were at the heart of this new concern, because, for merely one long-standing example, essentially all naval communications with the fleet, and among fleet units, had to be relayed through satellites. However, by this time, military space operations also were increasingly perceived as inherently joint-service activities, and one approach to improving joint-service space operations was to realign supporting space activities among the services. Accepting that premise, the question was, which services were best suited for which supporting space activities?

If this reasoning seems nebulous today, it seemed nebulous at the time as well. In hindsight, it is easier to see that the discussion was really a justification for a dedicated Space Force, on par with the traditional military services, but fostering the experience, expertise, and military doctrine suitable for operating

---

[86]Those two famous naval astronauts were present for the opening ceremony. In fact, all naval astronauts detailed to NASA reported periodically to the commander of NAVSPACECOM as their senior officer.

in the global space domain, rather than in the more territorial land, sea, and air domains. However, that idea was too far beyond more immediate defense concerns for it to be considered seriously. What happened at the time was that the space surveillance mission was consolidated under the Air Force as the single service best able to man, train, and equip both staff and facilities for that particular mission. In a joint-service arrangement for space surveillance operations, the Navy could still depend on the space surveillance support that it needed, without the organizational and budgetary burdens that it had to bear from the beginning up to that date, so that it could focus on other space roles for the fleet.

Transfer of the naval space surveillance mission to the Air Force was announced in early January 2004, and the organizational turnover was completed on 1 October 2004. Thereafter, the NSSS was known as the Air Force Space Surveillance System (AFSSS). The remainder of the NNSOC staff and functions were merged with a new, higher-level organization, Naval Network Warfare Command (NETWARCOM), and moved to that new headquarters in Little Creek, Virginia, near the headquarters for the Atlantic Fleet. Only the civilian space surveillance staff, with the operation center, computers, and communications facilities to support the Air Force space surveillance mission, remained in Dahlgren, now as Air Force employees.

The AFSSS operation continued with its normal success for the next decade. However, on 1 August 2013, partly because of budget problems in the federal government during 2012 and 2013, such as the congressional sequestration of funds in the middle of the fiscal year, Air Force Space Command decided to decommission all the AFSSS field sites by 1 October 2013. The operations center in Dahlgren would continue to function as the backup center for the primary space surveillance center, then at Vandenberg Air Force Base in California, and continue to support certain Navy-specific functions such as overflight warning for the fleet. At the time, the Air Force's hope was that contract work on the new, more powerful Space Fence would begin soon, and that changes to operational procedures at the other space surveillance sensors could make up for the missing NSSS tracking data until that new Space Fence came online several years later. Those operational changes at the other sensors were successful in that the total number of space surveillance observations for catalog maintenance did not decline significantly. However, some coverage at high altitude that the NSSS had provided could not be compensated completely, given the fixed locations of the other sensors. As it turned out, the new Space Fence program was delayed, partly for the same budgetary problems that led to the AFSSS closure. Besides that, locations for the new radar were debated for an extended time. However, eventually, the program went forward. On 27 March 2020, the new radar system was declared operational on Kwajalein Atoll in the Marshall Islands, midway between Hawaii and Australia. As the most powerful system of its type ever built, and in a location more commensurate with the immediate surveillance needs of the time, the Space Fence opened the door to a new phase of the space surveillance mission.

References

1 Altshuler, E. E., *The Rise and Fall of Air Force Cambridge Research Laboratories*, self-published via CreateSpace Independent Publishing Platform (Amazon), 2 January 2013.

2 Sturdevant, R. W., "From Satellite Tracking to Space Situational Awareness: The USAF and Space Surveillance, 1957–2007," *Air Power History*, Winter 2008, pp. 4–23.

3 Mansfield, R. L., "Hans Beat Wackernagel (1931–1992)," *Bulletin of the American Astronomical Society*, Vol. 29, No. 4, 1 January 1997, p. 1496, https://baas.aas.org/pub/hans-beat-wackernagel-1931-1992/release/2 [retrieved 22 July 2024].

4 Miczaika, G. R., and Curtis, H. O., "Evaluation of Satellite Observing Network for Project Space Track," Air Force Surveys in Geophysics, No. 101, TN-58-416, Geophysics Research Directorate, Air Force Cambridge Research Center, Air Research and Development Command, U.S. Air Force, Bedford, MA, June 1958. Scanned copy from the Univ. of Michigan available online at the Hathi Trust website, https://catalog.hathitrust.org/Record/102326270. The second author was H. O. Curtis from MIT Lincoln Lab., which was located in Lexington, MA, adjacent to Hanscom Field. Curtis later became the director of the operational element of Project Space Track at Hanscom, the NSSCC.

5 "Preliminary Design of an Experimental World-Circling Spaceship," Douglas Aircraft Co. Project RAND SM-11827, 2 May 1946, https://www.rand.org/pubs/special_memoranda/SM11827.html [retrieved 28 Feb. 2024].

6 Whipple, F. L., "Possible Hazards to a Satellite Vehicle from Meteorites," Douglas Aircraft Co., 1 Sept. 1946, Project RAND Supplement to Rept. No. SM-11827, pp. 61–66. The latter report SM-11827 had been published on 2 May 1946 as "Preliminary Design of an Experimental World-Circling Spaceship," and is cited in Chapter 1. It is available online at https://www.rand.org/pubs/special_memoranda/SM11827.html [retrieved 28 Feb. 2024].

7 Hayes, E. N., *Trackers of the Skies*, Howard A. Doyle Publishing Co., Cambridge, MA, 1968. Hayes had been the senior editor for the Smithsonian Astrophysical Observatory (SAO). This book contains one of the most complete accounts of the Moonwatch program organized by SAO and of the development of the Baker-Nunn cameras used in the early days of space surveillance.

8 E. Nelson Hayes, three papers in the series Annual Report of the Board of Regents of the Smithsonian Institution, U.S. Government Printing Office, Washington, D.C., https://ia801608.us.archive.org/29/items/annualreportofboy1961smit/, and similarly for the other two years:

Publication 4478, for Year Ended 30 June 1961|, "The Smithsonian's Satellite Tracking Program: Its History and Organization," pp. 275–322. Describes the program up to October 1957.

Publication 4530, For Year Ended 30 June 1963, "The Smithsonian's Satellite Tracking Program: Its History and Organization, Part 2," pp. 331–357. Describes the program from October 1957 to early 1958.

Publication 4613, For the Year Ended 30 June 1964, "The Smithsonian's Satellite Tracking Program: Its History and Organization, Part 3," pp. 315–350. Describes the program from early 1958 through mid-1959.

9 Henize, K. G., "Tracking Artificial Satellites and Space Vehicles," *Advances in Space Science*, edited by F. Ordway, Vol. 2, pp. 117–142, Academic Press, New York, 1960.

10 Brandenberger, A. J., "The Use of Baker-Nunn Cameras for Tracking of Artificial Earth Satellites," paper presented at the 1961 Fall Meeting of the American Society of Photogrammetry at the University of Notre Dame, South Bend, IN. Published in *Journal of Photogrammetric Engineering*, Vol. XXVII, No. 5, Nov. 1962, pp. 727–735. Available online at the American Society of Photogrammetry and Remote Sensing, https://www.asprs.org/wp-content/uploads/pers/1962journal/nov/1962_nov_727-735.pdf [retrieved 25 July 2024].

11 Green, C. M., and Lomask, M., *Vanguard: A History*, NASA SP-4202, NASA Historical Series, 1970 (foreword by Charles A. Lindbergh). https://www.nasa.gov/wp-content/uploads/2023/03/sp-4202.pdf. Another edition of this document was issued in 1971 by Smithsonian Institution Press. In 2009, a slightly corrected edition of NASA SP-4202 was issued as *Project Vanguard: The NASA History* by Dover Publications, Inc., Mineola, NY, with a new Introduction by Paul Dickson. See especially Chap. 9, "The Tracking Systems," pp. 145–163 in the Dover edition.

12 Hynek, J. A., "Memorandum to Satellite Tracking Program Staff," 3 Oct. 1958. Smithsonian Institution Archives as part of Accession 04-183, Fred Lawrence Whipple Papers. Available online at https://siarchives.si.edu/blog/right-places-right-time [retrieved 1 Sept. 2024]. The complete text of the memorandum is reprinted in Appendix A.

13 Henize, K. G., "The Baker-Nunn Satellite-Tracking Camera," *Sky and Telescope*, Vol. XVI, No. 3, Jan. 1957, pp. 3–7.

14 Fitzpatrick, P. M., and Findley, G. B., "The Tracking Operation at the National Space Surveillance Center (NSSCC)," Special Studies Sec., Directorate of Aerospace, Air Proving Ground Center, Eglin AFB, FL, 2 Sept. 1960. The cover page of this report records the following: "This work was conducted during a temporary assignment of the authors to the NSSCC, L.G. Hanscom Field, Bedford, Massachusetts, as part of Project 1772 (Project Space Track), Task 17721."

15 Sridharan, R., and Pensa, A. F. (eds.), *Perspectives in Space Surveillance*, MIT Lincoln Laboratory Series, MIT Press, Cambridge, MA, and London, England, 2017.

16 Stone, M. L., and Banner, G. P., "Radars for the Detection and Tracking of Ballistic Missiles, Satellites, and Planets," *Lincoln Laboratory Journal*, Vol. 12, No. 2, 2000, pp. 217–244.

17 Lovell, B., *Story of Jodrell Bank*, Oxford University Press, 1968.

18 Grometstein, A. A. (ed.), *MIT Lincoln Laboratory: Technology in Support of National Security*, Lincoln Laboratory, Massachusetts Institute of Technology, Lexington, MA, 2011.

19 Wahl, E. W., "Program Development in Orbital Computation at the U.S. National Space Surveillance Control Center," *Rockets and Astronautics: Proceedings of the Second International Symposium*, edited by F. Tamaki, Yokendo, Bunkyo-Ku, Tokyo, 1961, p. 190.

20 Wackernagel, H. B., "Empirical Parameters Determining the Lifetime of Artificial Earth Satellites," Project Space Track Report: Air Force Cambridge Research Center TN-455, May 1959.

21 Cuthbert, L. W., private communication, "Ballbuster in Orbit: The Official History of Spacetrack," Spacetrack Operations Report, circa 1966–1967. Best copy available, 1982. (Humor and satire.)

22 Corliss, W. R., "Histories of the Space Tracking and Data Acquisition Network (STADAN), the Manned Space Flight Network (MSFN), and the NASA Communications Network (NASCOM)," NASA CR-140390, June 1974. Scanned copy available online at https://ntrs.nasa.gov/api/citations/19750002909/downloads/19750002909.pdf. This

NASA report incorporates all the material from an earlier document by Corliss, "Evolution of the Satellite Tracking and Data Acquisition Network (STADAN)," NASA TM X-55658, Goddard Historical Note No. 3, NASA Goddard Space Flight Center, Greenbelt, MD, January 1967.

23 Naval Research Laboratory Memorandum Rept. 466, "A Scientific Satellite Program," 13 April 1955. A scanned copy is available online at the Defense Technical Information Center as Accession No. ADA 605755 at https://discover.dtic.mil, or directly at https://apps.dtic.mil/sti/tr/pdf/ADA605755.pdf. This is the Project Vanguard proposal that the Navy offered for the IGY.

24 Neufeld, M. J., "Orbiter, Overflight, and the First Satellite: New Light on the Vanguard Decision," *Reconsidering Sputnik: Forty Years Since the Soviet Satellite*, edited by R. D. Launius, J. M. Logsdon, and R. W. Smith, Vol. 11, Routledge Studies in the History of Science, Technology and Medicine, Harwood Academic Publishers, Amsterdam, The Netherlands, 2000, Chap. 8, pp. 231–257. A scanned copy of Chap. 8 is available online from the Smithsonian Institution at https://repository.si.edu/discover, searching "Neufeld" [retrieved 2 Aug. 2024].

25 President's Scientific Advisory Committee, "An Introduction to Outer Space: An Explanatory Statement," U.S. Government Printing Office, Washington, D.C., 26 March 1958. This 15-page report includes an opening statement signed by President Eisenhower, describing his purpose in releasing the document to the public and his general policy of fostering freedom of space and peaceful uses of space. Copies of the document are available online at a wide variety of websites, including, for example, http://www.plosin.com/beatbegins/pdf/IntroductionToOuterSpace.pdf [retrieved 7 Dec. 2022] and https://lunar.colorado.edu/jaburns/publicfiles/IntroOuterSpace.pdf [retrieved 7 Dec. 2022].

26 Ludwig, G. H., "The First Explorer Satellites" (prepared for Van Allen's 90th birthday celebration at the Univ. of Iowa), 9 Oct. 2004. Available online at the University of Iowa Space Physics website, https://space.physics.uiowa.edu/plasma-wave/van90/ExplorerSatellites_LudwigOct2004.pdf [retrieved 3 Sept. 2024].

27 Stuhlinger, E., and Ordway III, F. I., *Wernher von Braun, Crusader for Space: A Biographical Memoir*, Krieger Publishing Co., Malabar, FL, 1994 (reissued 1996 with corrections).

28 Neufeld, M. J., *Von Braun: Dreamer of Space, Engineer of War*, Alfred A. Knopf, New York, 2007. Published in association with the National Air and Space Museum and the Smithsonian Institution, Washington, D.C.

29 Easton, R. D., and Frazier, E. F., *GPS Declassified: From Smart Bombs to Smartphones* (foreword by R. W. Sturdevant, Deputy Director of History, Air Force Space Command). Potomac Books, Univ. of Nebraska Press, Lincoln, NE, 2013. https://www.nebraskapress.unl.edu/potomac/.

30 Mengel, J. T., "Tracking the Earth Satellite, and Data Transmission, by Radio," *Proceedings of the IRE*, Vol. 44, No. 6, June 1956, pp. 755–760. https://doi.org/10.1109/JRPROC.1956.274968. Also republished later as Mengel, J., "Tracking the Earth Satellite and Data Transmission by Radio," *1958 IRE International Convention Record*, Vol. 4, 1966, pp. 112–118. https://doi.org/10.1109/IRECON.1956.1150449.

31 Easton, R. L., and Fleming, J. J., "The Navy Space Surveillance System," *Proceedings of the IRE* (renamed *Proceedings of the IEEE* in 1962), Vol. 48, No. 4, April 1960, pp. 663–669. https://doi.org/10.1109/JRPROC.1960.287444.

32 Mengel, J. T., and Herget, P., "Tracking Satellites by Radio," *Scientific American*, Vol. 198, No. 1, Jan. 1958, pp. 23–29. https://doi.org/10.1038/scientificamerican0158-23. This paper describes for a scientifically literate general audience the Minitrack system as it was first used to detect and track Earth satellites.

33 Smith, R., private communication with one of the authors, dated 12 Jan. 2016, regarding the Minitrack stations as precursors to the Naval Space Surveillance System.

34 Easton, R. L., "The Mark II Minitrack System," Vanguard Rept. No. 21, Minitrack Rept. No. 2, Naval Research Laboratory Rept. No. 5035, 12 Sept. 1957. Available online at Defense Technical Information Center, Accession No. AD0142782, https://doi.org/10.21236/AD0142782. At this writing, also available in hardcopy via the print-on-demand service at https://www.lulu.com/, Lulu Press, Durham, NC.

35 Hayden, C. C., "A Golden Anniversary for the Space Fence: A Personal Reflection," private communication to a number of former NAVSPASUR staff including one of the authors, dated 1 Feb. 2011.

36 "Space Surveillance System Technical Summary Report No. 1," Naval Research Laboratory Memorandum Rept. 896, 31 Dec. 1958. Available online at the Defense Technical Information Center as Accession No. ADA459673 at https://discover.dtic.mil, or directly at https://apps.dtic.mil/sti/pdfs/ADA459673.pdf. This is the Naval Research Laboratory's first summary progress report on the project to build the Navy Space Surveillance System, funded under ARPA Order No. 7-58.

37 "Space Surveillance System Technical Summary Report No. 5," Naval Research Laboratory Rept. No. 5756, Applications Research Division, Naval Research Laboratory, Washington, D.C., 1 March 1962. Available online at Defense Technical Information Center as Accession No. AD0328442 at https://discover.dtic.mil, or directly at https://apps.dtic.mil/sti/pdfs/AD0328442.pdf.

38 Sturdevant, R. W., "Tracing Connections—Vanguard to NAVSPASUR to GPS: An Interview with Roger Lee Easton, Sr.," *High Frontier*, Vol. 4, No. 3, May 2008, pp. 52–55. Air Force Space Command, Peterson AFB, CO.

39 Cleeton, C. E., "The Space Surveillance System: A Summary Report," Naval Research Laboratory Rept. No. 5575, Applications Research Division, Washington, D.C., 31 Oct. 1960. Available online at Defense Technical Information Center as Accession No. AD0319956 at https://discover.dtic.mil, or directly at https://apps.dtic.mil/sti/pdfs/AD0319956.pdf.

40 "Ranging and Velocity Experimental Equipment," Naval Research Laboratory Rept. 6633, Space Applications Branch, Washington, D.C., 18 April 1968.

41 Gardner, I., Silverman, I., and Weaver, C., "Timation I Ground Station," Naval Research Laboratory Rept. 6876, Washington, D.C., 3 April 1969.

42 Niemann, R. A., "Dahlgren's Participation in the Development of Computer Technology," NSWC MP 81-416, Naval Surface Weapons Center, Dahlgren, VA, Sept. 1982.

43 Rife, J. P., and Carlisle, R. P., *The Sound of Freedom: Naval Weapons Technology at Dahlgren, Virginia 1918–2006*. NSWCDD MP-06/46, Naval Surface Warfare Center (Dahlgren Division), U.S. Dept. of the Navy, U.S. Government Printing Office, Washington, D.C., ca. 2006.

44 Junkins, J. L., "In Memoriam Samuel Herrick 1911–1974," *Acta Astronautica*, Vol. 3, 1976, pp. 53–54.

45 Osterbrock, D. E., and Seidelmann, P. Kenneth, "Paul Herget 1908–1981," *Biographical Memoirs of the National Academy of Sciences*, Vol. 57, 1987, pp. 59–86. A primary source on Herget's life and work, containing much information not available elsewhere. Both authors grew up in Cincinnati, Ohio, and credited Herget with inspiring them to become astronomers. Seidelmann received his doctorate under Herget, while Osterbrock received his doctorate under the famous astrophysicist Subrahmanyan Chandrasekhar.

46 Swift, E. R., "History of Satellite Orbit Determination at NSWCDD," Naval Surface Warfare Center–Dahlgren Div. TR-17/229, Dahlgren, VA, Jan. 2018.

Chapter 3

THE MATHEMATICAL MODELS FOR ORBIT PREDICTION

The foundation of satellite catalog maintenance is based on the astrodynamic models. In earliest times, models of satellite motion were still being developed, and orbit predictions were very rudimentary. Later, modeling capability exceeded computational resources. This forced trades between accuracy and system capacity that continued until the late 1990s. Much of this section is based on a 2007 history paper [1] by the authors.

I. EARLY METHODS: 1957–1963

The first formalized effort to catalog satellites occurred at the NSSCC located at Hanscom Field in Bedford, Massachusetts. The procedures used at the NSSCC were first reported in 1959 by Eberhard Wahl [2], who was the technical director of the NSSCC. In 1960, under Project SPACETRACK, Philip Fitzpatrick and George Findley [3] developed detailed documentation of the procedures used at the NSSCC. The following description is based on that historic documentation.

Observation of satellites was performed at more than 150 individual sites. Contributions came from radars, Baker-Nunn cameras, telescopes, radio receivers, and the Moonwatch participants who took observations on satellites by visual means. Table 3.1 describes the various observation types and sources.

The observations were transferred to the NSSCC by teletype, telephone, mail, and personal messenger. There a duty analyst reduced the data, determining corrections that should be made to the orbital elements before they were used for further prediction. After his analysis, these corrections were fed into an IBM-709 computer that computed the updated orbital data. These updated orbital data were then used in another phase of the same computer program to yield the geocentric ephemeris. From the geocentric

TABLE 3.1 SPACE SURVEILLANCE OBSERVATION DESCRIPTION

Observation type	Content	Source
Type 1	Two angles and slant range	Radars
Type 2	Two angles	Baker-Nunn cameras, telescopes, binoculars, visual sightings
Type 3	Azimuth	Direction finders
Type 4	Time of closest approach (Doppler)	Radars, radio receivers (for transmitting satellites)

ephemeris, three different products were computed and sent back to the observing stations for their planning of future observing opportunities.

The first, a bulletin, was a listing of the updated orbital elements and was the forerunner of today's TLE set. Additionally, it contained a table of pertinent data for a given satellite that provided an observing station with a simple tool for determining a geocentric ephemeris. The essential content of the bulletin was a table of revolution number, time of passage through the ascending node, and longitude of ascending node for three to seven days in advance. Additionally, there was a grid containing latitude, longitude, and height for a revolution near the middle of the time period for which the bulletin was valid. This information could be used in conjunction with the other data to determine an ephemeris for a given observing station.

The second product used the ephemeris in the General Look Angles Program (GLAP) to produce a tabulation of all satellite passes observable by a specific ground site. The tabulation contained one or more time points for each pass as well as the azimuth, elevation, and slant range at that time point.

The third product, the Fence Look Angles Program (FLAP), is a modified form of the GLAP, which was sent to the stations that comprised the U.S. Navy and U.S. Army observation fences. Instead of a sequence of look angles for a given pass, the FLAP program gave a time and point of intersection of the satellite orbit with the vertical plane containing the sensing beam. The orbit model employed in the NSSCC made use of the following empirical and theoretical equations to predict the time T of passage through the ascending node, the right ascension Ω of ascending node, and the argument of perigee ω all at revolution number N. All times are in days, and all angles are in degrees:

$$T_N = T_0 + P_0(N - N_0) + c(N - N_0)^2 + d(N - N_0)^3$$

$$\Omega_N = \Omega_0 + \dot{\Omega}_0(T_N - T_0) + \frac{1}{2}\ddot{\Omega}_0(T_N - T_0)^2$$

$$\omega_N = \omega_0 + \dot{\omega}_0(T_N - T_0) + \frac{1}{2}\ddot{\omega}_0(T_N - T_0)^2$$

with

$$\dot{\Omega}_0 = K \frac{\cos i_0}{a_0^{7/2}(1-e_0^2)^2}$$

$$\ddot{\Omega}_0 / 2 = -\dot{a}_0(7-e_0)\frac{\dot{\Omega}_0}{2a_0(1+e_0)}$$

$$\dot{\omega}_0 = -(5\cos^2 i_0 - 1)\frac{\dot{\Omega}_0}{2\cos i_0}$$

$$\frac{\ddot{\omega}_0}{2} = (5\cos^2 i_0 - 1)\frac{\ddot{\Omega}_0}{4\cos i_0}$$

$$K = -9.96^0/\text{day}$$

where a single or double dot above a term indicates the first- or second-time rate of change of that term, a subscript 0 indicates the value of that term at the epoch time, a is a unitless semimajor axis normalized by Earth radii, and P is the period of the satellite. The inclination i is assumed to be constant whereas the eccentricity e is determined based on the assumption that perigee height remains constant throughout the prediction span. The semimajor axis is related to the period by

$$P = 0.058672947\, a^{3/2}$$

where the period has units of days per revolution. The parameters c and d are the results of a least-squares fit to the residual differences between the predicted and observed times of passage through the ascending node. The parameter c may be expressed as

$$c = P_0 \frac{\dot{P}_0}{2}$$

which provides the time rate of change of the period P_0 from which the time rate of change of the semimajor axis a_0 may be computed.

II. THEORETICAL FOUNDATIONS: 1959–1969

In 1959, under Project SPACETRACK, Dirk Brouwer developed a solution [4] for the motion of a near-Earth satellite under the influence of the zonal harmonics J_2, J_3, J_4, and J_5. This work was later published in the *Astronomical Journal* [5]. In that same journal and on adjacent pages, Yoshide Kozai [6] published another solution to the same problem. Most analytical orbit prediction models in the U.S. Space Surveillance System today still have one of these two methods as their foundation.

In 1961, Brouwer and Hori [7] developed a modification to the 1959 Brouwer solution that included the effects of atmospheric drag. The atmospheric drag model was based on a static exponential representation for atmospheric density with a constant scale height. This choice of density model led to series expansions in the scale height. The complete model was much too extensive to be run on the computers of that time for numerous satellites, given the slow convergence of the series and given that numerous orbits had to be computed.

In the early 1960s the same group that had developed the original NSSCC documentation for Hanscom performed some pivotal work in atmospheric density modeling. Starting from the basic equations of hydrostatic equilibrium and assuming scale height varied linearly with altitude, they derived a density representation using power functions with integer exponents [8]. The importance of this work is that, when applied to an artificial satellite theory, it completely avoids series expansions that occur with exponential representations. This made possible the inclusion of drag in the Brouwer model in a more complete and compact manner. The resulting analytic orbit model was developed by Max Lane [9] in 1965 with further improvements by Lane and Cranford [10] in 1969. Fitzpatrick [11] gives a brief account of Lane's density modeling technique in his textbook.

A very important contribution to analytic satellite theory was made by R. H. Lyddane [12] in 1963. Lyddane showed that the Brouwer solution based on Delaunay variables could be reformulated in terms of Poincaré variables to avoid the small divisors of eccentricity and the sine of inclination while maintaining the first-order character of the theory.

III. OPERATIONAL IMPLEMENTATIONS: 1964–1979

The transition from paper to operational implementation took two paths. The tracking operation of the Naval Space Surveillance System (NAVSPASUR) adopted the entire 1959 solution of Brouwer with the modifications developed by Lyddane to avoid small divisors of eccentricity or the sine of inclination. This analytic satellite prediction model is now known as PPT3 (position and partials as functions of time). The equations of the original PPT model were implemented on an IBM 7090 computer in 1964 under the guidance of Richard H. Smith who also provided supplemental equations to account for atmospheric drag. At that time, the results of Brouwer and Hori could not be implemented operationally because of computer limitations, and the results of Lane and Cranford were not yet available. Smith adapted ideas from King-Hele [13] in a simple original model that is still in use. His semi-empirical drag model assumes that the effect of atmospheric drag on the mean motion can be represented as a quadratic time function. The linear and quadratic coefficients are treated as solved-for parameters during the orbit determination process. A time rate of change of the eccentricity is represented in terms of the mean motion rate by the following equations:

$$\dot{e}_0 = e_0(1 - e_0^2)\frac{\dot{a}_0}{a_0}, \quad \dot{a}_0 = -\frac{4}{3}\frac{a_0}{n_0}\left(\frac{\dot{n}_0}{2}\right)$$

The integral of the mean motion equation provides the model for the along-track drag effect.

PPT retains all long-periodic terms, including the ones with a zero divisor at the critical inclination. However, PPT handles these critical terms in a special way, as described by Hoots [1]. A special feature of PPT is that the "mean" mean motion is defined differently from Brouwer's quantity of the same name. Brouwer defined mean motion in terms of mean semimajor axis by essentially the Keplerian formula. However, for PPT, it was decided for computational reasons to define the mean motion as the entire coefficient of time in the linear term of the perturbed mean anomaly. That is, the PPT mean motion includes the zonal secular perturbation rate of mean anomaly that Brouwer derived. As a result, the expression for PPT mean motion explicitly contains perturbation parameters and functions of the other mean elements, like the definition adopted by Kozai [6]. Numerically, the PPT mean motion is closer to Kozai's value than to Brouwer's.

The other path from journal paper to operational implementation took place in Colorado Springs, Colorado. In 1961, the NSSCC was relocated to Colorado Springs and became known as the Space Detection and Tracking System (SPADATS) Center. The NSSCC algorithms were rehosted on a Philco Model 211 computer, and the group at Hanscom began to serve as the backup for the SPADATS Center. Following the rehosting in Colorado Springs, Geoff Hilton [14] provided updated documentation of the NSSCC algorithms. In 1960, Aeronutronic had begun developing the astrodynamics basis for a new system. The analytic orbit prediction model was based on the works of Brouwer and Kozai. To avoid small divisors of eccentricity or the sine of inclination, Arsenault, Chaffee, and Kuhlman [15] transformed the solution to a series in nonsingular parameters keeping only the most important terms. They included from Brouwer only those long- and short-period terms in position that do not contain eccentricity as a factor. They also adopted from Kozai the non-Keplerian convention relating mean motion to semimajor axis. The model is known as the simplified general perturbations (SGP) model. A complete documentation of SGP is provided by C. G. Hilton and J. R. Kuhlman [16]. Atmospheric drag was included in a similar manner to that of Smith except that the time rate of change of eccentricity was derived based on the assumption that perigee height remains constant as semimajor axis shrinks. In addition to becoming the principal analytic prediction model for centralized processing, SGP was also implemented at many of the satellite tracking sites around the world. In 1964, the SGP model became the primary orbital prediction model for the SPADATS system.

The improvement offered by an analytic density model rather than an empirical decay model led to a decision to implement the development of Lane and Cranford. However, by 1969 the number of satellites in the catalog had grown to a point that computers would not be able to manage the extensive terms in the model. Consequently, a simplified version of the Lane and Cranford work, known as SGP4, was developed and implemented operationally in 1970.

The simplifications leading to SGP4 were accomplished by retaining only the main terms that modeled the secular effect of drag. Similarly, the gravitational modeling was shortened by retaining from Brouwer only those long- and short-periodic terms in positions that do not contain eccentricity as a factor. The details of the derivation of SGP4 from the complete development of Lane and Cranford were documented in 1979 by Lane and Hoots [17]. The SGP4 model was used side by side with the SGP model until 1979 when it became the sole model for satellite catalog maintenance.

IV. Deep Space Modeling: 1965–1997

In 1965, the first highly eccentric, 12-hour-period satellite was launched. Soon it became apparent that a theory was needed that included terms to account for lunar and solar gravitation as well as the resonance effects of Earth tesseral harmonics. A semi-analytic treatment of this special class of orbits, which included lunar and solar gravity as well as geopotential resonance effects, was developed by Bruce Bowman [18] in 1967. By 1977, Dick Hujsak [19] had incorporated portions of Bowman's work in a new first-order model, which included all perturbations treated by Bowman and an extension to geosynchronous satellites. This new model was fully integrated with the SGP4 model for near-Earth satellites in 1979. This work completed the SGP4 model in use today. A comprehensive listing of the equations was provided by Hoots and Roehrich [20].

In 1997 the lunar, solar, and resonance terms from the SGP4 model were added to the Naval Space Command PPT model to provide improved prediction of higher-altitude satellites. This modified model became known as PPT3 and is documented in the work of Paul Schumacher and Bob Glover [21].

V. Model Accuracy Improvements: 2000–2024

Much of this section is based on a 2000 paper [22] by the authors. Although the satellite catalog had long been maintained with a moderately accurate (GP, or analytical) orbit model, it had been known for an equally long time that highly accurate (SP, or numerical) models were the preferred way to represent orbital motion. Interest in the SP catalog maintenance problem grew when, in August 1997, NASA published new requirements for space surveillance support of manned space flight operations [23]. These requirements implied nearly instant availability of SP orbit solutions for any object in the satellite catalog, on a permanent and continuous basis. The capability to produce and communicate these SP solutions was needed before the launch of the first element of the ISS in July 1998.

In response, the SP catalog development activities at NAVSPACECOM during 1997 culminated in an extended demonstration in December of that year [24]. Approximately 8800 satellite orbits were maintained for 30 days using distributed parallel processing on a 26-node subnetwork partitioned out of the NAVSPACECOM operational computer system. This demonstration system processed all the same observations as the operational GP-based system.

In a parallel effort, the Air Force was also pursuing implementation of catalog maintenance using SP. The target processing environment was a multiprocessor workstation. In 1998, the ASW became operational in Cheyenne Mountain and began supporting NASA ISS operations with collision assessment and covariance data. This operational support included maintaining a subset of the space catalog, about 700 satellites that pose a collision threat to the ISS. In September 1999, the ASW began operational processing of an SP space catalog of approximately 10,000 objects in Cheyenne Mountain.

The shift from GP-based processing to SP-based processing had several important implications besides the obvious gain in prediction accuracy for all cataloged objects. The shift was really from a model-limited system to a data-limited system. Using only GP orbit models, which may have kilometers of inherent prediction error, it was not possible to extract all the information present in the current space surveillance tracking data, most of which are precise to within hundreds of meters. No amount of high-quality data could improve the catalog accuracy beyond the limitations of the GP model itself. On the other hand, SP models have only meters of inherent error (except possibly for atmospheric drag effects), and so the data can be more fully exploited. With an SP model, the catalog accuracy was limited mainly by the amount and quality of the data available. The result was that maintaining an SP-based catalog encouraged the best use of the existing computer systems, sensor systems, and observational data. Moreover, the SP catalog encouraged (and still encourages) sensor system improvements that would not have been worthwhile if only GP accuracy were available to end users of the catalog.

Along with the gain in prediction accuracy, the SP catalog also enables a realistic assessment of prediction uncertainty for all cataloged objects. This is something that had not been within reach until that time. When the physical model of satellite motion has large inherent errors, as a GP model often does, the job of adequately characterizing the model uncertainty becomes too complicated to carry out in practice. With an SP model, this job, though still complicated, can be done, mainly because the inherent errors are much smaller. At the same time, regardless of the model used, space surveillance observational uncertainties can be characterized with highly accurate calibration reference orbits for a few satellites [25]. As a result, the prediction error covariance matrix, which depends on both the observational errors and the model errors, can be estimated with reasonable confidence for an SP orbit solution, whereas the same confidence cannot be achieved in practice for the less-accurate GP predictions. This fact implies that the SP catalog can enable realistic assessments of collision risk in close conjunctions, reliable tracking discrimination when cataloged objects are close together, and reliable acquisition or avoidance of any cataloged objects by laser systems. These are fundamentally new capabilities that have profound implications for both civil and military space control. They provide ample justification for trying to build the best possible SP cataloging system.

It is not enough to be able to develop an SP catalog. That information must be made available to users who will need to do their own calculations using the catalog information. For decades that information was provided using a standard TLE message that contained an epoch time, six orbital elements, and three parameters for modeling atmospheric drag. The perturbations of the GP model were fixed so all information needed was contained within the TLE. In contrast, the analyst could optimize the SP accuracy by selecting the most appropriate set of perturbations for each satellite.

To inform a user of exactly which perturbations were used, it was necessary to devise a new message called a vector covariance message (VCM). This not only contains the epoch and the state vector (position and velocity), but it also specifies exactly what perturbations and models were applied. To be a self-contained data set, it also provides timing constants and solar flux data. The VCM contains 32 lines of data and successfully conveys enough information to allow a user to approximately match predictions made by the central center.

A by-product of the high-accuracy catalog is an improvement in accuracy of the TLE catalog. The high-accuracy catalog is developed and maintained by fitting the latest observations of satellites. The epoch of the VCM is typically placed at the end of the data span. Once a new VCM is computed, it is used to generate an ephemeris extending for some number of days before and after the VCM epoch. By doing a least-squares fit of the TLE orbit model to the ephemeris, a set of TLE initial conditions is obtained. In this way, the TLE will not only be constrained to follow the observed trajectory before epoch but will also be constrained to follow the high-accuracy predicted trajectory after the epoch. In this way, the TLE users gain an immediate accuracy improvement with no effort on their part.

This is very important because TLEs are the "coin of the realm." They are recognized and used worldwide for many reasons. They are reliably available, require a small data transfer, and provide fast computation for moderate accuracy needs. They are used at all U.S. tracking facilities for tracking opportunity planning and data correlation.

During the first two decades of the 21st century, technology advancements in optical satellite tracking and processing resulted in measurement accuracy improvement by an order of magnitude or more. This technology improvement directly led to a need for a commensurate prediction model improvement. Previously the standard for correlating observed tracking with predicted position was governed by the uncertainty of the measurements. This uncertainty was reasonably in line with the prediction uncertainty of the SGP4 orbit model. But as optical technology advancements provided improved measurement accuracy, it became apparent that the prediction model accuracy must be improved to the level of measurement accuracy.

One might naturally conclude that a highly accurate numerical integration technique might be the answer to this need. However, some fundamental constraints prevent application of such an approach. The communications bandwidth and computational limitations at tracking sites cannot accommodate the workload of daily data exchange and track opportunity planning for today's large catalog. Furthermore, the standard data transmission messages are limited to the TLE data format. Thus, any upgrade to the satellite prediction model must satisfy the following constraints:

1. All necessary data required for model prediction are contained within the TLE.
2. The model must have a run time that is comparable to the existing SGP4 model.
3. The model must provide one or two orders of magnitude accuracy improvement.
    - More complete lunar–solar gravity modeling must be provided.
    - More complete geopotential resonance modeling must be provided.
    - Solar radiation pressure modeling must be included.
    - Orbital decay modeling should be improved.
4. The model must extend applicability to the cislunar region (orbits with multiday periods).

The new model is named SGP4-XP because it is a next step in the orbit model evolution of SGP and SGP4. The suffix "XP" connotes extended precision. In that spirit, the model is best described in comparison to the SGP4 model used in operations for over 50 years. The details of the new model are described by Hoots [26]:

1. Orbital elements: SGP4-XP employs mean orbital elements averaged over one orbital period, whereas SGP4 uses mean orbital elements averaged over both the orbital period and over the much longer period of the argument of perigee motion. The descriptive terminology is SGP4-XP uses singly averaged mean elements, and SGP4 uses doubly averaged mean elements.
2. Integration method: Both SGP4 and SGP4-XP are semi-analytic orbit models. This means that some portion of the theory is developed analytically. This is typically the averaging and the recovery of the periodics. The remaining terms on the right-hand side of the differential equations slowly vary and can be numerically integrated using a relatively large step size. SGP4 uses an Euler integrator, and SGP4-XP uses a Runge–Kutta integrator.
3. Earth gravity: SGP4-XP includes the complete solution of Brouwer that models the zonal harmonics J_2, J_3, J_4, and J_5. In contrast SGP4 does not include the effects of J_5. Furthermore, all periodics terms from Brouwer are truncated by dropping terms of order $O(eJ_2)$.
4. Lunar and solar gravity: Both SGP4-XP and SGP4 model the third-body perturbations as a series expansion in the ratio of the distance to the satellite divided by the distance to the perturbing body. The SGP4 model only considers the secular effect of the first term of the expansion. No periodics having the satellite period were developed. SGP4-XP carries the expansion to higher order and includes the satellite revolution periodics.
5. Geopotential resonance: SGP4 is specifically tailored for geosynchronous and Molniya orbits. Thus, the model is not applicable to the wide variety of deep space orbits used today. For example, the equations would not handle GPS-type orbits. Hence, the existing resonance code was discarded and replaced by a totally generalized approach valid for any eccentricity or inclination that is applicable to the following classes of resonance periods: 24 hour, 16 hour, 12 hour, and 8 hour.
6. Solar radiation pressure: In recent years, we have seen a dramatic increase in the number of satellites that are significantly perturbed by solar radiation pressure. These are referred to as HAMR objects. The effects are significant, and inclusion is necessary to gain the desired order of magnitude accuracy improvement. The new equations include both the average effect as well as the periodics over one revolution of the satellite. It should be noted that the new model properly accounts for orbits that are partially in the sun and partially in the shadow.
7. Atmospheric drag: The power-law static atmospheric density model of SGP4 was replaced with the Jacchia 70 atmospheric density model. This is the same model that forms the basis for the high-accuracy catalog density modeling. To avoid the requirement for a TLE recipient to obtain current solar flux values, a generic model was developed based on the behavior of solar flux averaged over multiple solar cycles. The model

simply uses the epoch of the TLE to determine where the elset falls in the generic cycle. It then uses that average value of solar flux as input to the Jacchia model. Usage of this model allows production of a meaningful ballistic coefficient rather than the B* term used in SGP4.

A few accommodations were necessary to provide the SGP4-XP initial conditions using the TLE format:

1. SGP4 is identified as "ephemeris type" 0, while SGP4-XP is identified as "ephemeris type" 4. This allows the user to know what model generated the TLE so it can be properly used.
2. SGP4 does not model radiation pressure. SGP4-XP does model radiation pressure, and the coefficient is provided in the space used for the second derivative of mean motion.

The government produces both an SGP4 and an SGP4-XP catalog of TLEs. Each is developed by fitting to the high-accuracy VCM catalog used for orbit maintenance. This allows users and tracking sites to continue their current use of SGP4 and to adopt the SGP4-XP model as their schedules may permit.

VI. CATALOG MAINTENANCE AND USER APPLICATIONS

Figure 3.1 provides an overview of the catalog maintenance process. Once a satellite is identified and added to the catalog, the orbit is maintained by a closed-loop process. The TLE is sent to the SSN with a tasking request for some level of tracking depending on the importance of the satellite. The sensors in the network respond by collecting the requested amount of data.

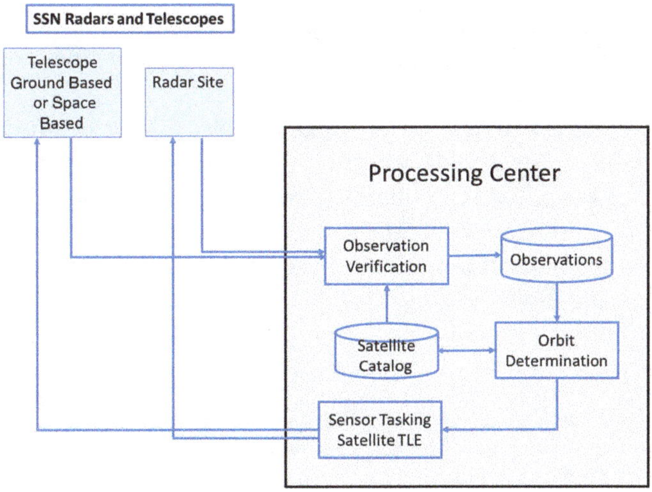

Fig. 3.1 Catalog maintenance overview.

The tracking data are sent to the central processing center where the tracking data are verified to belong to the requested object. If not, the track is declared an uncorrelated track (UCT). Verified observations are used in an orbit determination to produce an updated state for the satellite. This state (in the form of a TLE) is sent to the SSN for subsequent tracking, and the loop is closed.

The updated catalog is used for many important functions. The improved accuracy of the SGP4-XP model will directly translate to the accuracy and reliability of user support:

1. Observation planning: Sensor sites use the TLE data to determine the time windows when satellites will be in view of the site for daily operations. The efficiency of the TLE model facilitates accomplishing this large computation quickly. This preplanning allows instantaneous identification of penetrations of the surveillance fence.
2. Observation correlation: Sensors track objects and tag the observations based on the catalog they receive. When observations arrive at the processing center, the most recent catalog is used to verify the identity tag. If identity is not confirmed, then the remainder of the catalog is examined to find the proper identity. If no match is found, the data are declared an uncorrelated track.
3. Overflight prediction: A critical area of support is overflight warning. For satellites with an ability to collect electronic signals or take images, it is important to be able to warn ground resources of the times of overflight by these satellites.
4. Close approach determination: Each day screening must be performed to determine times at which pairs of satellites may have a physical close approach. Some entities do this for each satellite versus all other satellites. Others consider just active satellites versus all others. Regardless, this is a large computational load. A catalog of accurate TLE data can be used as a prefilter to reduce the satellite pairs to be considered using more accurate state vectors and covariance uncertainty metrics.
5. Laser clearing: There are various reasons why a laser beam is needed to aim at another satellite. It could be for communications, for precision tracking, or for other experiments. However, another satellite that passes through the laser beam could be inadvertently damaged. The United States provides a planning function to advise laser operators of clear times for their laser operations.

REFERENCES

1 Hoots, F. R., Schumacher, P. W., and DeVere, T. "50 Years of Satellite Tracking and Cataloging in the U.S.," AAS/AIAA Astrodynamics Specialist Conference, Paper 07-328, Mackinac Island, MI, Aug 2007.

2 Wahl, E., *Proceedings of the First Symposium (International) on Rockets and Astronautics*, Tokyo, 1959.

3 Fitzpatrick, P. M., and Findley, G. B., "The Tracking Operation at the National Space Surveillance Control Center," *Directorate of Aerospace, Air Proving Ground Center, Eglin AFB, FL*, 2 Sept. 1960.

4 Brouwer, D., "Solution of the Problem of Artificial Satellite Theory Without Drag," Air Force Cambridge Research Center, Geophysics Research Directorate, Bedford, MA, Oct. 1959.

5 Brouwer, D., "Solution of the Problem of Artificial Satellite Theory Without Drag," *Astronomical Journal*, Vol. 64, No. 1274, Nov. 1959, pp. 378–397.

6 Kozai, Y., "The Motion of a Close Earth Satellite," *Astronomical Journal*, Vol. 64, No. 1274, Nov. 1959, pp. 367–377.

7 Brouwer, D., and Hori, G., "Theoretical Evaluation of Atmospheric Drag Effects in the Motion of an Artificial Satellite," *Astronomical Journal*, Vol. 66, No. 5, June 1961, pp. 193–225, and "Appendix to Theoretical Evaluation of Atmospheric Drag Effects in the Motion of An Artificial Satellite," Vol. 66, Aug. 1961, pp. 264–265.

8 Lane, M. H., Fitzpatrick, P. M., and Murphy, J. J., "On the Representation of Air Density in Satellite Deceleration Equations by Power Functions with Integral Exponents," Deputy for Aerospace, Air Proving Ground Center, Eglin AFB, FL, March 1962.

9 Lane, M. H., "The Development of an Artificial Satellite Theory Using Power-Law Atmospheric Density Representation," AIAA Paper 65-35, Second Aerospace Sciences Meeting, New York, 25–27 Jan. 1965.

10 Lane, M. H., and Cranford, K. H., "An Improved Analytical Drag Theory for the Artificial Satellite Problem," AIAA Paper 69-925, AIAA/AAS Astrodynamics Conference, Princeton, NJ, 20–22 Aug. 1969.

11 Fitzpatrick, P. M., *Principles of Celestial Mechanics*, Academic Press, New York, 1970.

12 Lyddane, R. H., "Small Eccentricities or Inclinations in the Brouwer Theory of the Artificial Satellite," *Astronomical Journal*, Vol. 68, No. 8, Oct. 1963, pp. 555–558.

13 King-Hele, D., *Theory of Satellite Orbits in an Atmosphere*, Butterworths, London, 1964.

14 Hilton, C. G., "The SPADATS Mathematical Model," *ESD-TDR-63-427,* Aeronutronic Publication U-2202, 5 Aug. 1963.

15 Arsenault, J. L., Chaffee, L., and Kuhlman, J. R., "General Ephemeris Routine Formulation Document," ESD-TDR-64-522, Aeronutronic Publication U-2731, Aug. 1964.

16 Hilton, C. G., and Kuhlman, J. R., "Mathematical Models for the Space Defense Center," Philco-Ford Corp. Publication U-3871, Nov. 30, 1966, pp. 17–28.

17 Lane, M. H., and Hoots, F. R., "General Perturbations Theories Derived from the 1965 Lane Drag Theory," Project SPACETRACK Rept. 2, USAF Aerospace Defense Command, December 1979.

18 Bowman, B., "A First Order Semi-Analytic Perturbation Theory for Highly Eccentric 12 Hour Resonating Satellite Orbits," NORAD, Nov. 1971.

19 Hujsak, R. S., "A Restricted Four Body Solution for Resonating Satellites Without Drag," Project SPACETRACK Rept. 1, USAF Aerospace Defense Command, Nov. 1979.

20 Hoots, F. R., and Roehrich, R. L., "Models for Propagation of the NORAD Element Sets," Project SPACETRACK Rept. 3, USAF Aerospace Defense Command, Dec. 1980.

21 Schumacher, P. W., and Glover, R. A., "Analytical Orbit Model for U. S. Naval Space Surveillance: An Overview," AAS Paper 95-427, AAS/AIAA Astrodynamics Specialist Conference, Halifax, Nova Scotia, 14–17 Aug. 1995.

22 Schumacher, P. W., and Hoots, F. R., "Evolution of the NAVSPACECOM Catalog Processing System Using Special Perturbations," Fourth U.S./Russian Space Surveillance Workshop, Washington D.C., 23–27 Oct. 2000.

23 Letter from NASA Administrator (D. S. Goldin) to Commander in Chief, U.S. Space Command (Gen. Howell M. Estes, U.S. Air Force), 27 Aug. 1997.

24 Coffey, S. L., Neal, H. L., Visel, C. L., and Conolly, P., "Demonstration of a Special-Perturbations-Based Catalog in the Naval Space Command System," AAS Paper 98-113, AAS/AIAA Space Flight Mechanics Meeting, Monterey, CA, 9–11 Feb. 1998.

25 Schumacher, P. W., Gilbreath, G. C., Lydick, E. D., Davis, M. A., Seago, J. H., and Walters, S. G., "Design for Operational Calibration of the Naval Space Surveillance System," American Astronautical Society Paper 98-116, *Advances in the Astronautical Sciences*, Vol. 99, *Spaceflight Mechanics 1998*.

26 Hoots, F. R., "SPACETRACK Report No. 7: A New TLE Based Orbit Model," The Aerospace Corporation TOR-2024-00572, 15 June 2024.

Chapter 4

THE PROCESSING SYSTEMS

The science of astrodynamics as the estimation and prediction of the orbits of artificial space objects flowered in the early years of space surveillance. Moreover, as all students of that science in every generation soon come to realize, calculations of orbital motion tend to be long and involved, whether the method is analytical (typically via infinite series expansions) or purely numerical (as with numerical integration of systems of lengthy differential equations). Yet, the practical application of this growing body of knowledge to space surveillance was drastically constrained in the early days by the available computing facilities. Until the 1990s, the computer systems used for space surveillance, and for catalog maintenance, simply did not have the speed or capacity to host the most accurate astrodynamics methods known for Earth satellites, because of the sheer number of satellites that had to be tracked. The only recourse was to accept sometimes-severe approximations to shorten the mathematical formulations used to represent satellite motion. The result was the truncated analytical methods discussed in the previous chapter, notably SGP4 and PPT3, and purely numerical methods having less-than-complete representations of Earth's gravitational force, atmospheric drag force, and so on.[87] As one would expect, these simplified models of satellite motion were

[87]This basic situation of having to accept severe approximations has been understood by practitioners of celestial mechanics for more than two centuries. However, it was expressed particularly well, in a context not wholly different from space surveillance, by two scientists working at the Naval Weapons Laboratory in Dahlgren in the late 1960s and 1970s. Claus Oesterwinter and Charles J. Cohen, in a paper they published in 1972, famously wrote, "There is only so much algebra a man can do in forty years, so that all series have to be truncated somewhere" ([1], p. 317 in the journal paper.) They were speaking about the limitations in our knowledge of the orbital motion of the natural bodies in the solar system. Of course, the same principle had applied to the new regime of Earth satellite motion, beginning in the late1950s, regarding both the new mathematical formulations and the new computer implementations of those methods. Neither of these men was heavily involved in space surveillance, although both maintained interest in the mission and the associated technical problems until their retirements. Cohen worked in Dahlgren from 1944 to 1978 and was known mainly for his pioneering work in satellite geodesy. He also collaborated with Paul Herget on the software implementations at Dahlgren for the Naval Space Surveillance System (NSSS). Oesterwinter made his career

inherently less accurate than the better models that many persons had long known how to formulate and actually had implemented for other space applications involving only one or a few satellites at a time. It turned out that, historically, and very fortunately, improvements in computers kept pace with the growth of the space object population that had to be monitored, if those simplified orbit models were used. The effect was that the less-accurate models of satellite motion had to be used for many more years than anyone might otherwise have wanted: the growth of the catalog did not allow even the improved computer systems that became available over the years to accommodate the much longer calculations required by more accurate orbit models. The analytical models mentioned earlier became standards de facto, even with their known inadequacies, and were embedded in computer systems throughout the defense establishment in ways that were (and still are) expensive and cumbersome to change.

In the early 1990s, some relief from this computational bind appeared in the form of new computer technology, specifically, parallel processing. The basic idea is obvious, although a computer implementation of it has not always been so easy: some kinds of tasks can be divided up among many workers, each of whom can then do his part of the work simultaneously, that is, "in parallel," with the others, greatly reducing the time to finish the overall task. The most prominent computational task for space surveillance, maintaining the satellite catalog, turns out to be just such a task that can be easily parallelized: the same lengthy calculation program needs to be executed for each satellite, with the only difference being the input data for each satellite. For one picturesque example of this idea, it is known that in the very earliest days of the NSSS operation in Dahlgren, individual satellites were assigned to individual human analysts.[88] Each analyst was responsible for performing the following steps:

1. Know approximately when his satellite was expected to be detected by the Fence.
2. Read the raw antenna phase data off the strip chart recorders connected to the telephone lines.

at Dahlgren from 1959 to 1989, working mostly on applications of celestial mechanics to missile and satellite systems but occasionally serving as a consultant to NAVSPASUR staff on orbit-modeling questions.

[88]This information derives from private communications between one of the authors and Emory Hall Bales in late 1988 and early 1989 at the Naval Space Surveillance System (NAVSPASUR) headquarters in Dahlgren [2]. At that time, Bales was head of the Applications Section in the Software and Analysis Department, in charge of Fence data reduction. He had joined the NSSS project in Dahlgren in 1960, and, like most of the staff at the beginning, had taken his turn as one of those duty analysts. Throughout his career, Bales made substantial contributions to the Dahlgren operation, extracting the maximum accuracy from the Fence data and serving as the most authoritative source of expertise on Fence data characteristics. He sometimes confounded Naval Research Laboratory (NRL) scientists who had firsthand knowledge of the NSSS, with his original interferometric methods that were adapted to the details of the Fence design, and which were faster computing than more standard radar-processing methods. That role continued until he retired from the Naval Space Surveillance operation (by then part of Naval Space Command) in 1997.

3. Reduce the phase data into direction cosines using a special 4-ft-long, one-of-a-kind slide rule and a desktop calculator.
4. Prepare the input direction-cosine data for the main computer to update the orbit of the satellite.
5. Run the update program to predict the times and positions of the next few passes through the Fence.
6. Record the predictions for his use later when the satellite made its next pass through the Fence.

This operational concept for catalog maintenance became unworkable when the number of cataloged satellites outnumbered the duty analysts by only a modest factor: with the data-processing equipment available in 1960, a well-practiced analyst might require half an hour to complete these steps for one satellite, for each pass through the Fence. By June 1961, about fifty satellites were being tracked, so that much better automation and computation had already become mandatory. However, because of the limited computer technology available in the first three decades of space surveillance, the idea of parallel processing for catalog maintenance lay dormant for most of that time.

Two different technical innovations for parallel computing emerged almost together in the early 1990s, in forms that could be adapted for the space surveillance mission and for catalog maintenance. Although the new computing architectures did not enable the theoretical nicety of one processor for each satellite, as the early NSSS experience had suggested, drastic reductions in processing time had become possible, so that more accurate orbit models could be used. It is interesting from our perspective nowadays that the Air Force chose one of the two innovations and the Navy chose the other, resulting in differences of implementation that persisted almost up to this writing. In fact, the individuals working on the parallel-processing upgrades in each service in the early 1990s strove to be continually aware of the progress and plans of both teams. However, the different services had to start from different operational conditions at each center, with different acquisition processes, different budgets, and different schedules, not to mention somewhat different motivations. As described in the next chapter on the different operations centers, a constant goal was to maintain a "functional equivalence" between the Air Force cataloging center in Colorado Springs and the Navy's cataloging center in Dahlgren. This equivalence for the more accurate catalog maintenance was achieved early in the catalog upgrade process during 1997 and 1998 and was maintained thereafter, although the extensive coordination needed between the two service development teams proved to be a complicated matter. In any case, the result by the late 1990s was the implementation by both services of the most accurate orbit model ever used for space surveillance up to that time: a full-scale numerical integration of the differential equations of motion for every prediction for each satellite, using a force model that called for evaluating a few thousand trigonometric terms for gravity, besides table-lookups for the atmospheric density model, at each time step. The so-called high-accuracy or SP catalog of orbital data is still in use at this writing, with only relatively minor adjustments in the implementations at each catalog center over the years. Without any exaggeration, we can say that, given the availability of sensors and communications (a significant story, as we have seen in other chapters), the inception of the SP catalog was the single most important step taken in space surveillance to date, for reasons that are discussed later in this chapter.

One of the two technical innovations in computing was the commercial advent of multicore chips and multiprocessor electronics boards, with the associated operating-system software that could automatically manage the data flow to and from multiple internal processors on the common data bus with good efficiency. Although the internal details were mostly hidden from the user, so that it was not always termed "parallel processing" explicitly, such an arrangement allowed several then-new computer platforms to take advantage of the fundamental idea within a quite short development time and with minimal changes in the existing application software. The Air Force chose this path to implementing the SP catalog. Even though the efficiency of the parallelization was not always ideal internally in the machine, because of limitations in the operating-system software, the overall speed-up of processing for the catalog update turned out to be more than adequate for maintaining the SP catalog indefinitely. The Air Force's implementation at the Cheyenne Mountain Complex took a little less than two years from project start, including design, installation of new multiprocessor computers to supplement the existing computer system, and testing. The operational capability was available in the second half of 1998 at a cost that was modest compared to other work going on at the Mountain at the time.

Meanwhile, the Navy had a different opportunity to implement parallel processing for the catalog. As part of a somewhat overdue computer technology upgrade, the mission processing system at Naval Space Command in Dahlgren was changed from a mainframe-based design to a distributed-processing design, implemented operationally in a new building during 1995. Navy researchers at NRL proposed to adapt a then-new parallel-processing technique to allow the Dahlgren 240-node mission processing system to emulate a multiprocessor supercomputer. The idea was that special software could be installed on each computing node in the system to coordinate data transfers and computation to allow parallel processing, just as happens in a real supercomputer. The Navy's development of their SP catalog took about one year from the initial demonstration of the concept, producing an initial operational capability by the end of 1998, also at a modest cost compared to other work going on at Naval Space Command at the time.

In the remainder of this chapter, we seek to understand better both SP catalog developments just described, by briefly examining the computational capabilities available for space surveillance over the years. The very brief summaries offered later are naturally hard for present-day readers to appreciate because most of us have come to take advanced computer performance and capacity completely for granted in our everyday lives. However, inventing ways to use the very limited computing resources available in the first four decades of space surveillance demanded extreme ingenuity in some instances, the real history of which has yet to be written. For the present purpose, we set the computer-performance bar for comparisons deliberately low by contemporary standards: an unexceptional laptop computer, as it is being used to compose and edit these words, may be only about three years old, but already the main processor electronics are considered out of date by the industry and are no longer manufactured. Most of its performance specifications are no longer advertised or ranked in the current market, even by the manufacturer. This laptop computer weighs less than 5 lb with its 5-hour battery and consumes 28 W (peak) of electric power, with the display, keyboard, and one or two peripheral devices connected. The main processor board has four computational cores that can access the eight billion characters (bytes) of common main memory independently of each other, besides an additional 220 billion characters of common solid-state "disk" memory independently. The operating-system software is the 64-bit kind, the default choice offered by the manufacturer. The speed of memory access is around 3000 mega-transfers per second, fast enough to call up the *Encyclopedia Britannica* in a few seconds if it were stored locally on the laptop (and it could be). The speed of arithmetic operations for this machine is hard to estimate, since it is no longer advertised, but it is so high as to be essentially irrelevant for routine applications such as word processing. In fact, this machine is so ordinary that it can be underutilized quite economically for nothing but word processing and Internet-style communication. The clock speed is just over 3 GHz, meaning that the number of floating-point operations per second, using all four internal processors, is probably somewhere in the range of hundreds of millions. Given this mundane baseline for comparison purposes, let us consider a few examples of computer systems that our forebears in space surveillance have had to use for a mission that, from its very beginning, has had national importance.[89]

[89]The present authors owe a particular debt in this chapter to the authors Brian Weedon and Paul Cefola [3]. Their paper was a pioneering one in the subject of processing systems for space surveillance, in fact, essentially the only one up to this writing. The following discussion owes much to their choice of topics and to the questions they raise, as the reader of both narratives will realize. However, it would be too complicated to cite and interact with their discussion in detail in a more general work such as the present one. Here in our chapter introduction, we gratefully acknowledge their contribution.

I. THE AIR FORCE COMPUTER

1. PROJECT SPACE TRACK

Initially, in late 1957 and early 1958, Project Space Track, being merely another activity at AFCRL at Hanscom Field, shared access with everyone else at AFCRL to the laboratories' mainframe computer, an IBM 650.[90] The 650 was IBM's first mass-produced computer and had turned out to be more in demand than the company had expected. It had been introduced to the market in 1954, originally intended for scientific and technical applications. However, it soon became popular with businesses and even universities. The rotating-drum main memory contained up to 4000 10-digit words and spun at more than 12,000 rpm. That rate allowed about 40 instructions to be executed in 1 s. The vacuum-tube electronics in the arithmetic unit could perform additions in less than 2 ms and multiplications in about 13 ms. Input and output were done via magnetic tapes and punched-card readers. Even though the 650 required a special power supply delivering about 20 kilowatts (kVA), the popularity of the machine also guaranteed that a wide variety of system-software helps (translators, assemblers, primitive compilers) were developed over the years, and a wide selection of peripheral devices became available. In short, AFCRL could claim to be in the main wave of commercial computing capability at the time.

In August 1958, Space Track added its own computer, an IBM 610, dedicated to its mission. Introduced in 1957, the 610 was still relatively primitive compared to the 650. It used a magnetic drum memory holding 2600 digits, besides 6000 digits on a temporary tape drive. The vacuum-tube circuits for arithmetic were notably slow: some users wrote that it could take up to 20 s to calculate the value of a sine function, an awkward fact for analysts doing orbit-related calculations. A keyboard allowed the user to program the system by producing punched paper tape for input and output. The user could monitor the input and output data, as needed, by directing data to an 18-character-per-second electric typewriter. However, unlike the 3-ton 650 model (with all its peripheral equipment), the 610 weighed a mere 800 lb, fit comfortably in an office, required no special power supply or cooling, and could be operated by a single person.[91] In practice, it was used much like a desktop calculator, except that sequences of calculation steps could be stored or copied onto paper tape, and repeated with different data, and so it could be quickly mastered by anyone familiar with calculators.

[90] The IBM history accessible through the website https://www.ibm.com/history/650 [retrieved 10 December 2024], contains a variety of information about this model. Donald Knuth famously dedicated his classic work *The Art of Computer Programming* to the IBM 650 computer [4].

[91] Altshuler [5], Chapter 4, p. 55; and Wahl [6].

When Project Space Track became the operational NSSCC at Hanscom Field in December 1959, not only did the staff move to a new building, but also they obtained a new computer dedicated to the mission, an IBM 709.[92] This 1-ton machine required a special power supply delivering 100 kilowatts (kVA), plus special cooling. Peripheral equipment added to that total load. However, the magnetic core memory contained 32,000 36-bit words, and the arithmetic electronics could perform additions in 24 μs, despite being based on vacuum-tube circuits.

2. Space Defense Center

IBM had introduced the 709 in 1957, but rather soon, by December 1959, they introduced a transistorized version of the same computer, the more famous 7090 model, which was advertised as at least five times faster than the 709, processing about 100,000 floating-point operations per second, and requiring only about one-third as much power. The urgency of that development by IBM was not due to the needs of space surveillance but to a closely related activity: the BMEWS program wanted to replace its 709 computers with transistorized models that would be more reliable than the vacuum tube–based 709. In fact, BMEWS did achieve its full operational capability in 1961 with IBM 7090 computers. Nevertheless, for reasons that are not completely clear to the authors at this writing, the Space Defense Center in Colorado Springs, Colorado, in 1961 used a different computer, the Philco 2000 series Model 211, replaced by a Model 212 sometime in 1962. Whatever the reasons, these models turned out to be a good technical choice at the time, although maintenance and obsolescence eventually became problematic, as noted later. The 2000 series represented the first all-transistor computers, pre-dating the IBM 7090 by about two years.[93] The Model 212 had 65,000 48-bit words of memory and required only about 15 kilowatts (kVA) of power.

When the Space Defense Center moved from Ent Air Force Base in Colorado Springs into the Cheyenne Mountain Complex south of town, the Philco computers were also transferred. A triplet of Philco 2000 series computers were the core of the "Delta I" processing system, as the acquisition

[92] Wahl [6]. Fitzpatrick and Findley [7] reported that the NSSCC had a 709 "with other smaller computers."

[93] Sturdevant [8], p. 14, one of the very few sources on this point, mentions, but does not discuss, the Philco 2000 series in this context. Some additional, but still incomplete, background can be assembled at this writing, on the choice of the first computer system acquired by the Air Force for the space surveillance operation in Colorado Springs. Philco had been using transistors in small computers since 1955, mainly for military applications in airplanes. By 1957, their transistors were in several ways superior to those used by IBM, for high-speed digital applications. The 2000 series of computers, introduced in 1957, was Philco's effort to capture a share of the mainframe market from IBM, and that series appears to have been the first commercially available all-transistor computers. Some of this information is from a Philco brochure and related postings, courtesy of the Computer History Museum [9].

agent Electronic Systems Division of Air Force Systems Command called it, and they were used until 1979.[94] In any case, well before 1979 it had become clear that the satellite catalog would soon outgrow this workhorse computer system. Even worse, the Philco computers themselves were becoming hard to maintain, because Philco had gone bankrupt and had been purchased by Ford Motor Company in December 1961. In 1963, long before the Space Defense Center moved into the Mountain, the "Philco-Ford" brand no longer included computers. Despite the worsening supply-chain problems from that time on, an Air Force program to replace the Delta I system began only in 1972. Named the NORAD Cheyenne Mountain Complex Improvements Program (427M Improvements Program, using the budget-document nomenclature for short) by Electronic Systems Command, the computer upgrade was scheduled for completion in about 27 months after the contract start. However, several external factors combined to delay initial operation of the 427M system until 1979, by which time the catalog had grown to 5400 objects, more than 10 times larger than the catalog had been in late 1961.

3. The 427M System: Space Defense Operations Center

The 427M Improvements Program combined space surveillance functions with several other related defense activities at the Cheyenne Mountain Complex in a common facility known as the Space Defense Operations Center (SPADOC) supported by the SPADOC Computational Center (SCC). In 1972, several computer systems more advanced than the 212s were commercially available, and it must have seemed that the necessary computer upgrades to replace the Philco 2000 series were well within reach. However, one of the external circumstances that delayed and constrained the program, and which had a particularly deleterious effect on space surveillance, was, in fact, the choice of computer systems. Some background is necessary to understand why.[95]

Immediately after the Cuban Missile Crisis in October 1961, the DoD sought to improve electronic communications throughout the whole defense establishment and to provide better ways to inform the president and the secretary of defense about military actions in progress and about strategically important situations. The better-integrated communications network was called the World-Wide Military Command and Control System (WWMCCS,

[94] Del Papa and Warner [10], p. 19. The present authors also gratefully acknowledge private communications over the years with Lawrence (Larry) Grimaldi in Colorado Springs, about Cheyenne Mountain processing upgrade programs. All those communications are now impossible to date, but all of them were valuable for the present summary. Whatever errors we may have inadvertently introduced in our narrative are certainly not due to discussions with him but have happened despite them.

[95] Some of the following information is summarized in the graduate thesis by Smith [11], pp. 101–113, who helpfully put in perspective the complicated facts about the 427M program and other programs of the same type. Pearson [12] gives a very complete description of the problems with WWMCCS and their ramifications, including the case of the attack on the USS *Liberty* (pp. 71–78).

pronounced "Wimex" by most users). Although the system did make noticeable improvements in communications, it really consisted of a patchwork of different networks and software systems that had been in place already for many years in different organizations, departments, and regions of the world. The results of simply tying them together at the interfaces were not always good. For example, messages sometimes could be seriously delayed as they were being routed through a maze of potentially dozens of servers and systems, each with different software systems, capacities, and throughput speeds. This shortcoming of WWMCCS unfortunately manifested itself in the attack on the naval reconnaissance ship U.S.S. *Liberty* off the coast of Israel in June 1967 during hostilities between Israel and Egypt. Multiple orders from the Joint Chiefs of Staff for the *Liberty* to move away from the coast were delayed in the WWMCCS network for many hours. None of those orders reached the *Liberty* in time to avoid the Israeli Defense Forces' attack on the unidentified ship in their waters, resulting in the deaths of 34 Americans and heavy damage to the vessel. The blistering review of that incident, and others related to WWMCCS, in Congress afterward in 1971 led the joint chiefs to take decisive, if not perfectly informed, action to improve WWMCCS. Air Force Systems Command's Electronic Systems Division promptly purchased a few dozen Honeywell 6000 series computers to be installed at various critical nodes of the WWMCCS network. The idea was to reduce or eliminate unnecessary routing delays by having common hardware and software along key (high-priority or busier) message routes. One of the many network nodes that came under the joint chiefs' mandate was the SPADOC.

The Honeywell 6080 computers installed for the SPADOC turned out to have 256,000 36-bit words of memory and could execute almost one million instructions per second, several times better than the Philco computers. On the other hand, the catalog was also 10 times larger than it had been at the end of 1961. The 6080s turned out to be barely adequate for the computationally intensive space surveillance mission as soon as they were installed. Although slightly improved Honeywell models became available over the next few years, the catalog also grew somewhat faster than in previous years. Already by 1983, some kind of replacement for the 427M system was needed.

As a side note, we should mention that the kinds of difficulties involving delay and system performance that plagued the 427M program were to be repeated to some extent on other programs for space surveillance, over the next several decades. The reasons for this long-term trend of delays and performance shortfalls are complicated and evidently incompletely understood even by the military services. The problems were pervasive enough to have been not only endlessly debated within the services but also occasionally subjected to think-tank and academic studies. The graduate thesis by Smith [11] is merely one example of such an analysis. That author had been directly involved with aspects of defense acquisition and program management for several Cheyenne Mountain systems discussed in our narrative. He concluded[96] that most of the problems that delayed or otherwise compromised the end results of these programs were technical in nature and not the result of faulty program management. The technical problems arose partly because any delay in a program meant that ongoing technological progress in industry, which could be very rapid because of the competition between companies, would create mismatches between subsystems procured in different parts of an overall program. In particular, the space surveillance enterprise is a classic example of a "system of systems" that involves complicated integration of the different aspects of the mission, each of which calls for different expertise and must be engineered separately: tracking, communication, processing, threat detection and assessment, and other aspects. Incompatibilities among the systems that implement these very different aspects of the mission often lead to still further delays at any stage of the overall system integration. Naturally, if the cumulative delay in a program becomes long enough, the resulting overall system upgrade can be nearly obsolete by the time it is finished. Another reason for technical problems is illustrated earlier in the joint chiefs' mandate for common computing systems across different organizations regardless of different functions in each organization: high-level policy decisions, made with little, if any, technical insight, can soon become specifications for an acquisition program, which often turn out to be unworkable or unnecessarily constraining. Whether or not Smith's conclusions were definitive, the future space surveillance mission can be expected to provide ample opportunity to learn from the system limitations that surely will be encountered as the catalog and the scope of the mission grow, and to avoid both pitfalls just mentioned.

4. SPADOC 4: A PHASED PROGRAM OF UPGRADES

Efforts to replace the 427M system began in 1983. By that time, the technical needs were acute with the catalog having grown to unprecedented size, and a different approach to the upgrade was implemented. The idea was to divide the desired overall increment in space surveillance capability into smaller steps, each of which could be planned and engineered more easily, implemented more quickly, and scheduled in a particular priority order that addressed the more urgent needs earlier. Each increment of the program would seek to make the best use of new technology that would become available at that time. Fortunately, one of the most important upgrades, sufficiently large and fast computers for catalog maintenance, was within

[96]Smith [11], p. 332, in an outline of key findings. Smith discussed at length many specific examples from Cheyenne Mountain programs, and analyzed several different survey results, to arrive at his assessment.

reach for the first phase of the program, because computers had continued to improve steadily in the decade since 1972.

The specific content of each of the planned three phases (A, B, and C) is beyond the scope of our discussion because every aspect of all the missions in the Cheyenne Mountain Complex was involved. For our purpose, we note that an essential feature of the planned SPADOC 4 upgrade was that the program would be completed in 4.5 years from contract award in 1983, with the three planned phases being implemented at roughly one-year intervals, ending in 1987. What happened was that the development period for each of the phases turned out to be about six years, so that the A phase was implemented in about 1989, and the different phases were implemented at roughly two-year intervals thereafter. As it turned out, the C phase had to be split into two different phases, C1 and C2, so that the developmental period for the SPADOC 4-C2 phase lasted until 1995, for a total of 12 years to full implementation instead of the originally planned 4.5 years. Moreover, the upgrade contract had to be extended three years beyond the implementation of C2 to "clean up" (the term used informally at the time) some complicated system-integration problems. In any case, the upgraded SPADOC 4 system was in a stable state by 1998.

The initial delays in the SPADOC 4 upgrade program did have one positive effect: computers continued to improve in the meantime. The machines finally selected for the SPADOC 4 upgrade were the IBM 3090 series Model 200 mainframe computers. The 3090s had been announced in early 1985 and introduced on the market in 1986, in time for the SPADOC-4A development. These dual-processor computers were the most powerful in IBM's history up to that time, rated at 18 million operations per second and with memory that could be expanded to 256 MB. This was a fortunate circumstance, because the satellite catalog grew to about 8200 objects during the development periods for the A and B phases.

One negative effect of the rapid improvement in computers was already well known, and even expected, in the computer business: more rapid obsolescence. In 1990, only four years after its introduction, IBM discontinued the 3090 series in favor of a family of yet-more-powerful machines, the ES/9000 series, which were never used for space surveillance. The company did continue to service the 3090s for several years after that. Predictably, however, supply-chain problems and hardware maintenance difficulties with the 3090s began to appear by the end of the 1990s, although the machines continued to be adequate for catalog maintenance well past the year 2000.

5. 1ST COMMAND AND CONTROL SQUADRON AND CAVENET

Coinciding with the full SPADOC 4C inception in 1995, the space surveillance operational staff in the Mountain made an innocent-sounding decision that turned out to be quite important for the whole space surveillance mission for the next several decades. The staff was part of the 1st Command and Control Squadron (1CACS), as it was then called. 1CACS decided that the crews, working around the clock in three shifts, needed some kind of auxiliary computing facility in their offices so that they could do extra proficiency training for procedures and analysis methods in odd moments of free time. The crews also wanted to be able to do regular orbital analysis offline from the SPADOC system, since the higher-priority mission processing always happening on the 3090 mainframes often led to random delays in their lower-priority analyses, breaking up their attention and making it difficult to do problem-solving. Examples of such offline analyses might be matching up duplicate element sets or identifying observations not associated with any cataloged orbital element set or studying the orbital behavior of some space object over long periods of time. The solution for the crews' needs was for 1CACS to buy several Silicon Graphics workstations and set them up in-house, using their organizational budget for office equipment without any formal SPADOC program support or formal contractor assistance. By all personal testimonies, and judging by later results, the idea worked extremely well. The workstations turned out to have plenty of processing power and memory capacity to support all the crews' offline work. It was not feasible (or, for security and data-integrity reasons, proper) to connect the unofficial workstations to the same network that accessed the mainframe systems. However, catalog data, including orbital element sets and tracking observations, could be downloaded onto tape cartridges and hand-carried to the 1CACS workstations. Although data could not be uploaded in the same way (again, for security and data-integrity reasons), when some analysis problem had been solved offline, it was usually a straightforward job for the analyst to reproduce that solution on the operational system because no experimentation or other research was involved. Over time, the 1CACS crews produced an impressive library of offline software applications, many of them using the graphics displays available with the workstations, which would not have been feasible to implement in any manner on the operational system. They also set up a simple, but extensive, file-based database that allowed them to follow the motion of cataloged satellites over past years and to examine the observation histories for those objects. It was not unusual for the 1CACS crews to find errors in the catalog, such as the occasional observation associated with the wrong satellite, first using their offline system, and later to correct the errors on the operational system.

As workstations were added over the next several years, it became desirable to connect them together with an isolated local area network. That workstation network came to be known as CAVENet.[97] As the analysts continued to make ever-better use of CAVENet, eventually a somewhat sophisticated secure-one-way network connection to the SPADOC mainframes was allowed so that the analysts could solve problems offline using the most current catalog data. More important, as things would turn out, CAVENet would make possible the rapid introduction of the high-accuracy SP catalog maintenance in the Mountain, a processing load that the SPADOC mainframes would not have been able to handle.

6. The Air Force High-Accuracy Special Perturbations Catalog

It is helpful to understand why the SP catalog became a relatively sudden need for the military. In the natural evolution of things, the military eventually would have developed a high-accuracy catalog, because improving computer performance eventually would have encouraged it. Better accuracy allows more reliable tracking and identification of space objects, besides more reliable acquisition or avoidance of satellites by laser beams. More prominently as time goes on and as more objects orbit the Earth, better orbit accuracy also allows more reliable assessment of the collision risk when space objects happen to pass near each other. However, all such needs for high accuracy had, from the beginning, been met without maintaining the entire catalog with an SP orbit model. The reason was that the needs for high accuracy had always appeared infrequently and never for many satellites at once. All the computers up until the mid-1990s had been able to produce high accuracy for one or a few satellites at a time, in addition to maintaining the catalog with lower-accuracy orbit models. However, in 1995, preliminary discussions between the Air Force and NASA alerted many persons that the situation would suddenly be different with the launch of the ISS several years hence (the first module was launched on 20 November 1998). Previously, when shuttle missions were in orbit for a week or two at a time, the extra load of an occasional high-accuracy conjunction assessment between the shuttle and other satellites could be handled without much trouble. Now, with the ISS in orbit indefinitely, the extra load of high-accuracy conjunction assessments would be a permanent addition to the space surveillance mission. In fact, the military had always supported NASA on an informal basis of best effort not to interfere with the primary mission. Nevertheless, in anticipation of the ISS, a Memorandum of Agreement [13] between the two agencies was put in place in April 1996, which helped the military justify expending resources to support NASA on a more regular basis. When NASA's specific

[97] Correlation, Analysis and Verification of Ephemerides Network. The acronym was inevitable, given the location, but it took a little thought at 1CACS to decide what it meant.

list of space surveillance support requirements[98] for the ISS arrived in August 1997, it was clear that a high-accuracy catalog would be needed, at least for those satellites whose perigee height (lowest point on the orbit) was near or below the ISS altitude. Two items in the list could not be answered at all without an SP catalog of some kind. First, NASA requested that conjunction assessments between the ISS and any other object they specified be made available on demand within one hour, a timeline that could not be met reliably unless the high-accuracy orbit estimates already existed when the call came in. Second, NASA also wanted to calculate a probability of collision for each conjunction, not merely a miss distance, which meant that a whole set of statistical parameters (covariances) for the orbit estimate had to be reported along with the orbital information and on the same timeline. Reliable values of those parameters could not be generated from the lower-accuracy orbit estimates used in the regular catalog but could be generated from a high-accuracy orbit estimate. Many other technical considerations entered the discussion, but the goal was clear. At the time, about 2000 objects in the catalog had perigee altitudes near or below the ISS altitude. Given a facility that could maintain a database of that many objects with a high-accuracy SP orbit model, the entire catalog of more than 9000 objects at that time could be maintained at relatively little extra cost [15]. The question was how to implement it.

The answer was CAVENet. Once the Air Force's contractor on this project, Omitron Corporation, proposed to use a new 16-core multiprocessor workstation for the SP catalog, everyone close to the mission knew how it could be hosted most easily in the Mountain. Adding another workstation to CAVENet was as close to simplicity as computer networking gets. The ASW, as it was called, became a permanent addition to CAVENet. Sensor observations were downloaded from the mainframe systems in the SPADOC Computation Center, and the SP catalog was updated, three times per day. The initial operational capability for the ASW in mid-1998 was a list of about 700 satellites that posed the most risk to the ISS. By September 1999, the ASW's capacity had been expanded to the entire catalog of almost 10,000 satellites [16].

This arrangement for the Air Force's SP catalog remained in place until 2007. In that year the whole space surveillance operation was moved from the Cheyenne Mountain Complex to Vandenberg Air Force Base in California as part of a reorganization of the SPADOC. By that time, CAVENet, long considered indispensable, and starting to require a normal cycle of maintenance and upgrades, had been declared a formal program of its own,

[98]The list arrived with a letter from the NASA Administrator to the Commander in Chief of U.S. Space Command [14]. The whole list was extensive and beyond our present discussion. However, the handful of items that pointed most clearly to the need for an SP catalog were outlined in Schumacher and Hoots [15], and two of those have been summarized in the present discussion.

complementing SPADOC-4. The entire local area network, including the ASW, had to be moved along with the 3090 mainframes to support a new Joint Space Operations Center established at Vandenberg. Shortly after its arrival there, a concerted effort was mounted to upgrade the whole suite of CAVENet machines to include servers and more advanced workstations, since it was now no longer considered an "offline" system in any sense and was in use around the clock. This pattern would be continued almost up to this writing: it was always much easier to upgrade CAVENet a little at a time, so that it could take on progressively more of the routine catalog maintenance functions, than to upgrade the mainframe computer systems. In another 10 years, this situation was beginning to cause some consternation in parts of the space surveillance community, because it became progressively harder to keep even the CAVENet computers up to date. The extended paper by Weeden [17] and the editorial by Hitchens [18] are samples of that discussion.

II. THE NAVY COMPUTERS

1. SPACE SURVEILLANCE OPERATIONS CENTER

We described in Chapter 2 how the NORC came to be installed at the Naval Proving Ground in Dahlgren in 1955, and how it was the primary computational support for the NSSS for almost two years, from mid-1959 to early 1961. It was the only machine of its type, and its manufacturer IBM openly called it a supercomputer because it was thought to be the fastest computer in the world in 1955 and for several years after. The NORC electronics depended on 9000 vacuum tubes all functioning correctly for extended periods of time. The original main memory was a now-little-known set of cathode ray tubes (Williams tubes) providing 2000 words of memory, The NORC could run 15,000 arithmetic operations per second.[99] For the NSSS Fence and catalog computation, Easton and Fleming[100] were especially pleased with the NORC's high-precision floating-point arithmetic (13 decimal digits and a check digit, with sign and exponent), with the fact that the main memory was, in early 1960, being upgraded to 20,000 words of magnetic core memory, and with the processing center's eight magnetic tape drives, each one able to read and write 70,000 characters per second.

The NORC took about 15 minutes to update one orbit, even using the highly simplified orbit model available at the beginning, and extra time beyond that was needed to reduce the Fence observations.[101] In this connection, it is

[99] Rife and Carlisle [19], Chapter 4, p. 102.

[100] Easton and Fleming [20], p. 667.

[101] Swift [21], pp. 2–3. Little specific information is available about the orbit model used at the beginning of the NSSS project, but it must have been very much simpler and much less accurate than the PPT2 model

interesting to note that the use of a reasonably accurate SP orbit model was considered for the NORC implementation of the catalog. However, the idea had to be abandoned long before the NSSS processing was moved to Dahlgren, when it was found that the NORC would require several hours to update a single orbit on that approach.[102]

We also mentioned previously that a newly arrived commander of the Proving Ground, Captain M. H. Simons, Jr., had in late 1958 invited the NRL scientists and engineers working on the NSSS to move their processing center from that laboratory to the Proving Ground and to use the NORC, a machine that was far better by every measure than the computers being used at NRL at that time.[103] Less well known is that Captain Simons's invitation did not come merely as the sudden flash of a good idea. A group of senior scientists at the Proving Ground, led by Russell H. Lyddane, head of one of the departments there, had been seeking a tenant organization for the base.[104] Their idea was to improve their base management statistics, mainly to increase the utilization rate of equipment and facilities at the base and to share the relatively high base overhead costs with some other organization. Both issues were perennial ones for the research-oriented Proving Ground (and probably for every other research organization in the DoD over the years). The NORC was excellent bait, especially since NRL needed the extra computing power for the NSSS processing. A further incentive was that the science of satellite geodesy, sparked by the IGY and boosted by the local efforts of Charles J. Cohen, Richard Anderle, and Lyddane, was becoming an important specialty at the Proving Ground, because of its crucial applications in the Navy's Polaris Missile Program. This meant that the Dahlgren staff already included several experts in celestial mechanics and high-accuracy orbit calculations. NRL accepted Simons's invitation without much debate. On 20 February 1959, a few Dahlgren staff members began collaborating with NRL's NSSS data-processing team to learn the new methods being developed for the new space surveillance mission. On 24 May 1959, the SSOC was opened on a 24-hour-per-day basis at the NORC facility, Building 111, at the Proving Ground. The initial staffing included both NRL and Dahlgren base personnel. That same year, the Proving Ground was renamed the NWL, more correctly showing that

used by the mid-1960s. Probably, it was comparable to an early model used at the NSSCC, as discussed in Chapter 3 and in Hoots et al. [22], pp. 174–175.

[102]Private communication from former NAVSPASUR Analysis and Software Department Head Robert Cox to Gary R. Van Horn, 23 June 1998, in Dahlgren, Virginia. Also noted in Hoots et al. [22].

[103]It is not clear at this writing what computers were being used at NRL's NSSS processing center. After March 1956, the Vanguard Program was using IBM 704 mainframe computers at the Vanguard Computing Center in downtown Washington, D.C., operated by IBM. However, IBM built their own backup processing center at Poughkeepsie, New York, while NRL continued to use the computers already available to them, necessarily older and presumably less-capable ones than the 704, for Vanguard research and development at the laboratory. Green and Lomask [23], Chapter 9, pp. 159–161.

[104]Much of the following information is gleaned from the accounts in Rife and Carlisle [19], Chapter 4; Swift [21], pp. 2–3; Nieman [24], pp. 28ff.; and Hughey [25].

its main activity was defense-related applied research, complementing the wider ranging scientific research at NRL in Washington. In future years, the staff at NWL and its successor organizations became less directly involved with the space surveillance mission but continued to serve as a valuable technical consulting resource for the space surveillance operation during the entire half-century service life of the NSSS and cataloging operation.

At this writing, it is also interesting to reflect that this first host–tenant arrangement at NWL turned into a trend: the Dahlgren base attracted several other major tenant commands over the years, not all of them for the same reasons but all of them more operational in nature than the host activity, currently the Naval Surface Warfare Center–Dahlgren Division. The whole collection of colocated organizations is now referred to as the Naval Support Facility Dahlgren. At this writing, the direct descendant of the original 1959 space surveillance activity, the Space Force's 19th Space Defense Squadron, occupies the same building in Dahlgren that was constructed in 1994–1995 to house the Naval Space Command headquarters and the NSSS processing center.

2. Naval Space Surveillance System (NAVSPASUR)

IBM used many of the NORC design features in the later (and popular) 7090 series. One important example was automatic error checking on the NORC, without simply duplicating the hardware and comparing results from two computers running the same program side by side. Built-in error-checking was an unprecedented feature that Dahlgren scientists had insisted on (at extra cost at the time) for the NORC, and which became a standard offering on all subsequent IBM computers.[105] This feature, besides the fact that it was advertised as six times faster than the vacuum tube–based 709 mainframe that it replaced in the market, made the transistorized 7090 attractive for the extended calculations required at both NWL and NAVSPASUR.

Immediately after the NAVSPASUR command was established on 1 February 1961, the NSSS catalog processing was rehosted on an IBM 7090 computer, a step that immediately reduced the processing time to less than five minutes per satellite (plus the Fence data reduction time). New programming techniques and data-handling efficiencies were developed over the next several months, which further reduced the processing time to about one minute. This improvement came just in time to avert what would otherwise have been a computational catastrophe created by the first breakup of an object in orbit around the Earth [22], satellite 1961-Omicron, on 29 June 1961. That satellite, the upper stage of a booster that had deployed the Navy's Transit 4A satellite less than one orbital revolution before, exploded into several

[105] Rife and Carlisle [19], Chapter 4, p. 102; and Nieman [24], pp. 26–27.

hundred trackable pieces at an altitude of about 600 miles, more than quadrupling the number of detectable space objects and creating thousands of NSSS Fence observations that could not be associated with any object already cataloged. The NORC would have been hopelessly inadequate for the data association and orbit update tasks, but the IBM 7090 (aided by heroic human efforts[106]) processed the backlogged data in a matter of days. Although larger breakups in space have happened since then, no comparable event has happened in the history of space surveillance, in which the total number of cataloged space objects had to be increased severalfold so quickly.[107] The lessons were the kind to be taken to heart by everyone involved in space surveillance. The experience demonstrated, above all, the necessity of high computational power. Even though the routine catalog maintenance process may not always require extremely large capacity and speed in the computer system, events can occur that suddenly call for one or two orders of magnitude more computing power than usual for some limited time to maintain the catalog in an up-to-date, usable form.

In 1972, NWL began installing new computers from the Control Data Corporation (CDC) Cyber series. The first was a Model 6700. During the 1970s and 1980s, CDC machines were considered by many experts as the top choice for scientific and technical computing, thanks largely to the efforts of Seymour Cray in the 1960s and 1970s. Cray is sometimes remembered as the "father of supercomputing," an informal title he earned after he left CDC and founded Cray Research.[108] NAVSPASUR followed NWL's lead in this

[106]E. H. Bales [2] told stories of NAVSPASUR analysts unrolling long strip charts of incoming Fence analog data on the floor in the basement hallway of the processing center to survey the unprecedented number of detections and better match them between the different receiver stations.

[107]Naturally, the Air Force processing center was also faced with unprecedented numbers of tracking observations, and similar problems of maintaining the catalog, although the difficulties were acute at the Navy center because of the nature of the NSSS Fence data being processed there. In fact, the complete cataloging process for this event took a very long time at both centers, for reasons that we will come to appreciate better in the next chapter. Moreover, at 600 miles altitude, atmospheric drag is almost negligible and does not remove satellites from space very quickly: decades or centuries may be required for an object to reenter, depending on the satellite body characteristics. According to Anz-Meador et al. [26], the NASA *History of On-Orbit Satellite Fragmentations* (16th edition), p. 24, 317 objects were eventually added to the catalog from that upper-stage explosion. By 1 May 2022, the date of this NASA assessment, 193 of those objects were still in orbit. Consider also that John Gabbard from NORAD headquarters at Peterson Air Force Base in Colorado Springs reported that by 1 July 1982, 256 objects from the breakup had been cataloged, of which 195 were still in orbit [27]. In other words, many space objects, presumably all debris fragments, continued to be discovered and associated with the breakup for many years after the event. The same drawn-out cataloging process can happen nowadays, although more sensitive sensors and better computers allow a greater proportion of fragments from a breakup to be discovered much more quickly than in the early days of space surveillance. All the figures quoted here should be understood in relation to the size of the catalog at the time of the breakup: on 28 June 1961, the day before the Transit 4A upper stage exploded, the catalog contained 115 objects, of which 50 were still in orbit.

[108]Some of this information is gleaned from postings at Cray-Cyber.org (at https://cray-cyber.org/old/documentation/cdcmodels.php [retrieved 24 December 2024] and elsewhere on that website). It could be counted as an accident of history that Cray's famous supercomputers were never used for space surveillance, even though they were used for the U.S. government's Strategic Defense Initiative in the 1980s specifically to simulate the simultaneous tracking of many thousands of intercontinental ballistic missiles. The use of parallel processing for the satellite catalog was not a material concern until the more-or-less sudden

upgrade as soon as budget allowed. The CDC Cyber series Model 174 with dual processors was introduced to the market in 1973, and a pair of these were installed at NAVSPASUR in late 1974. A few years later, in 1979, CDC introduced the Cyber series 170 Model 760, and NAVSPASUR was able to install a pair of these machines around 1980. The command kept the 174s and used them for NSSS Fence data processing, dedicating the 760s to the growing astrodynamics-related processing load, including catalog maintenance. This suite of computers supported the NAVSPASUR operation until a completely redesigned mission processing system was brought online in 1995. By that time, the NAVSPASUR command had been disestablished, and the space surveillance operation had become part of Naval Space Command (NAVSPACECOM) in Dahlgren.

3. NAVAL SPACE COMMAND (NAVSPACECOM)

In the early 1990s, researchers at NRL began investigating how to implement space surveillance applications on then-new computing architectures, the Connection Machine CM-2 and CM-5 models available at the NRL. This work was done by a group led by Shannon Coffey in the Mathematics and Orbit Dynamics Section of the Naval Center for Space Technology at NRL, in collaboration with several space surveillance staff at NAVSPACECOM. The first application considered was the prediction of close approaches among satellites [28]. The Connection Machine allowed a straightforward and practical implementation of searching for close approaches among all satellites in the catalog, not merely the comparison of one satellite against all other satellites as was the standard approach in real operations. Despite this impressive advance, the cost and specialized nature of the Connection Machine models discouraged their incorporation into the NAVSPACECOM processing system.

Coffey's team then considered an alternative approach to parallel processing that was found in an open-source software package known as Parallel Virtual Machine (PVM). This software had been developed first at Oak Ridge National Laboratory, where the first version was released in 1989. Later versions were improved at the University of Tennessee and released in 1993.[109] The idea with PVM was to use an existing computer network as a single parallel-processing device, by means of specialized software installed on each computing node in the network to coordinate the transfer of data among the nodes. In essence, PVM allowed a user to design and implement his own

---

requirement for a high-accuracy catalog emerged in about 1995. By that time, other less-expensive options for parallel processing were available, as described in this chapter.

[109] Some of this information can be gleaned from Internet postings at https://web.archive.org/web/20070321103208/ and http://www.parawiki.org/index.php?title=Parallel_Virtual_Machine&redirect=no [retrieved 26 December 2024].

parallel computing system, using whatever computers happened to be available on any common network. Naturally, the speed of data transfer among the computing nodes would be limited by the speed of the existing network, unlike the high-speed characteristic of the extremely fast interprocessor connections in a real supercomputer. However, if the processing application did not require extensive communications between the individual computing nodes, then a substantial speed-up of the overall computation could be achieved with PVM, without any extra hardware procurement or major software development. Fortunately, many space surveillance applications, including catalog maintenance, proved to be good candidates for this kind of implementation.

Almost at the same time in the early 1990s, NAVSPACECOM had begun to redesign their entire processing system as part of a planned transition to an entirely new building and operations center on the Dahlgren base. The existing processing system, long overdue to be replaced, consisted of the CDC mainframe-type computers as mentioned earlier, with a few dozen remote-access terminals supporting operations, software development, and communications. The new space surveillance mission processing system was based on distributed processing, using 240 IBM RISC System 6000 workstations, plus certain function-specific larger RISC System 6000 processors, all interconnected with a high-reliability, high-bandwidth local area network. Each individual node would have significant processing capability of its own,[110] in a technology upgrade intended to last as many years as possible without any redesign being required. The idea was that individual processors could be upgraded to avoid obsolescence while maintaining the same network layout. This new mission processing system came online in Building 1700 at Dahlgren in 1995.

Coffey's NRL team, collaborating with several NAVSPACECOM operations staff, was able to implement a notable advance in processing NSSS Fence observations that could not be immediately associated with cataloged orbits [30]. This "uncorrelated target" (UCT) processing was one of the most computationally demanding applications among all the space surveillance applications. The existing method used in operations at that time required several days to process even a relatively small batch of five days of unassociated Fence observations, usually a few hundred. However, a 60-day database, typically containing several thousand of those unassociated observations, was always kept current. For their initial development, Coffey's team used an IBM SP2 computer at the DoD's Maui High Performance Computing Center. Advertised by IBM as a genuine supercomputer, the SP2 used an array of RISC System 6000 processors very similar to the processors

[110]Schumacher [29] noted that each workstation in that distributed system had computing power equivalent to 3000 NORC machines. Of course, in terms of sheer corporate computing power, the NAVSPACECOM mission processing system was not remarkable in its time, by any measure.

installed at NAVSPACECOM, making the eventual transfer of the new software to NAVSPACECOM relatively easy. As it turned out, the parallel version of this UCT processing was able to process the entire 60-day set of observations in about three hours, using 128 processing nodes on the IBM SP2. This result was so impressive that the NAVSPACECOM orbital analysts decided to start using NRL's implementation on the SP2 in Maui by remote log-in, more or less immediately, for real UCT processing to generate candidate orbits for the operational system. That practice continued until NAVSPACECOM later installed IBM SP2 machines locally in their mission processing system, allowing the Fence UCT processing to be done in-house.

4. THE NAVY'S HIGH-ACCURACY SPECIAL PERTURBATIONS CATALOG

The continuing collaboration between NRL and NAVSPACECOM then led, in December 1997, to the first demonstration of its kind: a month-long maintenance of a high-accuracy SP catalog of 8845 satellites, nearly the entire catalog at the time [31]. The demonstration used a 24-node partition[111] of the 240-node mission processing system, with two extra workstations on loan from NRL for user-control and database for the test. Meanwhile, the remainder of the mission processing system maintained the regular catalog as usual and supported all the other functions of the command. This 30-day demonstration proved the parallelizability and necessary speed-up of the catalog maintenance process without any design change of the mission processing system or additional hardware procurements. The demonstration did not continue beyond the month of December, somewhat to the chagrin of the operations staff, because it had been proposed and funded as a limited research project. However, despite delays in the follow-on funding for a permanent implementation, an initial SP catalog, updated continuously as observational data came in, was being maintained on a 3400-object subset of the entire catalog of about 9000 satellites, on the operational mission system at NAVSPACECOM by the end of 1998. When an IBM SP2 computer was later installed in the mission system, the high-accuracy catalog processing was moved to that platform, and the full SP catalog was maintained thereafter.

III. THE IMPORTANCE OF THE HIGH-ACCURACY SPECIAL PERTURBATIONS CATALOG

In the introduction to this chapter, we claimed that the inception of the SP catalog maintenance was the single most important step in space surveillance, after the building of the sensors and necessary communication systems for

[111] The partition included mostly workstations in the administrative offices of the command, the idea being that those users would be the least inconvenienced by extra processing going on in the background, invisibly to them. In fact, these users never noticed anything to remark on.

collecting the tracking data. The reason for this claim is not merely the gain in the accuracy of predicted satellite locations, although that is extremely important. For example, when two satellites linger near each other in nearby orbits, it can be difficult to distinguish them correctly. Very commonly in that case, observations of one satellite are attributed to a different one nearby when the simplified orbit models are being used, so that the future predictions of both objects are degraded until the mistake is detected and corrected. Using the high-accuracy SP orbit model greatly reduces this type of error. Other space surveillance applications, such as the conjunction analysis and collision avoidance that we have already mentioned, depend on the high-accuracy orbit model. We also mentioned the acquisition or avoidance of satellites by laser beams, a job that essentially cannot be done reliably with the simplified orbit models. However, all these applications point to a further aspect of the processing improvement available with the SP catalog, namely, our ability to quantify the uncertainty in every satellite prediction. NASA wanted to use the predictions from the catalog to calculate the probability that some space object would collide with the ISS, as part of their ongoing and highly formalized mission risk analysis for that program. However, they could not do that unless the uncertainty in the predicted positions of the objects in question could be quantified reliably. This example follows a general principle, which says that a probabilistic method to gauge the success or failure of some action is always preferrable for managing that action, if the probabilistic method can be based on reliable data and analysis.

It turns out that it can be very difficult to quantify the uncertainty in a prediction using the simplified orbit models. The reason is that, for many years, the tracking data used for space surveillance has been more accurate, on average, than the predictions available from the simplified orbit models. In round terms, we can say that, given perfect observational data, the simplified orbit models would often produce a calculated satellite position with about 1 km of error, predicting about one day ahead. Yet, the radar data being used to estimate the orbit would typically have a range error of only several hundred meters, and sometimes much less, at the time of each observation. Even worse for the statistical analysis involved, the prediction errors of the simplified orbit models are often of the systematic type, rather than being purely random, because those models have entirely omitted some force-effects from consideration. In such a case, no amount of extremely accurate tracking data will allow the cataloged orbit estimates to have less error than some minimum determined by the content of the orbit model itself. The satellite catalog was model limited, so sensor improvements could not benefit every object they observed, but only those few space objects that could be analyzed with a high-accuracy SP orbit model. When the SP catalog processing was instituted, suddenly the catalog was no longer model limited. Given perfect observational data, the SP orbit model could produce predicted positions with only a few meters of error in a one-day prediction,[112] often an order of magnitude lower than the errors in most of the sensor data used for space surveillance. In other words, since the full implementation of the SP catalog, the cataloging system has been data limited rather than model limited. This situation brings at least two benefits. First, because the surveillance sensor errors can be calibrated by periodically tracking a few special satellites whose orbits are known to have high accuracy independently of the cataloging system, a scheme mentioned in passing in Chapter 2, and the model's own errors are almost negligible by comparison, the uncertainty in the SP orbit model's predictions can be quantified reliably. Second, the accuracy of the SP orbit model encourages the best use of all sensor data (including the burden to calibrate it reliably) and encourages the addition of new, more accurate sensors as time goes on. Such improvements now benefit the cataloging of every satellite observed by those sensors [15]. These benefits to the cataloging system, and many other benefits in related space surveillance applications, stem directly from the parallel processing that enables the SP catalog to be maintained.

IV. ANOTHER NOTE ON CATALOG PROCESSING

One circumstance that has always helped space surveillance analysts keep pace with the growing catalog is that, even though it has become large, the trackable space population had grown relatively slowly since the first satellite was launched. Before the advent of megaconstellations in low Earth orbit since about 2016, but even allowing for occasional breakups causing sudden spikes in the count over the years, the number of trackable objects in space had increased by only a handful per day, on average. It turns out that discovering and adding new satellites to the catalog calls for much more computation per object than does simply updating the orbits of already-known satellites, as discussed in more detail in the next chapter, but only a few new objects had to be added each day. The result is that a large catalog of space objects has never had to be created from scratch in a short amount of time.[113] It is doubtful that, even at this writing, the computational capacity is in place to attempt such a feat in practice. In this sense, the catalog of Earth satellites could be considered a national treasure: if it were lost (quite

[112]This was, and still is, true, except in the case of satellites experiencing atmospheric drag. This particular force is governed by so many complicated variables that it remains the most difficult to model of all forces acting on a satellite. For example, one of the main variables, the atmospheric density, is no easier to predict than is the weather, when the satellite descends below about 200 miles of altitude. Above about 500 miles, the atmospheric density, and hence the drag force, is quite small compared to other effects. In that case, it does not matter so much that the drag force is being poorly modeled. This is why the SP cataloging process has always needed to incorporate somewhat elaborate special procedures, such as the Dynamic Calibration of the Atmosphere (DCA) for the High-Accuracy Satellite Drag Model (HASDM), to try to compensate for some of the drag mismodeling for low-altitude satellites.

[113]The problem was described several decades ago [32]. Of course, the hypothetical problem of reconstituting the catalog from scratch had been understood by many persons before and after the date of that paper; this reference is merely one of the more readily accessible sources on the subject.

unlikely, because multiple backups of the current database are generated continually and mirrored at different locations), then it might turn out to be irreplaceable, at least for a longer period than the space surveillance mission requires.

However, the advent of new highly sensitive surveillance sensors, both radar and optical, has at this writing, suddenly increased the number of detectable space objects by as much as an order of magnitude. This presents a problem almost as large as reconstituting the entire catalog from scratch. The reason for this sudden increase is not that vast numbers of new objects have somehow been put into space; rather, the new sensors can detect extremely small objects that have long been there but until recently were invisible to all other space surveillance sensors. Suddenly, we can see them, and we must decide how to handle them. One approach is to handle them at the sensor site by not reporting extremely dim (and therefore very small or very distant) objects, thereby preventing the catalog processing system from being overwhelmed. Of course, this stop-gap approach does not answer the real space surveillance need. At this point, it is worth pondering whether the problem of managing this dramatic increase in the potential size of the catalog will be permanently constrained by the available computing facilities, analogously to the way that the cataloging operation was constrained until the 1990s, or whether scalable parallel computing will eventually provide a way ahead.

REFERENCES

1 Oesterwinter, C., and Cohen, C. J., "New Orbital Elements for Moon and Planets," *Celestial Mechanics*, Vol. 5, 1972, pp. 317–395. Also published as Oesterwinter, C., and Cohen, C. J., "New Orbital Elements for Moon and Planets," Naval Weapons Laboratory TR-2693, Dahlgren, VA, March 1972. Report available online at Defense Technical Information Center, https://discover.dtic.mil/, as Accession No. AD-755164, or directly as https://apps.dtic.mil/sti/tr/pdf/AD0755164.pdf [retrieved 18 Dec. 2024].

2 Emory Hall Bales, late 1988 and early 1989, private communications with one of the authors about early NAVSPASUR Fence operations.

3 Weeden, B. C., and Cefola, P. J., "Computer Systems and Algorithms for Space Situational Awareness: History and Future Development," *Advances in the Astronautical Sciences, American Astronautical Society Paper AAS 10-420*, Vol. 138, *Applications of Space Technology for Humanity, Proceedings of the 12th International Conference of Pacific-Basin Societies (ISCOPS)*, edited by P. M. Bainum, A. K. Misra, Y. Morita, and Z. Chi, Quebec, Canada, 2011, pp. 205–226. Available online through Secure World Foundation at https://swfound.org/media/205395/computer-systems-and-algorithms-for-space-situational-awareness-history-and-future-development.pdf [retrieved 28 Feb. 2024].

4 Knuth, D. E., "The IBM 650: An Appreciation from the Field," *Annals of the History of Computing*, Vol. 8, No. 1, Jan. 1986, pp. 50–55.

5 Altshuler, E. E., *The Rise and Fall of Air Force Cambridge Research Laboratories*, self-published via CreateSpace Independent Publishing Platform (Amazon), 2 Jan. 2013.

6 Wahl, E. W., "Remarks on the Programming System of Project Space Track," Seminar Proceedings, Tracking Programs and Orbit Determination, Jet Propulsion Laboratory,

California Institute of Technology, Pasadena, CA, 23–26 Feb. 1960, pp. 161–164. Proceedings produced under NASA Contract No. NASw-6. Available at https://ntrs.nasa.gov/api/citations/19690071283/downloads/19690071283.pdf [retrieved 6 Dec. 2024].

7 Fitzpatrick, P. M., and Findley, G. B., "The Tracking Operation at the National Space Surveillance Center (NSSCC)," Special Studies Section, Directorate of Aerospace, Air Proving Ground Center, Eglin AFB, FL, 2 Sept. 1960. The cover page of this report records the following: "This work was conducted during a temporary assignment of the authors to the NSSCC, L.G. Hanscom Field, Bedford, Massachusetts, as part of Project 1772 (Project Space Track), Task 17721."

8 Sturdevant, R. W., "From Satellite Tracking to Space Situational Awareness: The USAF and Space Surveillance, 1957–2007," *Air Power History, Winter 2008*, pp. 4–23.

9 Computer History Museum, https://www.computerhistory.org/brochures/m-p/philco-corporation/[retrieved 10 Dec. 2024].

10 Del Papa, E. M., and Warner, M. P., "A Historical Chronology of the Electronic Systems Division 1947–1986," Electronic Systems Division ESD-TR-88-276, Air Force Systems Command, Hanscom AFB, MA, Oct. 1987. Available online from Defense Technical Information Center as Accession No. AD-A201708, or directly at https://apps.dtic.mil/sti/pdfs/ADA201708.pdf [retrieved 12 Dec. 2024].

11 Smith, D. R., "Evolving the Mountain: Defense Acquisition Management of Strategic Command and Control System Procurements," Ph.D. Dissertation, Univ. of Denver, 19 Aug. 1991.

12 Pearson, D. E., *The World Wide Military Command and Control System: Evolution and Effectiveness*, Air University Press, Maxwell Air Force Base, AL, June 2000. Available online at https://www.govinfo.gov/content/pkg/GOVPUB-D301-PURL-LPS20450/pdf/GOVPUB-D301-PURL-LPS20450.pdf [retrieved 2 Dec. 2024].

13 "Memorandum of Agreement Between United States Space Command and Johnson Space Center for Space Control Operations Relationship, Space Shuttle Program Support and International Space Station Program Support," 10 April 1996, by Roger de Kok, Director of Operations for U.S. Space Command and George W. S. Abbey, Director, Lyndon B. Johnson Space Center.

14 Letter from NASA Administrator (Daniel S. Goldin) to Commander in Chief, U.S. Space Command (Gen. Howell M. Estes, U.S. Air Force), 27 Aug. 1997.

15 Schumacher, P. W., Jr., and Hoots, F. R., "Evolution of the NAVSPACECOM Catalog Processing System Using Special Perturbations," *Proceedings of the Fourth U.S./Russian Space Surveillance Workshop*, U.S. Naval Observatory, Washington, D.C., 23–27 Oct. 2000, pp. 293–307.

16 Hoots, F. R., Schumacher, P. W., and DeVere, T., "50 Years of Satellite Tracking and Cataloging in the U.S.," AAS Paper 07-328, AAS/AIAA Astrodynamics Specialist Conference, *Advances in the Astronautical Sciences*, Vol. 129, Part 2, Univelt, San Diego, CA, 2008, pp. 1251–1278.

17 Weeden, B., "Going Blind: Why America Is on the Verge of Losing Its Situational Awareness in Space and What Can Be Done About It," Secure World Foundation, 10 Sept. 2012. Available online at https://swfound.org/, searching on "Going Blind" [retrieved 19 Dec. 2024].

18 Hitchens, T., "Key Space Monitoring Sensors Still Rely on Outdated CAVENet Computer System," *Breaking Defense*, 5 April 2022. https://breakingdefense.com/2022/04/key-space-monitoring-sensors-still-rely-on-outdated-cavenet-computer-system/[retrieved 19 Dec. 2024].

19 Rife, J. P., and Carlisle, R. P., *The Sound of Freedom: Naval Weapons Technology at Dahlgren, Virginia 1918–2006*, U.S. Department of the Navy.

20 Easton, R. L., and Fleming, J. J., "The Navy Space Surveillance System," *Proceedings of the IRE* (renamed *Proceedings of the IEEE* in 1962), Vol. 48, No. 4, April 1960, pp. 663–669. https://doi.org/10.1109/JRPROC.1960.287444.

21 Swift, E. R., "History of Satellite Orbit Determination at NSWCDD," Naval Surface Warfare Center, Dahlgren Division, TR-17/229, Dahlgren, VA, Jan. 2018.

22 Hoots, F. R., Schumacher, P. W., Jr., and Glover, R. A., "History of Analytical Orbit Modeling in the U.S. Space Surveillance System," *Journal of Guidance, Control and Dynamics*, Vol. 27, No. 2, March–April 2004, pp. 174–185.

23 Green, C. M., and Lomask, M., *Vanguard: A History*, NASA SP-4202, NASA Historical Series, Washington, D.C., 1970 (foreword by Charles A. Lindbergh). https://www.nasa.gov/wp-content/uploads/2023/03/sp-4202.pdf. Another edition of this document was issued in 1971 by Smithsonian Institution Press. In 2009, a slightly corrected edition of NASA SP-4202 was issued as *Project Vanguard: The NASA History* by Dover Publications, Mineola, New York, with a new introduction by Paul Dickson. See especially Chap. 9, "The Tracking Systems," pp. 145–163, in the Dover edition.

24 Nieman, R. A., "Dahlgren's Participation in the Development of Computer Technology," NSWC MP 81-416, Strategic Systems Department, Naval Surface Weapons Center, Dahlgren, VA, Sept. 1982.

25 Hughey, R. H., "History of Mathematics and Computing Technology at the Dahlgren Laboratory," Technical Digest, Naval Surface Warfare Center (Dahlgren Division), Dahlgren, VA, 1995, pp. 8–21.

26 Anz-Meador, P., Opiela, J., and Liou, J.-C., "History of On-Orbit Satellite Fragmentations," 16th ed., NASA TP-20220019160, Orbital Debris Program Office, Lyndon B. Johnson Space Center, Houston, TX, Dec. 2022. This report is also available in electronic form at http://www.sti.nasa.gov/ and http://ntrs.nasa.gov.

27 Gabbard, J., "History of Satellite Breakups in Space," *Orbital Debris*, compiled by D. J. Kessler and S.-Y. Su, NASA Conference Publication 2360, 1985, pp. 30–38. https://ntrs.nasa.gov/citations/19850012880 [retrieved 16 Sept. 2023].

28 Healy, L. M., "Close Conjunction Detection on Parallel Computer," *Journal of Guidance, Control and Dynamics*, Vol. 18, No. 1, July–Aug. 1995, pp. 824–829.

29 Schumacher, P. W., Jr., "U.S. Naval Space Surveillance Upgrade Program 1999–2003," *Proceedings of the Seventh U.S./Russian Space Surveillance Workshop*, edited by K. Seidelmann and F. Hoots, Naval Postgraduate School, Monterey, CA, 29 Oct.–2 Nov. 2007. Also published in *Proceedings of the Fifth European Conference on Space Debris*, edited by H. Lacoste, European Space Operations Center, Darmstadt, Germany, 30 March–2 April 2009.

30 Coffey, S. L., Jenkins, E., Neal, H. L., and Reynolds, H., "Parallel Processing of Uncorrelated Observations into Satellite Orbits," AAS Paper 96-146, *Proceedings of the AAS/AIAA Space Flight Mechanics Meeting*,. 12-15 Feb. 1996, Austin, TX, Advances in the Astronautical Sciences, edited by G. Edward Powell, R. H. Bishop, J. B. Lundberg, and R. H. Smith, Vol. 93, Part 1, pp. 739–756. Univelt, Inc., San Diego, CA, 1996.

31 Coffey, S. L., Neal, H. L., Visel, C. L., and Conolly, P., "Demonstration of a Special-Perturbations-Based Catalog in the Naval Space Command System," AAS Paper 98-113, *Proceedings of the AAS/AIAA Space Flight Mechanics Meeting*, 9–11 Feb. 1998, Monterey, CA, edited by J. W. Middour, L. L. Sackett, L. D'Amario, and D. V. Byrnes, Advances in the Astronautical Sciences, Vol. 99, Part 1, pp. 227 – 247. Univelt, Inc., San Diego, CA, 1998.

32 Schumacher, P. W., Jr., "Prospects for Improving the Space Catalog," AIAA Paper 1996-4290, *Proceedings of the American Institute of Aeronautics and Astronautics Space Programs and Technologies Conference*, Huntsville, AL, 24-26 Sept. 1996. https://doi.org/10.2514/6.1996-4290 or https://arc.aiaa.org/doi/epdf/10.2514/6.1996-4290.

Chapter 5

The Operations Centers

At present, maintenance of the U.S. satellite catalog is the largest continuous tracking enterprise on Earth. On any day, the number of satellites being tracked in near-real-time routinely by the U.S. space surveillance system is larger than the number of airplanes being tracked by the entire international air traffic control system. Moreover, from the very beginning, command and control of the U.S. SSN has presented certain fundamental challenges besides the sheer scale of the job. Many of the important sensors have (and always have had) a primary mission other than space surveillance, so that the primary and backup surveillance operations centers have implemented a unique two-tiered central-level-plus-sensor-level tasking scheme, which is described in this chapter. In addition, the locations and organizations of the operations centers over time have reflected the different approaches that the two services have taken in pursuing the space surveillance mission, in ways that are discussed.

The main aspects of the catalog maintenance process have been basically the same at all space surveillance centers, given that progressively more of the process has been automated over time. Nevertheless, the details of the implementations differed between Air Force and Navy centers, sometimes markedly. We mentioned some of the computer-related differences in Chapter 4, and some other aspects of the centers were mentioned in previous chapters. A few differences of staffing and software management have also been important to how the space surveillance mission was conducted over the years. Hence, although the same basic problem of maintaining the catalog had to be solved in every operations center, whether primary or backup center, some interesting differences in how the problem was approached will also emerge in the discussion. To begin, let us summarize a broad view of the developments described in previous chapters, as they affected the operations centers in both services, and add a few organizational details.

I. THE AIR FORCE SPACE SURVEILLANCE CENTERS

Project Space Track, the first organizational response of the American military to the launch of Sputnik, was initially housed in borrowed office space in the Geophysics Directorate of Air Force Cambridge Research Laboratories at Hanscom Field in late 1957. Over the next two years, 1957–1959, the main problems of how to keep a catalog of space objects were addressed. The fact that optical sensors, mainly the Baker-Nunn cameras, figured so heavily in the space surveillance job at first, created an outlook from the very beginning that all satellites in space had to be tracked to maintain a catalog. It was not possible to be selective about which satellites the Baker–Nunns would detect in their wide field of view. Unless all satellites were tracked, misidentification of satellites would be inevitable, defeating even the scientific purposes of the IGY.

In 1959, Space Track moved into a new dedicated facility at Hanscom and was renamed the NSSCC. This first operations center (as we would understand the term) had a new computer and conducted routine 24/7 operations for almost two years, 1959–1961. The same center also supported experimental and developmental work to improve the operation, continuing research work that Project Space Track had started.

In mid-1961, the NSSCC was moved to Ent Air Force Base in Colorado Springs, Colorado. The space surveillance staff at Hanscom was split, some of them moving to Colorado as part of the operational group, and the rest remaining at Hanscom to continue developmental work and to serve as a backup operations center for the new center. The new operations center was part of the NORAD command center complex at Ent, supporting missile warning. That space surveillance center operated during the period 1961–1966.

In 1966, the NORAD command center moved into the underground facility that came to be known as the CMC. The space surveillance operations center was located in the CMC until 2007. One of the major changes during this period was the establishment of the joint-service U.S. Space Command (USSPACECOM) in 1985. Before that time, the Air Force and Navy space surveillance centers had collaborated and shared tracking data regularly, and the NSSS Fence was important to the space surveillance support of the missile-warning mission. Under the joint-service arrangement beginning in 1985, the Air Force operations center in Colorado Springs was designated as the primary Space Surveillance Center, and the Navy operations center in Dahlgren, Virginia, was designated as the backup, officially named the Alternate Space Surveillance Center. The names changed over the years, but these two centers operated in this same relationship until 2004, when the DoD transferred the Navy's portion of the space surveillance mission to the Air Force.

Starting in 2007, as part of a reorganization that separated the space surveillance operations center geographically, but not functionally, from the missile-warning operation (possible because of improved communication systems), Vandenberg Air Force Base in California hosted the Air Force primary center. That arrangement continued until the Space Force was established as a separate branch of service in December 2019, at which time Vandenberg became a Space Force Base, continuing to host the primary space surveillance center.

II. The Navy Space Surveillance Centers

NRL in Washington, D.C., had been chosen to build, launch, and operate America's first satellites as part of the IGY. (Of course, as mentioned in Chapter 2, that history got much more interesting than anyone had anticipated, after Sputnik launched first.) The Minitrack system for their Project Vanguard was designed to do only cooperative tracking. However, Sputnik forced a change of outlook. In a matter of months, NRL had adapted the Minitrack design to become the NSSS, also known as the Fence, which enabled noncooperative tracking on all satellites it could detect. The operations center for the NSSS was set up first at NRL in mid-1958 but by mid-1959 had moved to the Naval Proving Ground (soon renamed Naval Weapons Laboratory) in Dahlgren. The reason for the move was specifically to provide better computational support to the space surveillance mission. The developmental center at NRL had not operated 24/7, but the Dahlgren SSOC did so from the day it opened.

In 1960, the Department of the Navy took over funding of the NSSS and SSOC operations as the development funding from the ARPA was phased out. In early 1961, the Naval Space Surveillance Center (NAVSPASUR) was established on the same base in Dahlgren, with operational control by NORAD, and the operations center was moved to the same building on base to colocate it with the NAVSPASUR headquarters. NAVSPASUR would remain in that multipurpose building until 1993.

In 1983, the new Naval Space Command (NAVSPACECOM) was colocated with NAVSPASUR. When USSPACECOM was established in 1985, NAVSPACECOM became the naval component of the joint-service command, while the backup Alternate Space Surveillance Center continued as part of NAVSPASUR. Then, in 1993, NAVSPASUR was disestablished as a separate command, and the SSOC, with its joint-service functions, was merged organizationally into NAVSPACECOM.

A major upgrade to the naval SSOC came in 1995, when NAVSPACECOM moved into a new purpose-built facility on the Dahlgren base. The computer system was changed from mainframes to a distributed-processing system that allowed an efficient and adaptable type of parallel processing for space surveillance.

NAVSPACECOM continued as the naval component of USSPACECOM, until the latter was replaced by U.S. Strategic Command (USSTRATCOM) in

2002. At that time NAVSPACECOM was replaced by NNSOC in Dahlgren, while a new command, Naval Network Warfare Command (NAVNETWARCOM), located in Little Creek, Virginia, near the Atlantic Fleet headquarters, became the naval component command representing the Navy's space interests at USSTRATCOM. Once the Space Force was established in December 2019, the Navy again began to reorganize its space-related activities to align better with the new branch of service.

NNSOC continued to operate the NSSS Fence and SSOC until the DoD transferred the entire space surveillance mission to the Air Force in 2004. After that time, the Air Force took over the Dahlgren space surveillance facilities and continued to operate the NSSS Fence, renaming it the AFSSS. Most of the Navy's space surveillance operations staff remained at the center, as Air Force employees, after 1 October 2004.

The AFSSS Fence service life ended in 2013, and the system was decommissioned. The SSOC continued in its same facility in Dahlgren, becoming part of the Space Force in December 2019.

III. Air Force and Navy Approaches to Managing the Space Surveillance Operations Centers

A few differences can be noted in the two services' approach to operating space surveillance centers, some significant and deserving of further study beyond this discussion, and some casual, merely enriching the history.

1. The Air Force primary space surveillance center was moved across the continent from Boston to Colorado Springs to California during the service's tenure in the space surveillance mission. The Navy backup center moved across the base in Dahlgren, but no further. In the first two years of space surveillance, the Air Force center was located at a laboratory and benefited greatly from its interaction with the scientific and academic community. In its later locations, the primary center was geographically separated from the service laboratories, which may have inhibited regular interaction between space surveillance and scientific staff. In contrast, the backup center was located on a naval applied research and development base and was just over an hour's drive from the main installation of NRL. Staff at both laboratories were knowledgeable and interested in the space surveillance mission, and working-level interaction and collaboration with space surveillance staff were frequent.
2. The Air Force's management of the space surveillance operations centers was complicated by the service's close involvement in the missile-warning mission. Especially, the need to integrate space surveillance operations into missile warning, and to task missile-warning sensors for surveillance tracking, greatly complicated software development, computer hardware upgrades, and system integration for all aspects of space surveillance.

In comparison, the Dahlgren operation was relatively insulated from these problems, except for the need to remain "functionally equivalent" to the Air Force's primary space surveillance center.

3. Much of the technical expertise at the operations centers has always resided with the orbital analysts, the persons closest to the daily technicalities of the mission. After the NSSCC moved from Hanscom to Colorado Springs in 1961, the analysts on duty in the space surveillance center were mostly military members. In contrast, the orbital analysts at Dahlgren were mostly civil service staff. This was true from the earliest days at NRL in 1958 and continuing after the Air Force assumed the entire space surveillance mission in 2004, and even after the Space Force was established in 2019. This difference had the result that military members rotated into and out of the Air Force orbital analyst positions frequently, about every two years by military policy, while Dahlgren orbital analysts often had decades of experience at their tasks. It should be noted also that the Air Force military analysts working at Cheyenne Mountain relied heavily on key civilian experts at Air Force Space Command and contractor facilities.
4. The Navy space surveillance center always had direct control of the NSSS Fence operation and observation flow. This was a sensor that was especially simple to understand and use. It did not have to be tasked and was often the first sensor to detect breakups and maneuvers, in the same way that it had always provided cueing for optical sensors and narrow-beam radar trackers. The Navy operations center's status as backup to the Air Force primary center did not affect this direct control and use of the Fence data. NAVSPASUR analysts often made special use of NSSS Fence observations in their catalog processing, even though all space surveillance sensor observations were being shared with the Air Force center. This feature of the operation continued when the Air Force took over the Dahlgren space surveillance operation, until the Fence was decommissioned in 2013. In contrast, the Air Force primary center never had direct control of any sensor in the same way that the Navy center did, even though the Air Force center was the primary tasking agent for the whole SSN.

In the remainder of this chapter, we discuss aspects of the problem that every space surveillance operations center has had to solve, namely, how to maintain the satellite catalog in a timely, accurate, and comprehensive way. Although the details of implementation at the centers depended on the computers and communication systems available, the size of the catalog, and the types and locations of sensors available to do the tracking, all of which changed over the years, the form of the basic process has remained surprisingly constant once it was formulated at Project Space Track and NRL before 1960. Our purpose here is to describe these "invariant" features of the cataloging process.

IV. THE CATALOG MAINTENANCE PROCESS[114]

The basic problem of space surveillance and catalog maintenance is data association, namely, "tagging," or labeling, each observation as originating in the detection of a certain space object. This fact might seem hardly necessary to mention. Yet, space surveillance tracking is almost all the "noncooperative" type, a type of tracking that has some inherent difficulties and potential limitations. Said another way, in virtually every case, the object being tracked does not help us identify it, say, by emitting signals on a particular frequency.[115] In contrast, the operators of active payloads, including those for commercial programs, scientific missions, or manned space missions, usually do use some kind of "cooperative" tracking scheme to determine the orbits of their satellites. Equipment such as radio beacons, onboard sensors, laser retroreflectors, onboard GPS receivers, and other systems are commonplace for active payloads. Such systems provide accurate orbits very economically. Cooperative tracking also neatly identifies which satellite is being tracked in a constellation of several objects, because each payload can, for example, be assigned a unique frequency or access code. However, for space surveillance, noncooperative tracking is mandatory: the only information we can obtain routinely about an object in space is its position, in the case of a radar, or its direction, in the case of a telescope, at some known time. Often, we have a short sequence, or "track," of observations from the sensor, which we can expect all belong to the same object, whatever it is. However, we still must match that track with some orbit already contained in the catalog or else create a new catalog entry to account for the track. The latter step usually involves extensive and specialized computation, as discussed later.

The satellite catalog is commonly thought of as a database of orbital elements for all trackable space objects because that is the form in which

[114] Some of the material in this section has been gleaned from a 1996 conference paper by Schumacher [1] that is still readily available at this writing. Refer to https://doi.org/10.2514/6.1996-4290 [retrieved 27 December 2024] or https://arc.aiaa.org/doi/abs/10.2514/6.1996-4290 [retrieved 27 December 2024]. The experience recorded in that paper was specifically for the space surveillance operation at Naval Space Command in Dahlgren in 1996. The present discussion emphasizes those features of the cataloging system that were 1) common to both services at the time and 2) fundamental in the sense that they persisted for many years, at least up to the end of the Navy's space surveillance mission in 2004. For this reason, we often use the first-person present tense in the following discussion simply to avoid awkward grammar. Most of the same ideas apply to a cataloging operation anytime and anyplace, which is what the discussion intends to convey. The details of how these ideas apply, or are implemented, in the current cataloging system in the current operational centers are not discussed.

[115] To date, there is no system in place for satellites, which is analogous to the IFF (Identification Friend or Foe) transponders carried on board most military airplanes to help manage military operations around the world. Commercial airplanes also usually carry some similar kind of transponder to aid the civilian air traffic control system. Besides the fact that international agreements for a such a system in space have proved elusive, only a small fraction of the trackable space population consists of active satellites that would be able to radiate signals. Most space objects are inert, either empty rocket bodies or debris of some type. This feature of the satellite population is really a reason why space surveillance continues to be a military mission, as we mentioned before: neither benign nor hostile space objects are likely to provide us with helpful signals, but they always have to be distinguished very quickly, a job that can be done only with a complete and up-to-date catalog on hand.

catalog information is published. However, it would be more correct to think of the catalog as a database of correctly associated observations for all trackable space objects (which typically is not published, for various good military reasons[116]). The goal of catalog maintenance is then to obtain the most correct associations possible for all tracking data available for all satellites. The orbital elements, defined by some orbit model, are only a natural and efficient means by which the correct associations can be discovered, through applying the laws of motion for a satellite. Given the database of correctly associated observations, it is a straightforward job to derive, for any satellite or for all satellites, any orbit representation defined by any orbit model whatsoever, and to pursue any other analyses that may be required, such as collision avoidance, overflight warning, historical analysis of certain satellite motions, and so forth. In this sense, the high-accuracy SP catalog is more than just the database of accurate orbit representations. Rather, the SP catalog is the database of observations that have been associated using the highly accurate SP orbit model, so that the associations are correct in more instances than they would have been with a simplified orbit model.

At this stage, we should make explicit a fact that has been implicit in our discussion so far: no orbit model is perfect. The improvement in accuracy obtained by replacing the simplified orbit models with a (much more computation-intensive) high-accuracy SP orbit model certainly is significant, as we discussed in the previous chapter. However, the prediction error of even an SP model grows with the time span over which the prediction is made. The prediction error of one of the simplified orbit models necessarily grows faster, exhibiting the same level of prediction error as the SP model but over a much shorter time span. Hence, for any orbit model we use to build the catalog, the farther we must predict in time, the less likely it is that we can correctly associate an observation with a cataloged orbit. The situation is worse if we must start with some extra error at the initial time (the beginning of the prediction), because that error is added to the prediction error of the orbit model itself. Unfortunately, as all statisticians will appreciate, exactly this situation occurs when a cataloged orbit is estimated from observations: some error at initial time is unavoidable because neither the orbit model nor the observations are perfectly accurate. What this means in practice is that the catalog cannot serve its purposes if it is not "maintained," that is, if each orbit is not updated sooner than the prediction error grows enough to prevent us from associating tracking data reliably for that orbit. For the same reason,

[116]The military is, very logically, reticent to discuss the specific capabilities and limitations of any of its sensor systems, and the set of observations for any particular satellite reveals much of that type of information. For similar reasons, and to avoid creating unnecessary vulnerabilities for satellite systems, the military does not, at present, publish the high-accuracy orbit representations available from the SP catalog. However, the U.S. military has always published the standard element sets derived from the lower-accuracy simplified orbit models mentioned in previous chapters and continues to do so.

observations that are many days old are of little use in maintaining the catalog and including them may actually degrade the accuracy of an orbit update. Rather, one always needs a good collection of recent observations to estimate the satellite's current motion. All the elaborate infrastructure of space surveillance—sensors, operations centers, computers, communication systems—exists to keep observations arriving soon enough, and cataloged orbit updates occurring often enough, so that the likelihood of correct data association remains as high as possible for every orbit. If the association likelihood happens to get too low for some satellite, perhaps because the orbit updates for that satellite were delayed for some reason, then the space surveillance system can, effectively, "lose track of" that object. That problem will not fix itself; without some intervention, the chances that we can ever associate future observations of that satellite continue to become smaller and smaller as the prediction error from the estimated orbit on file in the catalog continues to grow larger and larger. Indeed, the fact that a satellite is "lost" as far as the catalog is concerned does not necessarily mean that the satellite is not being observed. Sensors may be tracking it from time to time, but they cannot tell that their observations belong with that old orbit estimate in the catalog, and so they will simply report the observations to the operations center as "unassociated," that is, "uncorrelated." Later in this discussion, we consider what intervention must occur to restore a lost satellite to the catalog.

Being able to correctly associate all observations on a particular satellite can call for extensive computation so that, by design, the space surveillance system takes every workable short-cut to getting the most reliable associations possible. Associating new observations of a satellite with its already-existing cataloged element set turns out to be an efficient process, and the update of that element set then follows routinely. However, correctly associating new observations, when none of those observations matches any existing cataloged element set, turns out to be the most computationally intensive part of the entire catalog maintenance job. We explain in more detail later why this type of association processing, also known as uncorrelated target (UCT) processing, grows disproportionately more difficult as the number of unassociated observations increases. It will also become evident why UCT processing is a potentially limiting factor in how well the catalog can be maintained and, therefore, in how well all the other aspects of the space surveillance mission can be accomplished.

The problems of managing the whole SSN of sensors in such a way that they can supply the necessary tracking data to the cataloging centers are what the military understands as problems of "command and control." Many years ago, Uhlmann [2] described essentially the same network-management problems in a related context, that of missile defense, by a simple but insightful analogy with baseball. Consider three outfielders facing a fly ball. Usually, they would be successful at catching the ball, at least with a little practice so they know how to work together, deciding which one ought to call for the catch. Now consider what happens if they are faced with three simultaneous fly balls. They would have noticeably more difficulty catching all three balls, but if they can communicate among themselves well enough, then they might succeed. However, the three-on-three problem is easy compared to the problem of 10 outfielders trying to catch 10 simultaneous fly balls. Of course, everyone would agree that it is hopeless to expect 100 outfielders to be able to catch 100 simultaneous fly balls. In this latter case, the difficulties of distinguishing and somehow labeling the individual balls, communicating that information among the players, and coordinating the actions of the players so that no two of them go after the same ball, have increased far out of proportion to the number of balls in the air. Following this same baseball analogy, we might compare the current SSN to about 20 outfielders (sensor sites) trying to catch about 45,000 fly balls (satellites). Fortunately, space surveillance tracking is easier than this analogy implies. Satellites orbit the Earth repeatedly, and each sensor has many chances to "catch" them, unlike the case with ballistic missile tracking. However, the actual difficulty of the space surveillance problem is substantial, even potentially overwhelming. Fortunately, a few mitigating circumstances have, to date, worked in favor of the cataloging system:

1. An essentially complete and up-to-date catalog database of all trackable space objects has been maintained at some operations center since the first satellite was launched. Moreover, the information in that database has always been available in some form to each of the sensors, so that each sensor has a good chance of correctly associating each of its new observations with a cataloged orbit. In other words, at no time have we had to generate the entire catalog ab initio at the operations center, but rather we have been able to grow the catalog at about the same rate as trackable objects were added to the space population.
2. Satellite orbital motion can be modeled and predicted much more accurately than can the motion of other high-interest objects such as missiles or aircraft, or even baseballs. Just as importantly, the timescale of orbital motion is such that an orbit update at the operations center within tens of minutes of the observation at the sensor, is adequate for practically all surveillance purposes, allowing opportunity for extensive computation.
3. The vast majority (at this writing, more than 90%) of currently cataloged objects do not maneuver, split into several distinct satellites, or use evasive measures to prevent detection.
4. The space environment provides a relatively low-clutter, low-noise background for most of the sensors, and so the number of false detections reported to the operations center is extremely low.
5. As large as the current catalog size is, the spatial density of trackable objects is still low enough to prevent frequent confusion of targets. As a result, each

sensor can reliably associate individual observations correctly into single-object tracks, whether the object is in the catalog or not. Naturally, the chances of confusion are higher in certain very popular, more crowded orbits, such as the circular GEOs. However, these cases are still rare enough that some extra attention to those regions by the operations staff can keep the cataloged orbits accurate enough for essentially all purposes.

6. Since the cataloging operation was started, computer capabilities have continued to grow faster than the catalog size. As a result, the operations center has usually had some processing performance margin and has usually responded well to stressing events such as satellite breakups.

All these factors combine to simplify the problem of data association enough for straightforward processing techniques to work almost all the time. In fact, the entire concept of operations for the space surveillance system depends on such simplifying factors, and the entire system infrastructure has evolved accordingly. If any of these fortuitous circumstances were to change for the worse in the long term, we would have to both rethink the current concept of operations and invest in major technical improvements throughout the system. Sometimes one or more of these circumstances does change temporarily, and in those cases, we do face special difficulties maintaining the catalog. For example, solar storms disturb the upper atmosphere at satellite altitudes, temporarily degrading our ability to model near-Earth orbits accurately. On one occasion (March 1989), a solar storm temporarily endangered the integrity of the catalog because of unmodeled atmospheric perturbations affecting an unprecedented number of orbits. Fortunately, our ability to model satellite orbits is not going to degrade in the long term. Also, some procedural changes at the operations centers shortly after that event have reduced the chances that a similar problem would recur, at least with the present catalog size and sensor network. For another example of temporarily changed circumstances, satellite breakups may instantly increase the trackable orbiting population by a few hundred objects. In that case, a substantial fraction of the objects detected by a sensor will not be in the catalog. Moreover, in the early stages of a breakup, the spatial density of objects can be high enough to cause mis-associations among different space objects. These effects must be overcome through extra association processing at the operations centers. Once the breakup pieces have drifted apart (as they do naturally), typically after a few hours, orbits for most of the trackable pieces can be determined within another few days or several weeks.

One consequence of the preceding listed mitigating factors, which has special practical importance even in routine, "nonstressing" operations, is that each individual sensor can, with high probability, associate a new observation with the correct satellite orbit before the data are forwarded to the operations centers. Only a small fraction of the associations needed for catalog maintenance must be made at the center itself. This circumstance avoids a serious bottleneck. Since the time that the catalog grew larger than a few hundred, perhaps a thousand, space objects, the operations centers have not had enough processing capacity to do all, or even a large fraction, of the data association for all the observations on all the satellites in the catalog. The association processing that is done at the centers consists mainly in verifying the satellite numbers supplied by the sensors, which may involve detecting and correcting the occasional mis-association, plus identifying the relatively small percentage of observations that the sensors could not associate or, occasionally, mis-associated. It turns out that most observations that the sensors could not associate with a cataloged orbit belong to cataloged objects after all. Sometimes the cause of a sensor's failure to associate it stems from the sensor not having the most up-to-date element set so that the observation really does not match that old element set but does match the more recent updated element set at the operations center. This identification process, which can involve a refined search of the entire catalog, but which can be largely automated, is the first phase of UCT processing. Once the verification and identification steps have been completed, the incoming newly associated observations for each satellite are added to its other associated observations, and the orbital elements of each of those satellites can be updated. Finally, the updated elements, along with sensor-specific tracking assignments (so-called tasking), are distributed to all the sensors for use in acquiring future observations.

This two-step sensor-level-plus-central-level approach to data association has always been a fundamental feature of the satellite cataloging operation. In general, it has been very successful. It permits the necessary autonomy and flexibility at remote sensor sites, many of which have a primary mission other than space surveillance (such as missile warning). It also allows the initial association processing at the sensor to be tailored to the characteristics of each individual sensor without increasing the central-level processing load. It turns out that in the current SSN under routine operating conditions, well over 90% of the observations are correctly associated by the sensors, making the central-level verification process relatively fast. Of the 10% or fewer remaining, the majority turn out to be already in the catalog after all, as described previously. The high association success rate at the sensor level is what makes it feasible to do the refined search for identification at the central level, because only a small fraction of the incoming observations must go through that identification step.

Of course, it may happen that new objects have appeared in space, and that one or more new element sets must be created and examined for inclusion in the permanent catalog. This is the computationally intensive second phase of UCT processing. It involves track-to-track association for an unknown number of new space objects, plus initial orbit determination, and is discussed in more detail later.

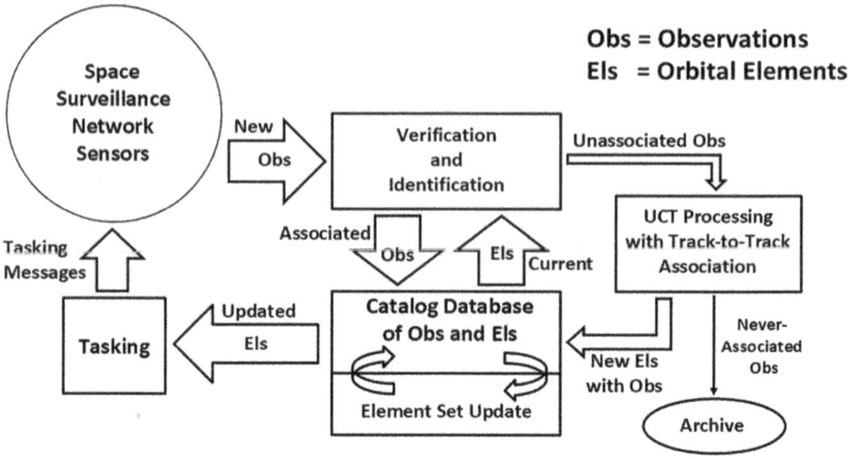

Fig. 5.1 Data flow for catalog maintenance.

Consider Fig. 5.1 that shows the main parts of the catalog maintenance process. The widths of the arrows in the illustration indicate relatively high, medium, and low amounts of data transferred. Relative amounts of computation required in each process block are not indicated.

The logical flow of data for catalog maintenance at the operations centers can be broken down into the following interrelated steps:

1. Verify the sensor-level data associations.
2. Identify unassociated or mis-associated observations with known satellites.
3. Update the database of element sets.
4. Generate element sets and associate tracking data for the remaining UCTs.
5. Formulate tracking assignments (tasking messages) and send them to each sensor.

1. *Verification*

As mentioned in the preceding text, all the sensor-level associations are verified at the operations centers as the observations arrive. This is a relatively fast and automatic process because only the single element set indicated by the sensor needs to be retrieved from the database and the satellite position predicted for the time of that one observation. Except in rare circumstances, such as soon after a breakup, a great majority of the sensor-level associations agree with the central-level verification and are judged to be correct.

2. *Identification*

The few sensor-level mis-associations, plus the observations that could not be associated at the sensor level, are passed to the identification process. Here some kind of comparison with the entire database of elements must be made for each observation. In practice, we can retrieve and propagate element sets for only those orbits that come near the observed altitude, drastically reducing the number of cataloged orbits that must be checked. However, the identification process is still much slower (per observation) than the verification process. Usually, most of the observations presented for identification can be associated with objects that are already in the database when this refined search is made, and so the extra effort is worthwhile. The observations that fail to associate with any cataloged orbit in the identification step are passed to the UCT processing step.

3. Updating the Cataloged Element Sets

As new observations become available after verification, and possibly identification, an attempt is made to update the corresponding element set. The automatic update is a least-squares differential-correction process, using the most recent element values as the prior estimate, to obtain a new estimate of the orbit that best fits the observations. Historically, on average, about 98.5% of the element database can be updated successfully without human intervention. The least-squares fitting process may fail or produce poor results for a variety of reasons, so that the orbital analysts at the operation center must examine about 1.5% of the database elements daily. A variety of software tools are available to them, some of which enable them to accept or reject individual observations, update only a subset of the complete element set, adjust the number of differential-correction iterations, and so on. Typically, new orbital analysts acquire proficiency with these software tools only after considerable learning time, so that human expertise turns out to be an essential part of the cataloging system.

As a sidenote, we can expect, based on history, that the need for highly trained and experienced analysts increases as the catalog grows, even with better automation and even with the improved SP orbit model. The reason is that more of the detailed orbital phenomena that are observable in the sensor data will be represented in the element sets and will call for analysis and explanation. It has long been the case that various kinds of hard-to-understand phenomena show up for individual satellites [3]. As the catalog grows and the automatic processing system is improved, a smaller percentage of the catalog will need human attention, on average, but the absolute number of cases needing some special analysis will still be significant, perhaps even rising, and naturally these cases will be the hardest cases to understand. Some of these analysis problems, such as apparently maneuvering debris, are of such a nature that no automatic system is likely to be able to handle them satisfactorily. In other words, the orbital analysts will always have to stay fully engaged with the entire cataloging process, no matter how good the models and the processing algorithms are.

4. UNCORRELATED TARGET PROCESSING

Ultimately, the orbital analysts must dispose of UCTs either by 1) associating them with a cataloged element set or a candidate element set created in previous UCT processing or 2) creating a new element set from the UCT data as a candidate for inclusion in the catalog. Option 1, not shown explicitly but implied in the Verification and Identification block in Fig. 5.1, is still possible after the automatic identification step. It may be that the most recent element set has been corrupted by some mishandling or that the element set has not been updated recently so it no longer provides accurate predictions. Of course, it may be that no association will ever be made for some observation, perhaps because the tracking data are corrupted or because the object has such low probability of detection by the sensors that no element set can be maintained. Historically, the practice was to retain all UCT data online and readily accessible for up to 60 days for high-altitude orbits and up to 30 days for near-Earth orbits. Beyond these time spans, the prediction accuracies available from the simplified orbit models were judged to have degraded so much that the chance of making correct associations even with known satellites is practically nil. These "old" UCT data would be relegated to an archive, because they might still prove useful in some future analysis.

Whenever the identification step, either automatic or manual, cannot associate observations with known satellites, orbital analysts must use special software tools to make the proper data association. Although the analyst work is inherently slower than real time, it is still essential to make the correct data associations as soon as possible. Human expertise built up through long experience has always been indispensable for this analysis. Moreover, satellite breakup events present special challenges because many unassociated observations and an unknown number of new satellites are involved. The number of association hypotheses that may have to be investigated always goes up much faster than the number of unassociated observations and rapidly becomes unmanageable without human expertise to identify the most probable associations a priori. Advanced computation and parallel-processing techniques have made it possible for the analysts to consider many more association hypotheses than in the past. Nevertheless, human expertise is still crucial in the overall management of the cataloging system.

When the automatic identification fails to make associations, the analysts usually try first a "manual identification" against existing element sets, namely, Option 1 mentioned earlier. This involves relaxing the association criteria in some systematic way to account for known, expected, or possible inaccuracies in the element sets and observations. The added payoff in this "manual" (but still computerized) identification step is that, for each observation, association can be checked, not only against the catalog but also against candidate elements sets that have not yet been judged reliable enough or accurate enough to add to the catalog. Those candidate orbits would have been generated in prior UCT processing. In fact, Option 2 is just that generation of new element sets.

The computational burden goes up dramatically if Option 2 must be exercised. If none of the observations under consideration at this stage of the processing can be associated with any orbit already known or postulated as a candidate, where is the analyst to start? The somewhat technical answer depends on a couple of methods from the science of astrodynamics:

1. An element set can be calculated if two or more positions of the satellite on the orbit are known, provided we also know the times of those positions.
2. An element set can be calculated if three or more lines of sight (directions) to the satellite on the orbit are known, provided we also know the times of those lines of sight.

Tracking radars typically provide short sequences ("tracks") of two or more measured positions of a satellite at measured times, and, in either case, having more observations in the track is usually better. Hence, Method 1 for generating an element set can be used for radar tracks. Similarly, telescopes typically provide tracks of three or more measured directions to a satellite at measured times, and so Method 2 for generating an element set can be used for optical tracks. In the best case, all the measurements in each track reported by the sensors belong to the same space object, even if that space object is not yet cataloged. This is one of the six favorable mitigating circumstances that we listed earlier in this discussion: if objects are not too close together in space, on average, then the sensor can reliably associate the individual observations into a single-object track to be reported. In that case, one element set can be generated for each track, and that element set becomes a candidate for eventual inclusion in the catalog.[117] The analyst must then assess the whole batch of single-track element sets to find out which ones describe the same orbit, despite small numerical differences in the values of the elements among the different tracks. This is a type of multidimensional statistical clustering or classification problem, the general version of which has received much theoretical attention over the years, so that quite sophisticated approaches now exist for identifying groups having similar characteristics within a large population. However, in the first decade of space surveillance, those methods

[117] The specific criteria for including a new element set in the catalog need not concern us, except for us to note that the new space object must be detectable by more than one sensor and has to have at least a certain minimum number of observations associated with it, within some span of time. These criteria are revised occasionally, depending on the number, type, and locations of the available sensors, the goal being to guarantee that every cataloged orbit is "maintainable" in the same sense to which we alluded earlier in this chapter. The effect of enforcing these criteria for inclusion in the catalog is that, at any time, there are likely to be some element sets that do not quite meet all the criteria but can still serve as candidate orbits in UCT processing. Depending on circumstances, such as having only one sensor that can detect them, these candidate orbits can remain in their provisional status for a long time and continue to have observations associated to them occasionally.

were not well developed. Instead, Lawrence Cuthbert, whom we mentioned in Chapter 2, devised a computational help for the human eye to do the clustering in a plot of orbital characteristics, based on single-track values for the orbital plane, the period of the orbit, and the natural precession rate of the orbital plane. The method was modified over the years by Robert (Bob) Morris beginning in the 1960s, with the result that some version of the Cuthbert–Morris algorithm for associating radar tracks is still one of the most-used tools for UCT processing at the primary operations center [4].

Method 1, handling radar tracks one at a time, does not always work well in practice. If the individual measured positions in the track happen to be too close together, for example, or the data are inadequate in other ways, the method produces inaccurate single-track element sets that cannot be used reliably for data association by clustering or any other approach. The obvious questions, "how close is too close?" or "how good do the data have to be?" turn out to be difficult to answer rigorously, for a variety of reasons beyond this discussion. In practice, orbital analysts at the operational centers usually answer the questions "operationally" by simply trying the single-track method on the UCT data they have in hand: poor quality and quantity of the resulting candidate element sets is the negative answer to the question! In that outcome, a different strategy is needed: the analyst can select pairs of tracks from the whole batch of unassociated radar tracks. Each selected pair then constitutes a data-association hypothesis that those two tracks belong to the same space object. By selecting a position from each track in the pair, the analyst can apply Method 1 and construct an element set that represents the hypothesis that those two tracks are associated. Then all these hypothetical element sets can be clustered by the Cuthbert–Morris algorithm or other track-comparison method.

The main difficulty with the two-at-a-time version of Method 1 is that it can require very large computation. Whatever the number of unassociated tracks that the analyst begins with, the number of track-pairs is almost proportional to the square of the number of tracks. For example, a hundred tracks will lead to nearly 5000 track-pairs, and a thousand tracks will lead to nearly half a million track-pairs, that must be checked for association. Yet, with a catalog as large as the present one, it may take only a few days to accumulate a thousand unassociated tracks. The situation is (and, historically, always has been) much worse in the case of a breakup. It turns out that exactly this problem of unavoidable computation-intensive UCT processing occurred for the NSSS Fence system. As mentioned in previous chapters, that multistation radar interferometer did not do extended tracking on any object that it detected. Rather, the line-of-sight data could be triangulated to produce one position per satellite, per pass through the Fence coverage. The only viable approach to using Fence UCT data to produce candidate element sets was just what we have described: pairs of triangulated positions were used to generate hypothetical element sets, and the association of later unassociated data was checked for association with all the hypothetical element sets. The need to do that large UCT computation quickly enough to keep up with the observation flow from the NSSS Fence is what motivated Shannon Coffey's team at NRL to propose and demonstrate parallel UCT processing on the distributed mission processing system in Dahlgren [5].

The case of optical telescope UCT tracks has (and, historically, always has had) some important differences from the radar case. If the optical tracks reported to the operations centers allow the single-track approach of Method 2 to be used, the computational burden, per track, is about the same as for single radar tracks. However, if the optical tracks are too short or otherwise inadequate for reliable single-track UCT processing, the analyst can consider hypothetical tracks constructed from the whole batch of reported unassociated tracks. On this approach, an unavoidable computational burden immediately appears: the hypothetical tracks are made up of reported tracks, selected three at a time. One hundred reported tracks imply more than 160,000 hypothetical tracks to check for association, while a thousand reported tracks imply more than 166 million hypothetical tracks. At this writing, a comprehensive approach to solving this kind of UCT processing problem still has not been developed and tested. The solution almost certainly lies in formulating some alternative to the "brute-force" Method 2 just described. Meanwhile, historically, optical-sensor UCT observations have been processed in large numbers only when the single-track Method 2 is sufficient to generate the candidate orbits. Otherwise, optical UCT data have been used, not to generate the candidate orbits directly but to be checked for association with candidate orbits generated some other way, adding to the statistical support for those candidates.

5. *TASKING*

The remaining part of the overall catalog maintenance process is known as tasking, that is, generating instructions for the sensors to track satellites and send that data to the operations centers. The general approach to tasking developed slowly in the first few years of space surveillance, because only the Baker–Nunn cameras were owned by the space surveillance organizations, and not many satellites had to be tracked. By the end of 1958, the NSSS Fence was available and dedicated entirely to space surveillance. Moreover, the Fence could not be, and did not need to be, tasked. The Fence simply detected whatever passed through its coverage (if the object was big enough to detect) and automatically sent the raw observational data to the operations center in Dahlgren in real time, without any prior information about satellites in orbit. In contrast, optical sensors and narrow-beam tracking radars needed to be cued somehow, and when those types of sensors became numerous, tasking methods began to be developed. These sensors need prior information about when and where to point, to be able to detect a specific object and record the observational data. Cued sensors inherently have a limited tracking capacity. That is, they can track only a limited number of objects in a limited time span, because they need to re-aim their apparatus between different tracking contacts and wait for the satellite to appear in the field of view. Compare these features with a typical missile-warning radar: these radars were designed specifically to detect and automatically track objects without any advance notice, and they can do it in all weather, day or night. These radars would be ideal space surveillance sensors, except that their mission is something else entirely. Nevertheless, they do detect many satellites, and, as we have seen in previous chapters, they do depend on the space surveillance system to enable them to distinguish satellites from missiles at a moment's notice. The job of tasking is to make use of all these different sensor types, in their existing locations, and during the time periods when they are available for space surveillance, to obtain enough tracking data to maintain the catalog. It is a notably complicated problem. One part of the practical solution for many years was to predict how many observations that the NSSS Fence would make during the next day. Then the rest of the SSN would be tasked for the remaining number of observations needed for each satellite.

One of the most important features of the tasking methods employed for space surveillance is that these methods do not attempt to schedule the sensors' operations [6, 7]. Instead, tasking messages to each sensor contain a list of satellites that will be visible, the priority for tracking each satellite, the most recently updated cataloged element set for each satellite, and the number of tracks needed on each satellite from that sensor during the next day. The sensor control system then must schedule the times when those tracks will be taken. Developing the schedule involves predicting when each satellite will be visible to the sensor (including lighting conditions, for an optical sensor) and determining a workable sequence in which all the requested tracks can be made. It is entirely possible, and in practice it occurs rather often, that the sensor cannot schedule all the tracks it has been tasked to make. That is why each satellite has a priority assigned; if not all tracks can be obtained, the sensor makes the higher-priority tracks in preference to the lower-priority tracks.

It has long been necessary in space surveillance to allow the sensors to do their own scheduling and to construct the tasking messages with the minimum information that the sensors need to do the scheduling. This strategy allows sensors whose primary mission is something other than space surveillance to contribute to the catalog without excessive contention over essential tasks in either mission. Perhaps more important, letting the sensors do their own scheduling circumvents the need for the operations centers to have to solve the global scheduling problem for the entire SSN, a colossal computational problem that almost certainly could not be solved satisfactorily with existing computers and the time available within the catalog update cycle.

The question of number of tracks needed for each orbit update is also a difficult one. Statistical studies are done periodically to determine, on average, how many tracks per day will produce specific levels of accuracy in the orbit estimates, for different categories of orbits. Then that total number of tracks is allocated among the sensors in some manner. Naturally, the way that an analyst chooses the orbit categories affects the number of tracks needed, and several different types of orbit category have been devised over the years. The earliest was invented and used by an orbital analyst named John Gabbard, working at Cheyenne Mountain, who became prominent in space surveillance later in his career. The Gabbard classes, as they were called, were several dozen groupings of orbits based on two parameters, perigee height and apogee height. His scheme was used for many years. Miller [7] describes how Gabbard's scheme was augmented and analyzed over the years to try to make it more rigorous in some sense. Eventually, in the late 1990s, when the SP catalog was being developed, it became apparent that Gabbard's classes of orbits, devised for use with the simplified orbit models of his time, were not the right means to task the sensors to enable the higher accuracies possible with the better SP orbit model. The most sensitive variable affecting the amount of tracking needed for low orbits turned out to be the rate at which atmospheric drag removed energy from the orbit. This is perhaps not too surprising in retrospect, because drag is perennially the force that is modeled least well, and extra measurements of the actual motion might help compensate for that mismodeling. In any case, a scheme based on the trend in energy dissipation rate across orbit classes is still used.

6. Conclusions

Before we leave the subject of catalog maintenance, we should reemphasize just how many assumptions are built into the processes we described. Earlier in this chapter, we listed six simplifying factors that have helped make it feasible to keep a catalog in the first place. We also noted that if any of those factors were to change for the worse, then we would have to redesign and replace many of the analytical and computational tools that we have learned to depend on for catalog maintenance. A small-scale example of this fact is the change to the tasking method that we just described. When the SP catalog became necessary, it was not only new computers and new software that had to be provided. It suddenly became necessary to revise all the assumptions that, for more than two decades, had guided the tasking of the SSN, a crucially important aspect of catalog maintenance. An entirely new method of tasking had to be invented if the SP catalog was to serve its purposes. We can expect to face the need for even more striking innovations as the domain of space activities expands outward from Earth orbit into the Earth–Moon system. One example of the impending need for innovation is already upon us. Since Sputnik was launched, we have grown comfortable with the familiar periodic elliptical orbits currently listed in the catalog. However, trajectories throughout the Earth–Moon system are radically complicated to describe, compared to our traditional families of orbits: even the very concept of orbital element sets no longer applies in that larger domain. Consequently, maintaining a catalog of space objects that are moving throughout the Earth–Moon system is bound to be not only quite difficult by current measures but also profoundly interesting.

References

1 Schumacher, P. W., Jr., "Prospects for Improving the Space Catalog", AIAA Paper 96-4290, *Proceedings of the American Institute of Aeronautics and Astronautics Space Programs and Technologies Conference*, Huntsville, AL, Sept. 1996. https://doi.org/10.2514/6.1996-4290 [retrieved 27 Dec. 2024].

2 Uhlmann, J. K., "Algorithms for Multiple-Target Tracking," *American Scientist*, Vol. 80, March–April 1992, pp. 128–141.

3 DeVere, G. T., and Teets, R. B., "An Examination of the Histories of Interesting Debris Satellites," AAS Paper 96–120, *Proceedings of the AAS/AIAA Space Flight Mechanics Meeting*, 12–15 Feb. 1996, Austin, TX, edited by G. Edward Powell, R. H. Bishop, J. B. Lundberg, and R. H. Smith, Advances in the Astronautical Sciences, Vol. 93, Part 1, pp. 289–305. Univelt, Inc., San Diego, CA, 1996.

4 Miller, J. G., "Special Perturbations Uncorrelated Track Processing," AAS Paper 07-228, *Proceedings of the AAS/AIAA Astrodynamics Specialists Conference*, 19–23 Aug. 2007, Mackinac Island, MI, edited by R. J. Proulx and T. F. Starchville, Jr., Advances in the astronautical Sciences, Vol. 129, Part 1, pp. 589–604. Univelt, Inc., San Diego, CA, 2008.

5 Coffey, S. L., Jenkins, E., Neal, H. L., and Reynolds, H., "Parallel Processing of Uncorrelated Observations into Satellite Orbits," AAS Paper 96-146, *Proceedings of the AAS/AIAA Space Flight Mechanics Meeting,* 12–15 Feb. 1996, Austin, TX, edited by. G. Edward Powell, R. H. Bishop, J. B. Lundberg, and R. H. Smith, Advances in the Astronautical Sciences, Vol. 93, Part 1, pp. 739–756. Univelt, Inc., San Diego, CA, 1996.

6 Miller, J. G., "A New Sensor Resource Allocation Algorithm for the Space Surveillance Network in Support of the Special Perturbations Satellite Catalog," AAS Paper 03-669, Proceedings of the AAS/AIAA Astrodynamics Specialists Conference, 3–7 Aug. 2003, Big Sky, MT, edited by J. de Lafontaine, A. J. Treder, M. T. Soyka and J. A. Sims Advances in the Astronautical Sciences, Part 3, pp. 2623–2644. Univelt, Inc., San Diego, CA, 2004.

7 Miller, J. G., "A New Sensor Allocation Algorithm for the Space Surveillance Network," *74th MORS Symposium Proceedings*, Military Operations Research Society–Working Group 5, Colorado Springs, CO, 28 Aug. 2006. Also published in *Military Operations Research*, Vol. 12, No. 1, 2007, pp. 57–70. https://www.jstor.org/stable/43941064 [retrieved 26 Dec. 2024]. Addressed sensor tasking for the SP catalog to meet new Capstone accuracy requirements.

Chapter 6

The Space Surveillance Network

Within the SSN, we get data from a variety of sources, and the measurement content of each depends on the technology employed. Consequently, we refer to the sources generically as sensors. Although some sensors also have a capability to characterize a satellite, our discussion primarily concentrates on the satellite location measurement data they provide. We describe the specific systems used in approximate historical order. Some of the equipment has a remarkable history of being used for tracking, deactivated, put in storage, and then reused for tracking in another location. Our focus is on the location of a tracking site and the type of technology used for observation collection at that site.

Sensors are further distinguished by the type of support they provide. A dedicated sensor has a primary mission of space surveillance. A collateral sensor performs space surveillance as a secondary or tertiary mission. A contributing sensor is a private or research and development asset that supports space surveillance under a contract.

Sensors are further categorized as radars that can be mechanical trackers, phased arrays, or detection fans. Telescopes are employed for both ground- and space-based observing. Passive radio frequency sensors are mechanical trackers that can only detect and track emitting satellites.

Before the launch of the first man-made satellite, there was expectation that a launch would occur and that it would be a U.S. satellite. The United States had been testing rockets for years, and some number of sites already existed for receiving telemetry.

I. Moonwatch Observers (1956–1975)

Volunteers who recorded visual sighting of satellites provided a little-known but important component of early tracking efforts. Project Moonwatch [1] was an amateur science program formally initiated by the SAO in 1956 and finally discontinued in 1975. The SAO organized Moonwatch as part of the IGY. Its initial goal was to enlist the aid of amateur astronomers and other citizens who would help professional scientists spot the first artificial satellites. Thousands of teenagers, housewives, amateur astronomers, schoolteachers, and other citizens served on Moonwatch teams around the globe. Using stopwatches and specially designed telescopes, hand-built or purchased from vendors like Radio Shack, scores of Moonwatchers nightly monitored the skies. Until professionally staffed optical tracking stations came online in 1958, this network of amateur scientists and other interested citizens played a critical role in providing crucial information regarding the world's first satellites.

II. BAKER–NUNN CAMERAS (1956–1992)

The Satellite Tracking Program (STP) [2] of SAO was a worldwide network of stations responsible for the optical tracking of satellites. The goal of STP was to obtain photographs of satellites against a star background in sufficient number and accuracy to allow the determination of highly precise orbits. Data derived from the orbits provided information concerning variations in the density and temperature of the upper atmosphere and helped construct new representations of the Earth's gravitational potential and geometrical figure. A special satellite tracking camera, designed by James C. Baker and Joseph Nunn, was installed at each station.

By June 1958, there were 12 tracking stations spread around the world. When primary operations were moved to the Colorado Springs location in 1961, five additional tracking stations were dedicated to satellite tracking and were operated by the Air Force. By the mid-1960s, that configuration had evolved to the sites and locations provided in Table 6.1 except for the St. Margarets site, which came online in the mid-1970s. Longitude is positive east longitude, negative values are west longitude, and the sensor number is the identity assigned in the tracking station database.

TABLE 6.1 BAKER–NUNN CAMERA LOCATIONS

Location	Country	Sensor Number	Latitude (deg)	Longitude (deg)
Pulmason	South Korea	25	35.16731	128.75559
St. Margarets	New Brunswick, Canada	27	46.89944	−65.21020
San Vito	Italy	28	40.64096	17.84558
Mt. John	New Zealand	29	−43.98728	170.46384
Edwards Air Force Base	California, United States	30	34.96299	−117.91449
Cold Lake	Alberta, Canada	32	54.43555	−110.28476

Fig. 6.1. Baker–Nunn camera [3].

The Baker–Nunn film was developed, and analysts used plastic overlays of star charts, a microscope, and a grease pencil to "field reduce" the observations and send them in, usually within 24 hours. The film itself was then shipped to central sites (there was one in Colorado Springs) and processed by a special digital scanning device to precisely determine the position of the satellite relative to the stars on the film. This processing is known as "precision reduced." This took a week or more and was not done very often for satellite catalog maintenance. SAO had their own site and did their own precision reduction. The Baker–Nunn cameras provided invaluable contributions to satellite catalog maintenance until 1992 when the last site in Cold Lake, Canada, was closed. A photograph of a Baker–Nunn camera is shown in Fig. 6.1.

III. SPACE TRACKING AND DATA ACQUISITION NETWORK (1957–1961)

The NRL Vanguard satellite project included a radio-tracking network known as Minitrack [4], which used radio interferometers and Yagi antennas to track satellites using signals emitted by onboard beacons to determine their positions and orbits. Sputnik and other Soviet satellites failed to conform to the beacon standard, illuminating a need to track nonradiating or noncooperative satellites. Consequently, a high-priority program was initiated to acquire the capability to track satellites by signals reflected off them. An experiment using a transmitter in Fort Monmouth, New Jersey, and Minitrack receivers demonstrated the concept, and from this experiment a larger and more elaborate system was developed by NRL. The new system was to be called the Naval Space Surveillance System (NAVSPASUR), and the first two stations were operating in July 1958, less than six weeks after NRL was tasked to construct the system.

IV. NAVSPASUR (1958–2013)

By 1961, NAVSPASUR [5] had reached initial operational capability. Two transmitter sites at Jordan Lake, Alabama, and Gila River, Arizona, produced a wall of VHF energy stretching across the United States along a great circle inclined at 33 deg to the equator. Four receiver stations were built: San Diego, California; Elephant Butte, New Mexico; Silver Lake, Mississippi; and Fort Stewart, Georgia. Later additions to the NAVSPASUR network included a two-mile-long transmitter at Lake Kickapoo, Texas, and two gap-filler receivers at Red River, Arkansas, and Hawkinsville, Georgia. These sites were built between 1961 and 1965, completing the system. This "fence" could detect basketball-sized objects in orbit as high as 26,000 miles above Earth.

The system sent to the NAVSPASUR processing facility at Dahlgren, Virginia, real-time raw signal phase and amplitude data measured on more than 12 miles of dipole arrays. The raw data were converted by interferometric methods into apparent direction cosines (as seen from the receivers) for use in updating the satellite catalog. Near-Earth satellites passed through the field of view of the Fence four to six times per day and were, on average, illuminated by two transmitters and detected by four receivers on each pass. The result was an abundant data set produced without any cueing or prior knowledge of the satellite population. More than 98.5% of the orbits visible to the Fence were updated without human intervention. By 1970, NAVSPASUR was one of the top five contributors of tracking data and continued to do so until its closure in 2013.

V. MILLSTONE HILL (1957 TO PRESENT)

The original Millstone Hill Radar [6] began operations in the fall of 1957. This was a UHF radar, operating at about 450 MHz. On 5 October 1957 it detected the Soviet Sputnik I satellite, just one day after its launch. On 11

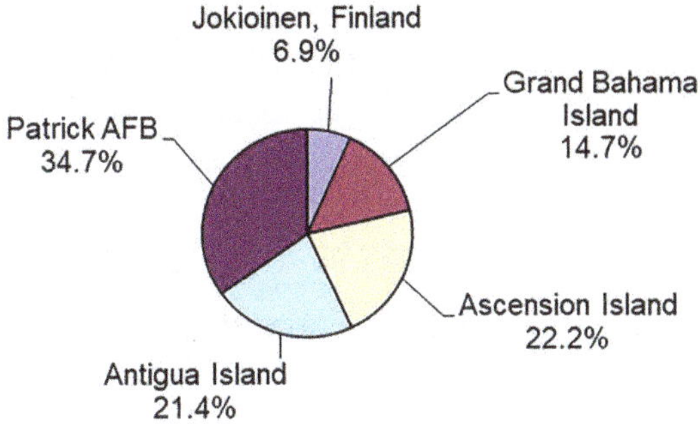

Fig. 6.2. Top five sensor contributions, 1960 [7].

April 1958 it became the first radar (at least the first U.S. radar) to track (not just detect) a satellite when it tracked Sputnik II from horizon to horizon. This original UHF version of the Millstone Hill Radar was the basis for the FPS-49 and FPS-92 missile-tracking radars in the BMEWS early warning system.

This original radar was upgraded to L-band in 1962, which involved installing a new antenna with the same diameter. The old antenna was moved to the BMEWS early warning radar site in Diyarbakir (Pirinçlik), Turkey, where it was designated as the AN/FPS-79 and continued to be used for space surveillance until about 1997. In 1963, Millstone successfully detected a satellite in geostationary orbit (likely the first time this was accomplished by radar), which involved noncoherent integration of about 40,000 pulses over about 45 minutes. Millstone is a contributing radar to the SSN.

In 1960, the top five sensors contributing data were primarily range support sensors developed to support space launch operations. These sensors supplied almost 25,000 total observations for the entire year, representative of the very small satellite population at the time. The relative contribution from the top five sensors is illustrated in Fig. 6.2. It should be noted in this and subsequent graphs that the contributions of other sensors not on the graphs were also very important. In many cases, the sensors with the fifth and sixth most observations were nearly equal. The graphs are meant simply to show how the relative contributions of the primary sensors have evolved over nearly 70 years.

VI. DIYARBAKIR (PIRINÇLIK) TURKEY (1962–1997)

Pirinçlik Air Base [8] was a joint U.S.–Turkish facility near Diyabakir, Turkey. It was the site of an important tracking capability for monitoring launches from the Soviet Union and normally provided the first radar tracking of Soviet launches from the Kapustin Yar launch facility. It had a detection radar (AN/FPS-17) and a tracking radar (AN/FPS-79). Together they were able to provide tracking of two objects simultaneously. Its location close to the southern

Soviet Union made it the only ground sensor capable of tracking actual deorbits of Soviet space hardware. In addition, the Pirinçlik radar was the only 24-hour-per-day Eastern Hemisphere deep space sensor. The site discontinued operation in 1997 because of U.S. base closure and realignment activities.

VII. SHEMYA (1963 TO PRESENT)

Shemya [9] is a western island in the Aleutian chain off the coast of Alaska. It is of strategic importance because of its proximity to the Soviet Union Kamchatka Peninsula where missile testing was conducted. In 1963, three AN/FPS-17 radar trackers were installed there. Observations from Shemya were usually the first radar reports of Soviet satellite launches from Tyuratam in the early days of satellite tracking. In 1977, the equipment was replaced by the FPS-108 Cobra Dane phased array radar. The Shemya site has been a major contributor to space surveillance both in the tracker era as well as the phased array era. It has been among the top five contributors for much of its existence. Cobra Dane is a collateral SSN site.

VIII. TRAVELING RADAR OPERATORS (1967–1996)

We briefly depart from our discussion of tracking sites and instead follow one operational organization as it was moved around the world for performing space surveillance. The 17th Space Surveillance Squadron [10] was first organized in Moorestown, New Jersey, and operated a radar sensor until it was deactivated in December 1969. The site was used for satellite tracking before the 17th taking over operations. It was one of the top five contributors in our 1965 snapshot.

The squadron was activated in 1971 at Ko Kha, Thailand. It tracked Chinese missile launches and southerly satellite launches from the Soviet Union until its deactivation in May 1976. The next stopover was a move to San Miguel, Philippines, in 1982. At San Miguel, the unit operated a GPS-10 radar in support of space surveillance. The site was deactivated in 1989. Finally, the Squadron was reactivated in Edzell, Scotland, and operated sensors for the Deep Space Surveillance Network until 1996.

IX. BALLISTIC MISSILE EARLY WARNING SYSTEM (1961 TO PRESENT)

The BMEWS [11] was a radar system built by the United States to provide early warning of Soviet Union intercontinental ballistic missiles. The sites were situated at far north latitudes watching for missiles coming over the North Pole. All the BMEWS sites have a collateral SSN role.

BMEWS consisted of two types of radars. The first type of radar had very large, fixed parabolic reflectors with two primary feed points. They produced two fan-shaped microwave beams that allowed them to detect targets across a very wide horizontal front at two narrow vertical angles. These were used to provide coverage of missiles rising above the horizon, and by tracking them at two points as they climbed, enough information to determine their rough trajectory was available.

The second type of radar was used for fine tracking of selected targets and consisted of a very large steerable parabolic reflector. These radars provided high-resolution angular and ranging information for rapid calculation of the probable impact points of the missile warheads.

The first site at Clear, Alaska, became operational in July 1961. It consisted of three fixed detection fans and one mechanical tracker. In April 1968, the site was upgraded to a phased array radar with two faces.

The site at Thule, Greenland, also became operational in early 1961. It consisted of four detection fans driven by three transmitters and one mechanical tracking radar. In June 1987, the site was upgraded to a phased array radar with two faces.

The original BMEWS system reached full capability in September 1963 when the site at Fylingdales [12] in North York Moors, England, was completed. Unlike the other two sites, it consisted of three mechanical trackers. Two of these were used to scan the horizon rather than the fixed detection fans used at the other two sites. In 1992, the site was upgraded to a phased array radar with three faces.

As the number of spacecraft being launched increased, the United States developed dedicated sensor systems. By 1965, the United States had built several new radar sensors devoted to the detection and tracking of ballistic missiles and satellites. Many of these used mechanical dish tracking radars coupled with large, grid-like detection radars capable of providing many more observations each day. The number of observations collected by the top five sensors increased dramatically, with over 3,100,000 observations collected in 1965, an increase by over two orders of magnitude from 1960. The relative contribution from the top five sensors is illustrated in Fig. 6.3.

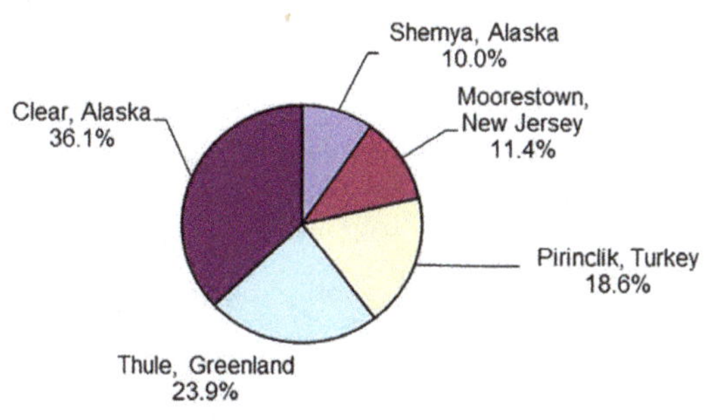

Fig. 6.3. Top five sensor contributions, 1965 [7].

X. EGLIN SPACE SURVEILLANCE RADAR (1969 TO PRESENT)

During the 1960s, the Soviet Union developed a system to deliver nuclear weapons with missiles in Earth orbit. The system was called fractional orbital Earth bombardment. Such a missile could orbit the Earth before beginning its reentry, allowing it to arrive from any direction. Such an attack would likely be undetected by the BMEWS system that looked only for attacks coming over the North Pole. The possibility of such a threat from space, as well as the increasing number of satellites in Earth orbit, convinced the United States that it needed to greatly expand its space tracking capabilities, and the AN/FPS-85 was designed for this mission. It is a dedicated SSN sensor.

It is significant that this was the first surveillance system built with a dedicated space surveillance mission. Previously, all radars were developed primarily for missile warning, and space surveillance was a secondary mission. The AN/FPS-85 at Eglin AFB [13] was the world's first large phased array radar. Not only did the phased array have much greater tracking capacity, but also the accuracy of the measurements provided a 100-fold improvement in the accuracy of the observations from any previous stations. In the early 1960s, the differential orbit corrections suffered from the effects of using very inaccurate observations from the BMEWS stations. This all changed in 1969 when Eglin became operational with extremely accurate observations on the great majority of the catalog. Also noteworthy, the Naval Space Surveillance System (NAVSPASUR) had begun to make substantial contributions. The total number of observations contributed by the top sensors increased only slightly from 1965 to just over 4,040,000 satellite positional measurements. The relative contribution from the top five sensors is illustrated in Fig. 6.4.

The only radar added to the space surveillance system between 1970 and 1975 was the radar at Kho Kha, Thailand. It was one of the top five contributors in the 1975 snapshot. The contributions of the Eglin phased array and NAVSPASUR radars became dominant by 1975, with more than 75% of nearly 3,620,000 total observations supplied by these two sensors. The relative contribution from the top five sensors is illustrated in Fig. 6.5.

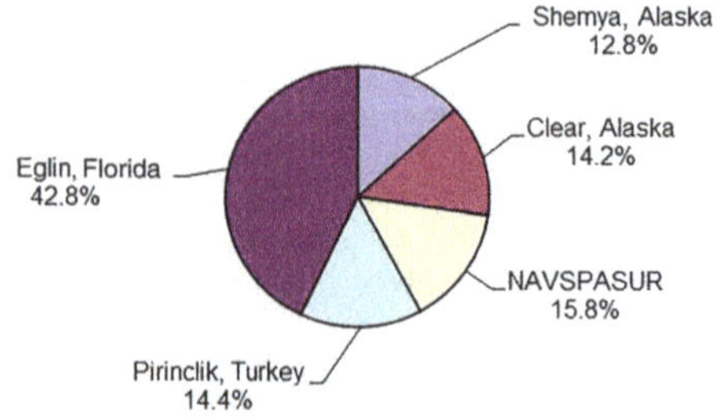

Fig. 6.4. Top five sensor contributions, 1970 [7].

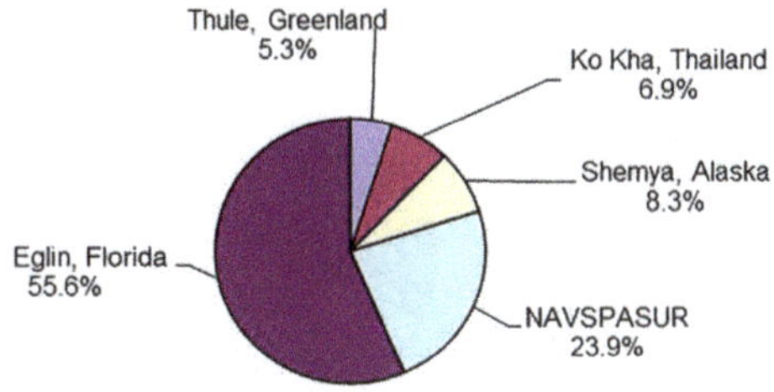

Fig. 6.5. Top five sensor contributions, 1975 [7].

XI. PERIMETER ACQUISITION RADAR ATTACK CHARACTERIZATION SYSTEM (1977 TO PRESENT)

The Perimeter Acquisition Radar Attack Characterization System (PARCS) [14] is located in North Dakota and became operational in 1975 as part of the U.S. Army's Safeguard antiballistic missile system. It is an AN/FPS-16 phased array with a single face pointed north and operates in the UHF band. In 1977, the site was transferred to the Air Force for missile-warning and space surveillance use. PARCS is also known as Cavalier, the closest town in North Dakota. It has a collateral SSN role.

XII. ASCENSION ISLAND (1978 TO PRESENT)

Ascension Island [15] is part of the British Overseas Territory and is located on the equator roughly halfway between South America and Africa. Initially, the site was developed to support long-range missile testing in the 1950s. In the 1960s, it supported the Gemini manned space program. In 1978, it began to support space surveillance as a collateral sensor. The current equipment is an FPQ-15 C-band radar.

XIII. PAVE PAWS (1980 TO PRESENT)

Beale Air Force Base in north-central California is home to a PAVE PAWS [16] early warning radar site that also supports the space surveillance mission. It is an AN/FPS-121 radar with two faces operating in the UHF band. It first began operations in August 1980. All the PAVE PAWS sites have a collateral role in the SSN.

Cape Cod Air Force Station in Massachusetts has a PAVE PAWS early warning radar site that also supports the space surveillance mission. It is an AN/FPS-121 radar with two faces operating in the UHF band. It first began operations in April 1980.

Robbins Air Force Base in Georgia has a PAVE PAWS early warning radar site that also supports the space surveillance mission. It is an AN/FPS-121 radar with two faces operating in the UHF band. It first began operations in 1986. It was closed in 1995 as part of Cold War drawdown.

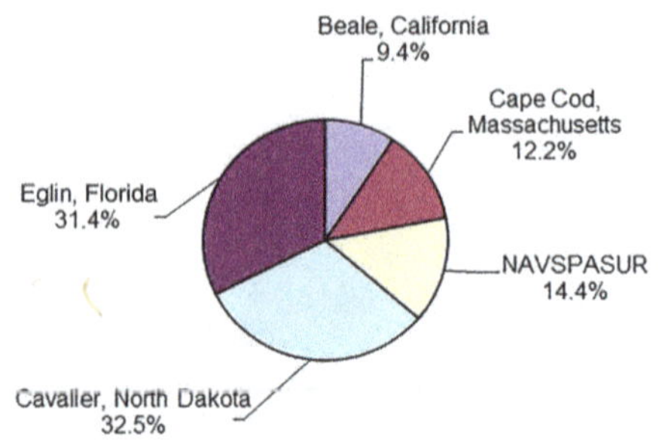

Fig. 6.6. Top five sensor contributions, 1980 [7].

El Dorado Air Force Station in Texas has a PAVE PAWS early warning radar site that also supports the space surveillance mission. It is an AN/FPS-121 radar with two faces operating in the UHF band. It first began operations in 1987. It was closed in 1995 as part of Cold War drawdown.

The 1980s marked the beginning of the phased array radar era, because four of the five most significant contributors were phased array systems. New radars at Beale, California, and Cape Cod, Massachusetts, began to make large contributions, and the converted Antiballistic Missile radar at Cavalier, North Dakota (commonly known as PARCS), also made significant contributions to over 5,410,000 observations sent in 1980. The relative contribution from the top five sensors is illustrated in Fig. 6.6.

One important set of systems not addressed by this summary includes those that provide data on satellites in deep space orbits, or those with orbital periods greater than 225 minutes. Because most satellites in orbit are in low-altitude orbits, it logically has followed that the most prolific SSN sensors have been those capable of tracking this orbit regime. However, with a growing number of satellites in deep space orbits, the ability of sensors to track these satellites was a significant requirement.

XIV. Ground-Based Electro-Optical Deep Space Surveillance (1980s to Present)

The Baker–Nunn cameras provided space surveillance of geosynchronous satellites for over three decades. But the need for automatic observation generation and greater throughput demanded a new technology replacement for the film-based Baker–Nunn system. The Ground-Based Electro-Optical Deep Space Surveillance (GEODSS) [17] system was designed to be that successor. The AN/FSQ-144 system employed an Ebsicon vacuum tube to detect and produce satellite position measurements. As the availability of solid-state charge-coupled devices (CCDs) moved forward, CCDs were refitted to the GEODSS cameras with dramatic gains in throughput and accuracy.

The initial plan for worldwide viewing included five sites: Diego Garcia in the central Indian Ocean; Taegu, South Korea; Maui, Hawaii; Socorro, New Mexico; and a site in Portugal. Plans for the Portugal site were dropped due to negotiation difficulties. The remaining four sites were fielded in the 1980s. The site in South Korea was closed in 1994 due to degraded operation from weather patterns.

XV. RONALD REAGAN TEST SITE (1980S TO PRESENT)

The Ronald Regan Test Site on Kwajalein Atoll near the equator in the mid-Pacific Ocean serves as an ideal site for live testing of missile interceptors. The mechanical tracking radars also provide support to space surveillance.

ALTAIR [18] is a mechanical tracker operating in both VHF and UHF bands. It began support to the SSN in 1982 as a contributing SSN sensor. It tracks both near-Earth and deep space satellites.

ALCOR [19] is a mechanical tracker operating in C-band. It began support to the SSN in 1985 as a contributing SSN sensor. It tracks both near-Earth and deep space satellites.

TRADEX [20] is a mechanical tracker operating in L-band. It began support to the SSN in 1998 as a contributing SSN sensor. It tracks both near-Earth and deep space satellites.

The dominance of phased array radar sensors continued in 1985. Notables are the increased contributions from the Cavalier, North Dakota, radar and the decrease in contributions from the Eglin radar. Eglin had been assigned an additional mission of providing missile-warning support that limited its satellite tracking contributions somewhat. Almost seven million observations were supplied by these prolific sensors. The relative contribution from the top five sensors is illustrated in Fig. 6.7.

The contributions in 1990 showed the emergence of the "Big Three" radar sensors: Eglin, Cavalier, and NAVSPASUR. These sensors were responsible for more than 75% of the data supplied for the next 17 years and were critical to the success of the satellite tracking mission. The top five sensors supplied almost nine million satellite measurements in 1990. The relative contribution from the top five sensors is illustrated in Fig. 6.8.

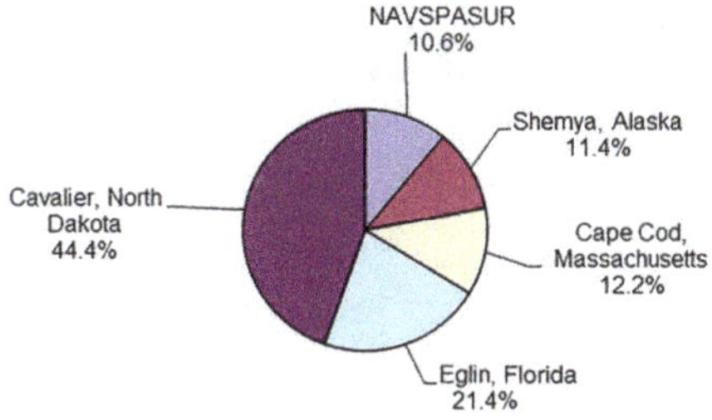

Fig. 6.7. Top five sensor contributions, 1985 [7].

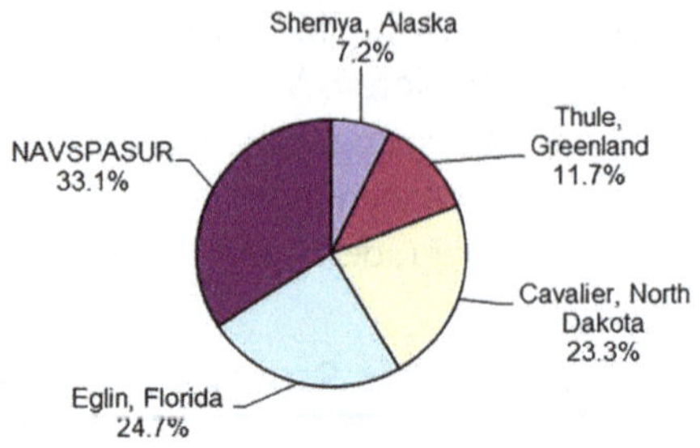

Fig. 6.8. Top five sensor contributions, 1990 [7].

Besides radars and telescopes, the 1990s saw the addition of a third technology for space surveillance. For active satellites, if their signals can be detected by a ground antenna, then the time and direction of the signal can be recorded as a satellite location observation. Using this technology, sites at Griffiss Air Force Base [21] in New York; Misawa [22] in Japan, and Edzell, Scotland, in the United Kingdom formed the Deep Space Surveillance Network in support of space surveillance. This support was discontinued in 1995.

As established in 1990, the importance of the "Big Three" sensors was maintained in 1995. The continued growth in the number of satellites was a driving factor in the collection of almost 25 million observations in 1995. In 1993, a major tasking change occurred, and this produced more observations on sparsely tracked objects, really helping the accuracy of the orbit determination, especially for deep space satellites. This also greatly reduced the lost list. The relative contribution from the top five sensors is illustrated in Fig. 6.9.

XVI. GLOBUS II

The Globus II [23] radar is an AN/FPS-129 mechanical tracker operating in X-band. It is in Vardo, Norway, and began support to the SSN in 1998 as a

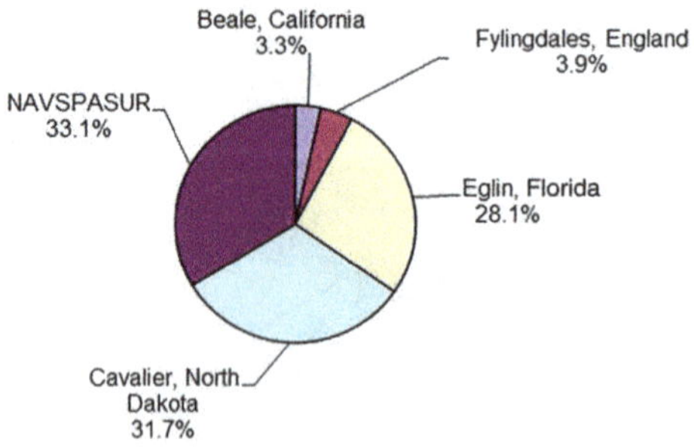

Fig. 6.9. Top five sensor contributions, 1995 [7].

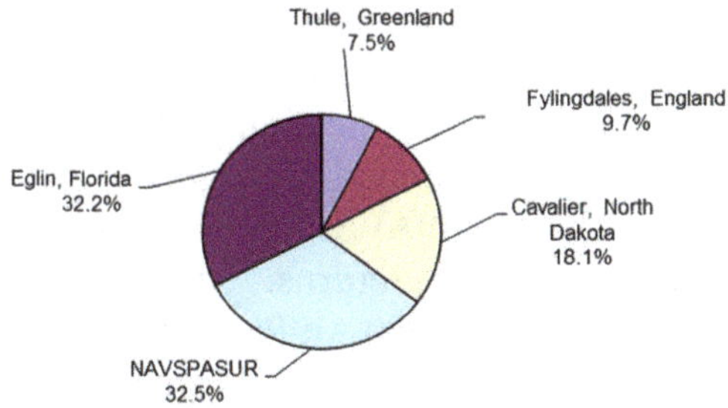

Fig. 6.10. Top five sensor contributions, 2000 [7].

dedicated SSN sensor. The Globus radar along with the Millstone Hill Radar in Massachusetts and the radars at Kwajalein Atoll in the Pacific supply valuable deep space tracking data around the equator.

By 2000, the number of observations collected by the SSN reached a new high, with almost 35 million observations processed, and the contributions of the Big Three sensors continued to be very important. The relative contribution from the top five sensors is illustrated in Fig. 6.10.

XVII. RAVEN

Raven-class telescope systems [24] are based on a paradigm where the goal and concept of operations of the mission, along with the desired data products, define the components that will be used in that system. One particular hardware and software configuration is optimized for deep space metrics and supports the SSN. The Raven telescope on Maui began support to the SSN in 2005.

Significant changes occurred in the SSN contributions by 2005. The advent of the SP Catalog required more data than at any time in the last 40 years, and the contributions of sensors reflected this increased demand. The relative contribution from the top five sensors is illustrated in Fig. 6.11.

Nearly 110 million observations were supplied in 2005, a factor of more than four times the number of observations produced in 2000. The difference

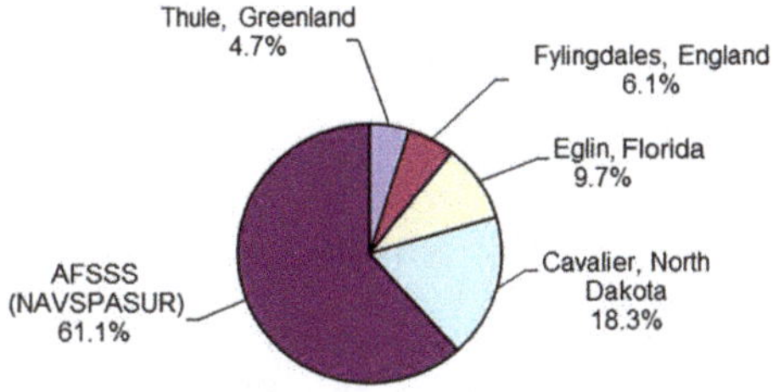

Fig. 6.11. Top five sensor contributions, 2005 [7].

between 2000 and 2005 is largely due to a single contributor, NAVSPASUR. In the past, NAVSPASUR acquired numerous single station observations but was required only to forward to Cheyenne Mountain multiple station observations that could be triangulated. The greater data needs of the SP catalog led to the requirement for NAVSPASUR to forward all observations, both single and multiple station returns. Of note was the transfer of NAVSPASUR from Navy control to the Air Force, which resulted in renaming it the AFSSS.

Telescopes play a critical role in providing SSA of the deep space domain. However, they have limitations requiring the satellite to be in sunlight while the ground-based observing site is in darkness. Furthermore, they need relatively cloud-free skies. A space-based orbiting telescope could largely overcome these limitations and fill ground network data gaps.

The Midcourse Space Experiment (MSX) satellite [25] was launched in April 1996 as a 15- to 18-month missile-tracking experiment. After such experiments were completed, the Air Force began using the spacecraft for SSA operations in 1998 as part of an Advanced Concept Technology Demonstration. The satellite operated more than seven years past its expected lifetime and has very successfully validated the concept of space-based observations.

Following the MSX pathfinder and proof of concept, we see this as the dawning of the era of space-based space surveillance. Five separate systems were launched and became operational between 2010 and 2020.

XVIII. SPACE-BASED SPACE SURVEILLANCE SYSTEM

The Space-Based Space Surveillance System [26] is an optical sensing satellite and follows the successful MSX concept demonstration. Its orbit is sun-synchronous at 630-km altitude. It began providing observations to the SSN in August 2012.

XIX. SPACE TRACKING AND SURVEILLANCE SYSTEM

The Space Tracking and Surveillance System [27] satellite was launched in May 2009 to demonstrate new technologies for missile detection. The satellite is in a near-Earth sun-synchronous orbit. Operational control was transferred to Air Force Space Command in Jan 2011 as a dedicated sensor for space surveillance.

XX. SAPPHIRE

Sapphire [28] is a Canadian space surveillance satellite. It was launched in 2013 and is a contributing sensor to the SSN. The satellite is in a sun-synchronous orbit 483 miles by 493 miles. The satellite tracks objects ranging from medium Earth orbit to geosynchronous. Before launch, the sensor was tested at a ground site in Canada and provided observations from August 2011 until the Sapphire launch in 2013.

XXI. GEOSYNCHRONOUS SPACE SITUATIONAL AWARENESS PROGRAM

Geosynchronous Space Situational Awareness Program [29] satellites provide a space-based capability operating in the near-geosynchronous orbit regime supporting U.S. Space Command space surveillance operations as a dedicated SSN sensor. Two satellites were launched in 2014 following by another two in 2016.

XXII. ORS-5

The Operationally Responsive Space–5 satellite (ORS-5) [30] was launched in August 2017 into a low inclination orbit at 599-km altitude. It is used to track satellites and debris in geosynchronous orbits. It reached initial operational capability in May 2018.

XXIII. S-BAND FENCE

The NAVSPASUR fence provided significant contributions to SSA for 55 years. The capability of a high-elevation fence for unqueued detection of satellites has proven to be invaluable. In 2009, the Air Force decided there was a need to upgrade the legacy system. The new Space Fence [31] was envisioned to be a system of two or three S-band ground-based radars designed to perform unqueued detection, tracking, and accurate measurement of orbiting space objects. The shorter wavelength of the S-band Space Fence will allow for the detection of much smaller satellites and debris. The first site is located on Kwajalein in the Pacific. It began operations in March 2020.

XXIV. HAROLD E. HOLT NAVAL COMMUNICATIONS STATION

The Harold E. Holt Naval Communications Station [32] located on the northwest coast of Australia is the home of a C-band mechanical tracker. Its mission is to provide Southern Hemisphere coverage of RSOs for catalog maintenance, SOI, and support for special events. It became operational in 2014 as a contributing sensor to the SSN.

XXV. SPACE SURVEILLANCE TELESCOPE

The Space Surveillance Telescope [33] is a high-sensitivity, wide field-of-view optical system with a 3.6-m aperture. It was first assembled at a facility

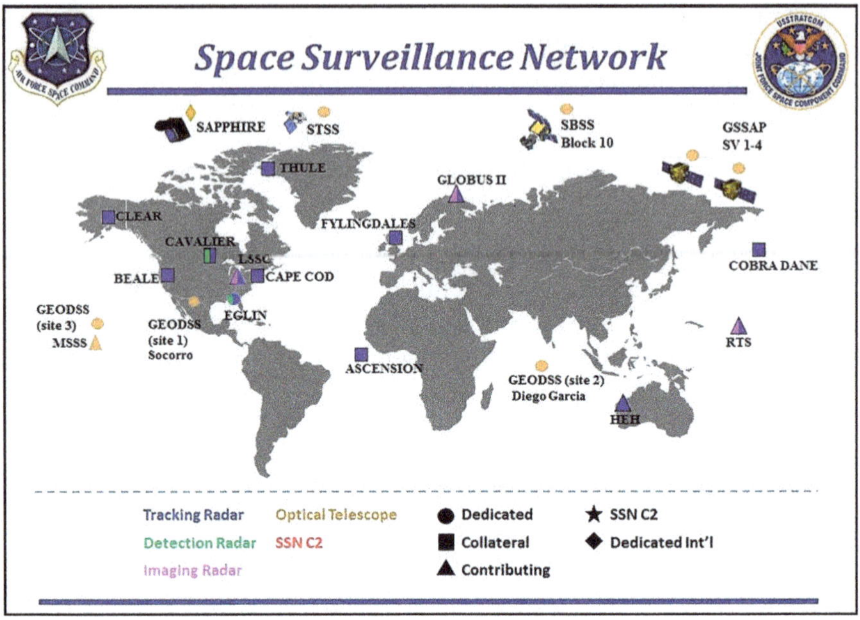

Fig. 6.12. U.S. space surveillance sites [34].

in White Sands Missile Range in New Mexico. Following construction and testing, the telescope and associated hardware and operational system were transported to the Harold E. Holt site in western Australia in 2017. The telescope began operations as a dedicated SSN sensor in 2020.

Figure 6.12 shows the global distribution of the U.S. SSN.

REFERENCES

1 "Operation Moonwatch," Wikipedia, https://en.wikipedia.org/wiki/Operation_Moonwatch

2 Smithsonian Astrophysical Observatory, Satellite Tracking Program, "Satellite Tracking Station Records, 1953–1968," Smithsonian Institution, Washington, D.C., https://siarchives.si.edu/collections/siris_arc_216832

3 "Baker Nunn Satellite Tracking Cameras," Boller & Chivens, https://bollerandchivens.com/?p=561

4 "Minitrack," Wikipedia, https://en.wikipedia.org/wiki/Minitrack

5 "Navy Space Surveillance System [NAVSPASUR]," GlobalSecurity, https://www.globalsecurity.org/space/systems/navspasur.htm

6 "Space Surveillance Sensors: The Millstone Hill Radar (May 5, 2012)," mostlymissiledefense, https://mostlymissiledefense.com/2012/05/05/space-surveillance-sensors-millstone-hill-radar/

7 Hoots, F. R., Schumacher, P. W., and DeVere, T., "50 Years of Satellite Tracking and Cataloging in the US," AAS/AIAA Astrodynamics Specialist Conference, Paper 07-328, Mackinac Island, MI, Aug. 2007.

8 "Pirinçlik Air Base," Wikipedia, https://en.wikipedia.org/wiki/Pirin%C3%A7lik_Air_Base

9 "Shemya," Wikipedia, https://en.wikipedia.org/wiki/Shemya

10 "17th Expeditionary Space Control Squadron," Wikipedia, https://en.wikipedia.org/wiki/17th_Expeditionary_Space_Control_Squadron

11 "Ballistic Missile Early Warning System," Wikipedia, https://en.wikipedia.org/wiki/Ballistic_Missile_Early_Warning_System

12 Smith, A., "Fylingdales Long Range Radar Station," Subterranea Britannica, 11 Oct. 2002, https://www.subbrit.org.uk/sites/fylingdales-long-range-radar-station/#:~:text=Fylingdales%20is%20a%20long-range%20radar%20station%2C%20which%20forms,two%20are%20at%20Thule%2C%20Greenland%20and%20Clear%2C%20Alaska.

13 "Eglin AFB Site C-6," Wikipedia, https://en.wikipedia.org/wiki/Eglin_AFB_Site_C-6

14 "AN/FPQ-16 PARCS," Wikipedia, https://en.wikipedia.org/wiki/AN/FPQ-16_PARCS

15 "RAF Ascension Island," Wikipedia, https://en.wikipedia.org/wiki/RAF_Ascension_Island

16 "PAVE PAWS," Wikipedia, https://en.wikipedia.org/wiki/PAVE_PAWS

17 "Space Surveillance Sensors: GEODSS (Ground-Based Electro-Optical Deep Space Surveillance) System (August 20, 2012)," mostlymissiledefense, https://mostlymissiledefense.com/2012/08/20/space-surveillance-sensors-geodss-ground-based-electro-optical-deep-space-surveillance-system-august-20-2012/

18 "ALTAIR (Radar)," Wikipedia, https://en.wikipedia.org/wiki/ALTAIR_(Radar)

19 "Space Surveillance Sensors: ALCOR Radar (May 17, 2012)," mostlymissiledefense, https://mostlymissiledefense.com/2012/05/17/space-surveillance-sensors-alcor-radar-may-17-2012/

20 Hall, T. D., Duff, G. F., and Maciel, L. J., "The Space Mission at Kwajalein," *Lincoln Laboratory Journal*, Vol. 19, No. 2, 2012, https://www.ll.mit.edu/sites/default/files/page/doc/2018-05/19_2_3_Hall.pdf. Retrieved 10 May 2025

21 "1st Space Surveillance Squadron," Wikipedia, https://en.wikipedia.org/wiki/1st_Space_Surveillance_Squadron

22 "Misawa Air Base," Wikipedia, https://en.wikipedia.org/wiki/Misawa_Air_Base

23 "GLOBUS," Wikipedia, https://en.wikipedia.org/wiki/GLOBUS

24 Kervin, P., Hoo, V. S., Nishimoto, D., and Liang, D., "Rapidly Deployable Raven-Class Systems: SSA Support in the Field," https://apps.dtic.mil/sti/pdfs/ADA531674.pdf

25 Singer, J., "Air Force Space Command Discontinues Use of MSX Satellite," SpaceNews, 19 June 2008, https://spacenews.com/air-force-space-command-discontinues-use-msx-satellite/

26 "Space Based Space Surveillance," Wikipedia, https://en.wikipedia.org/wiki/Space_Based_Space_Surveillance

27 "USA-205," Wikipedia, https://en.wikipedia.org/wiki/USA-205

28 "Sapphire (Satellite)," Wikipedia, https://en.wikipedia.org/wiki/Sapphire_(satellite)

29 "Fact Sheets: Geosynchronous Space Situational Awareness Program," United States Space Force, https://www.spaceforce.mil/About-Us/Fact-Sheets/Article/2197772/geosynchronous-space-situational-awareness-program/

30 "ORS-5 (Operationally Responsive Space-5)," Satellite Missions Catalogue, eoPortal, https://www.eoportal.org/satellite-missions/ors-5#eop-quick-facts-section

31 "Space Fence," Wikipedia, https://en.wikipedia.org/wiki/Space_Fence

32 "C-Band Radar," Air Force Association, https://raafa.org.au/c-band-radar/

33 Ryan, D., "The Space Surveillance Telescope in Western Australia Captures Its First Image," MIT Lincoln Laboratory, 2, Feb. 2021, https://www.ll.mit.edu/news/space-surveillance-telescope-western-australia-captures-its-first-image

34 "Space Situational Awareness—The Story So Far and an Open Way Forward," Libre Space Foundation, https://libre.space/2020/03/02/space-situational-awareness/

Chapter 7

THE U.S.–RUSSIAN SPACE SURVEILLANCE WORKSHOPS

In 1992, the National Research Council Aeronautics and Space Engineering Board Commission on Engineering and Technical Systems formed a committee to study the space debris issue. At one of the initial committee meetings, Dr. Stephen Knowles of the U.S. Naval Space Command and Dr. Stanislav Veniaminov, the Russian representative to the committee, were discussing how the two countries performed space surveillance and concluded that the scientists involved in space surveillance in the two countries should have a workshop to discuss the mathematical methods and processes used in space surveillance in the two countries. Maintaining knowledge of the orbits of all space objects was an important task for each country. In 1983, the United States had published a report [1] summarizing their algorithms for maintaining the space object catalog. The Russian approach was unknown but expected to be similar as the laws of physics were well known. Drs. Knowles and Veniaminov returned to their respective countries and obtained permission to have such a workshop. The 1st U.S.–Russian Space Surveillance Workshop was held at the U.S. Naval Observatory 13–15 July 1994. The workshop was deemed a success, and the two groups had a total of nine workshops over the next 18 years as detailed in Table 7.1. The last workshop, held at the beautiful Lake Baikal in Russia in 2012, was well attended and very successful, but by 2014, the international political situation had reached a point that prevented any further workshops.

In this chapter, we present the history of the workshops, the attendees and the papers presented, and a discussion of the highlights of each workshop. Additionally, we provide information about the setting for the workshops as well as highlights of entertainment and social events provided by the host countries. We hope that this record will provide useful background information for those who may pick up the mantle and carry on these meetings in the future.

TABLE 7.1 U.S.-RUSSIAN SPACE SURVEILLANCE WORKSHOP DATES AND LOCATION

Dates	Location
13–15 July 1994	U.S. Naval Observatory, Washington, D.C.
4–6 July 1998	Poznan, Poland
20–23 October 1998	U.S. Naval Observatory, Washington, D.C.
23–27 October 2000	U.S. Naval Observatory, Washington, D.C.
24–27 September 2003	Pulkovo Astronomical Observatory, St. Petersburg, Russia
22–26 August 2005	Pulkovo Astronomical Observatory, St. Petersburg, Russia
8–12 October 2007	Naval Postgraduate School, Monterey, CA
18–22 October 2010	Wailea Marriott Resort, Wailea, Maui, HI
27–30 August 2012	Krestovaya Pad, Listvyanka, Russia (Lake Baikal)

It is hoped that you will gain a sense that this 18-year journey was more than just an exchange of technical information. We dissolved mindsets that had resulted from decades of adversarial relationships between our countries. We have grown to know each other as fellow scientists trying to understand and master the mysteries of the motion of space objects. What began in 1994 as a sterile exchange of technical presentations quickly evolved into genuine friendships through our social events, our musical performances for one another, and our willingness to look at problems through another person's eyes.

I. FIRST U.S.–RUSSIAN SPACE SURVEILLANCE WORKSHOP

The first workshop was by invitation only, and each team consisted of senior scientists from each country who were involved in the research, development, and management of space surveillance. The workshop attendees are listed in Table 7.2. Each country selected the papers that would be presented during the workshop, and the papers are listed in Table 7.3. The attendees and papers presented at Workshops 2–7 are provided in Appendix B. Dr. P. Kenneth Seidelmann of the U.S. Naval Observatory was the general chair for the United States, and Dr. Stanislav Veniaminov of the Space Research Center (Kosmos) was the general chair for the Russian team. Dr. Felix Hoots was the technical chair for the United States and handled the organization of the technical presentations. We had professional translators, but they struggled at times with the technical jargon. Fortunately, Dr. Veniaminov spoke excellent English, which helped considerably the technical communication between the two teams. Each team consisted of individuals who had been raised during the Cold War between the two countries, and now they were gathering as colleagues to discuss how the respective countries performed space surveillance, and how they might cooperate in making space flight safer. The atmosphere from the beginning was one of potential cooperation.

TABLE 7.2 FIRST U.S.–RUSSIAN SPACE SURVEILLANCE WORKSHOP ATTENDEES

U.S. Attendees	
Terry Alfriend	General Research Corp.
Paul Cefola	Draper Laboratory
John Clark	HQ AFSPACECOM
Shannon Coffey	Naval Research Laboratory
Don Danielson	Naval Postgraduate School
Andre Deprit	NIST
Tom Eller	Kaman Sciences
Dan Fonte	USAF/ Air Force Research Lab (AFRL)
Felix Hoots	General Research Corp.
Kenneth Johnson	U.S. Naval Observatory
Nicholson Johnson	Kaman Sciences
Jeff Huth	Kaman Sciences
Bernard Kaufman	NRL
Stephen Knowles	Naval Space Command
Mark Matney	Lockheed EFC
P. Kenneth Seidelmann	US Naval Observatory
Paul Schumacher	Naval Space Command
David Vallado	AFRL
Russia Attendees	
Gennady Batyr	Space Research Center COSMOS
Vladimir Boikov	Vympel Corp
Sergei Y. Kamensky	Vympel Corp.
Zakhary N. Khutorovsky	Vympel Corp.
Alexander Menshikov	Vympel Corp.
Andrey Nazarenko	Center for Program Studies
Stanislav Veniaminov	Space Research Center COSMOS
Vasiliy Yurasov	Space Research Center COSMOS

The workshop started with a reception at the home of Dr. Knowles, and the highlight was watching the World Cup finals while enjoying Kentucky Fried Chicken. This was followed by three days of presentations of the papers listed in Table 7.3. The room was arranged with a long table with the two teams on opposite sides of the table and chairs behind those sitting at the table. A review of Table 7.3 shows that the papers presented at the workshop were primarily focused on the dynamic models used for catalog maintenance and some of the problems that occur in space surveillance, such as collision risk analysis. A key paper was a comparison of the catalogs of the two countries and proposals on sharing tracking data. Unfortunately, the sharing has yet to occur. Before the workshop, the U.S. analytic orbit model SGP4 [1] used for maintaining the U.S. Space Object Catalog was well known and documented, but the Russian model was not known. One key outcome of the workshop was the paper on the Russian analytic orbit model used for maintenance of their catalog, and it revealed their model was more precise than the U.S. model. Dr. Alfriend was editor-in-chief of the AIAA *Journal of Guidance, Control and Dynamics* (*JGCD*), and he and Dr. Hoots

TABLE 7.3 FIRST U.S.–RUSSIAN SPACE SURVEILLANCE WORKSHOP PAPERS

Title	Presenting Author
Principal Factors in Determining Space Surveillance Orbits	S. Knowles
Models of Motion Employed in Space Surveillance Activities: A Retrospective Analysis	A. Nazarenko
U.S. Naval Observatory Support of Space Surveillance	K. Seidelmann
Prediction Procedures Used in Satellite Catalog Maintenance	V. Boikov, A Testov
Analytical Orbit Model for Naval Space Surveillance	P. Schumacher
Some Preliminary Results of Russian Participation in ODERACS Experiment	G. Batyr
Correlation of U.S. and Russian Satellite Catalogs	N. Johnson
Correlation of Satellite Catalogs Maintained by Space Surveillance Systems of USA and Russia	G. Batyr
Transformation of Satellite Elements Between the US and Russian Space Surveillance Systems	A. Nazarenko
Advances in Orbit Determination and Collision Prediction Techniques for Naval Space Command	S. Coffey
Collision Risk Analysis for Satellites in LEO	Z. Khutorovsky
A Review of Joint Russian–American Activity in the Field of Space Surveillance, and Proposals on Sharing Satellite Tracking Data and the Exchange of Space Surveillance	S. Veniaminov

approached Dr. Khutorovsky with a proposal for him to write a survey paper on the Russian orbit modeling for publication in the *JGCD*. He agreed and said he would request permission to publish this proposed paper. He was successful in obtaining permission, and the paper was published in 1999 [2]. This led to several more publications [3, 4] in the *JGCD* on the Russian orbit models used for space surveillance. The last paper of the workshop by Drs. Batyr and Veniaminov was titled, "A Review of Joint Russian–American Activity in the Field of Space Surveillance, and Proposals on Sharing Satellite Tracking Data and the Exchange of Space Surveillance Specialists," which demonstrated the atmosphere of the workshop. A consolidated list of 14 potential areas of joint space surveillance cooperation was compiled shortly after the workshop by Dr. Veniaminov and Mr. Nick Johnson of NASA. These potential areas were:

1. Establish periodic (e.g., quarterly) exchange of unclassified satellite element sets directly between the U.S. DoD and the Russian Ministry of Defense. Data exchanges are currently conducted between NASA and the Russian Space Agency, which do not operate space surveillance systems and are not intimately familiar with space surveillance requirements and procedures.
2. Develop a correlation table of U.S. and Russian satellite numbers (International Designators) to facilitate communications and data exchanges during critical periods. Update the correlation table with exchange of each new American and Russian database (see item 1).

3. Establish procedures, communications protocols, and data formats for the exchange of data during critical periods (e.g., reentry of potentially hazardous satellites). This involves training of crew members on both sides as well as periodic exercises. Products of items 5–10 will contribute to this capability.
4. Develop procedures for performing joint debris collision avoidance calculations, specifically for the upcoming STS-Mir missions (i.e., an extension of normal STS COMBO operations and Mir collision avoidance calculations).
5. Develop and verify element set transformation routines for conversion from U.S. to Russian and from Russian to U.S. formats. Potentially, the use of a common algorithm would be advantageous, but the software employed by both sides need not be identical. Work in this area has already been performed by the Russian SSS support organizations and by the Center for Program Studies of the Russian Academy of Studies.
6. Investigate the feasibility of coordinating satellite breakup processing and subsequent hazard assessments. Informal notification and exchange of general satellite breakup information has already been conducted.
7. Compare pre-reentry orbit propagation and reentry prediction techniques, including comparison of historical data in low Earth orbit and in highly elliptical Earth orbits. The objective is to improve models on both sides, not to create a common model.
8. Compare techniques for compensating for solar activity influence on orbit propagation accuracies. The objective is to improve models on both sides, not to create a common model.
9. Compare effective ballistic coefficient calculation techniques and data, including attitude dependency before reentry. The objective is to improve models on both sides, not to create a common model.
10. Investigate uncontrolled LEO satellite attitude stability variations and trends. This work will complement item 9 due to ballistic coefficient dependencies on satellite attitude.
11. Compare satellite radar cross-sectional estimation techniques and data with initial emphasis on the UHF band (due to commonality of frequency use by the United States and Russia and data availability). This work will aid both sides in their long-term modeling of the near-Earth space environment and space safety programs. Comparisons of higher-frequency data could be undertaken if the UHF data exercise proves beneficial.
12. Compare satellite breakup records. Preliminary analyses indicate respective historical records contain some events not recognized by both systems. A more complete record of satellite fragmentations would aid both sides in developing satellite evolution models.
13. Develop methods for detecting and tracking small satellites, including optimum search theory models for narrow-angle and narrow-beam

facilities. These techniques would be applicable to both radar and optical facilities.

14. Prepare and exchange a series of lectures by leading space surveillance experts in the United States and Russia and exchange specialists and postgraduates in the field of space surveillance on a temporary basis.

The final statement by Drs. Batyr and Veniaminov in the last workshop paper was "One of main and paramount objectives of our cooperation we consider adjusting sound contacts between U.S. and Russian SSS both on the level of technical specialists and on the level of heads and leaders. And generally, it is a sheer abnormality and even indecency in the whole world there exist only two so powerful and unique space surveillance systems—yours and ours, and between them there is no interaction. This situation should be urgently corrected." This helped to start a period of cooperation between the two country's space surveillance scientists.

The workshop ended with a reception at the U.S. Naval Observatory on Wednesday evening and on Thursday evening a social hour followed by a dinner at the Gangplank restaurant in Washington, D.C.

The papers presented at this first workshop are listed in Table 7.3.

II. Second U.S.–Russian Space Surveillance Workshop

Both the U.S. and Russian teams thought the first workshop was a success and should be continued. It was decided to hold the next workshop in Poznan, Poland, 4–6 July 1996 in conjunction with the International Astronomical Union (IAU) Colloquium 165 on the Astrometry and Dynamics of Natural and Artificial Celestial Bodies that was being held 1–5 July 1996. The workshop had a common program with the IAU on 4 July and a separate program on 5 and 6 July. Holding the workshop in Poland helped to increase the Russian attendance. Both Dr. Seidelmann and Dr. Veniaminov continued their role as the general chairs for the U.S. and Russian teams, Bernard Kaufman of the U.S. Naval Research Laboratory served as the technical chair for the United States, and Dr. Iwona Wytrzyszczak of the Adam Mickiewicz University in Poznan was the Polish organizer. Translations for the workshop were handled by astronomers Dr. Alexander Krivov of St. Petersburg University and Dr. Sergei Rudenko of the Main Astronomical Observatory of Kiev. The excellent performance of these translators significantly contributed to the success of the workshop and proved to be a major improvement over the professional translators used at the first workshop in Washington, D.C.

The attendees and the papers presented are given in Appendix B. The number of papers presented increased to 14, and again, the focus was on mathematical methods for orbit determination and prediction with some focus on the effects of the atmosphere on orbits and approaches for predicting the atmospheric effects. The different approaches used by the two countries in addressing the same basic problems were interesting and generated much discussion.

In the closing discussion of the workshop, everyone concluded that the presentations were very useful. Both countries were seeking to solve the same problems, and the truth was unique; however, different approaches were being taken. The work by the Russians on modeling atmospheric effects and using observations to generate real-time corrections for atmospheric drag was found to be particularly interesting. Similarly, the use of workstations and parallel processing by Dr. Shannon Coffey of the Naval Research Laboratory was of special interest. This work by Dr. Coffey demonstrated that developing and maintaining a high-accuracy space object catalog using numerical integration methods (SP) was feasible. It was suggested that we might identify common problem areas as cases for future discussions and communication between workshops. These common problem areas could identify cases where both organizations have difficulty individually, or where cooperation could be immediately beneficial. The following items were identified for future attention as common problems arose:

1. Immediately answer concerning breakups or explosions in space.
2. Track breakups in GEO/GTO orbits.
3. Track multiple simultaneous or overlapping breakups.
4. Calculate dangerous conjunctions.
5. Note untrackable hazards to other satellites.
6. Compare methods of orbit predictions used by Naval Space Command, USAF Space Command, and Russian Scientific Research Center "Kosmos," and IAC "Vympel."
7. Note reentry time and location predictions.
8. Compare atmospheric density modeling and parameter values.
9. Test comparable atmospheric models.
10. Create an observational database for long time periods for geostationary satellites. SRC "Kosmos" and Institute of Theoretical Astronomy initiated this activity in Russia.
11. Test methods of evaluating observational accuracies. Some Russian satellites have laser reflectors that could be used for calibration.
12. Compare techniques of uncorrelated target processing.
13. Exchange observations of specific passive satellites.
14. Regularly compare catalogs and resolution of differences detected.
15. Follow criteria for including objects in the catalog, especially untrackable or intermittently observed objects.
16. Add to catalogs or establish a separate catalog containing more information about objects: history of objects, technical makeup of objects, subsystems.
17. Compare or develop models of breakups, circumstances that lead to breakups, breakup results analysis.

18. Exchange orbit model software for testing purposes.
19. Ensure close cooperation between the RSSS and the U.S. SSN in operational works on reentry of space objects (SO), SO in contingency, and other risk SO.

III. Third U.S.–Russian Space Surveillance Workshop

The third workshop returned to the United States and was held 20–23 October 1998 at the U.S. Naval Observatory. Drs. Seidelmann and Veniaminov continued as the general chairs from the United States and Russia, and Dr. Felix Hoots was the technical chair. Because the first two workshops were more general in subject matter, it was decided to concentrate this workshop on deep space, as opposed to low Earth objects. The intent of the workshop was to further the communication between the two communities and mutually learn from the knowledge, methods, and experiences of the other. The significant overlap in the people participating in the workshops contributed to the development of a friendly and cooperative atmosphere at the workshops. Three of the papers at the workshop were coauthored by people from both countries.

This workshop consisted of 23 papers, 11 U.S. papers, 9 Russian papers, and, for the first time, 3 jointly authored papers. These joint papers were an indication of the desire of both groups to begin cooperation between the two countries. There was considerable interest in the accuracy and time interval used in Russia for geosynchronous satellites orbit determination and prediction. The paper titled, "Near-Geostationary Orbit Model, Used in Satellite Catalog Maintenance" by Boikov, Testov, and Khutorovsky, was of significant interest as it showed that the Russian analytical model for near-geostationary space objects was more precise than the one used in the United States as it included the effects of the primary longitude-dependent gravitational perturbations. As a result of the conversation of Drs. Alfriend and Hoots with Dr. Khutorovsky at the first workshop about publishing their research in the *AIAA JGCD*, a paper on the mathematical model and method for near-geostationary was encouraged and eventually occurred. Also, the increase in the probability of collisions between satellites and space debris due to the steadily rising number of space objects inspired plans for cooperative research. There were no formal recommendations, but the opinion was that there should be 1) another workshop in a couple of years; 2) continued communication between the participants and the organizers; 3) better exchange of catalogs of the two countries, and the Russian catalog should be more generally available to the U.S. space surveillance community; and 4) cooperative research efforts between individuals and organizations.

The Russian technical chairman judged the workshop highly active based on the questions and discussion. He also felt the workshop was more saturated with scientific and technical ideas and information than all meetings he had attended in the past several years. It was pointed out that a NASA workshop participant had contacted a Russian participant concerning the puzzle of a North Korean satellite. With quick cooperation through email, the situation was clarified. He also mentioned that unification of the capabilities of the American and Russian Space Surveillance Systems would lead not just to simple summation but to a much more significant jump of their joint operational capabilities as a superadditive function of composition. This would allow establishing and solving many new actual tasks that could not be solved by either single system separately.

Both countries felt the workshop was very successful, and these workshops were leading to more cooperation between the countries and should be continued.

At the end of the first day, there was a reception held by the U.S. Naval Observatory 1730–1930 hrs, and this was followed by a tour of the observatory. Thursday evening after the technical sessions, there was a social hour from 1830 to 1930 hrs. This was followed immediately by a dinner. Both events were held at the Savoy Suites Hotel. On Saturday evening, there was a dinner at the home of Dr. Seidelmann, and Dr. Veniaminov entertained the group with a short piano recital.

IV. FOURTH U.S.–RUSSIAN SPACE SURVEILLANCE WORKSHOP

Continuing with having the workshop every two years, the fourth workshop was held again at the U.S. Naval Observatory 23–27 October 2000. Drs. Seidelmann and Dr. Veniaminov continued as the general chairs from the United States and Russia, and again Dr. Felix Hoots was the technical chair. There were 15 U.S. attendees and 11 Russian attendees, and the number of papers increased to 26. There was obviously increasing interest in both countries to continue the workshops. The attendees and list of papers presented are in Appendix B. Figure 7.1 is a picture of the attendees at the U.S. Naval Observatory (USNO).

Increasing cooperation between the scientists of the two countries was evident. There were seven papers with coauthors from the United States and Russia, of which one resulted from funding Dr. Cefola had obtained for joint research between Russia and the United States. The USAF Space Command decided to participate and presented two papers. In the late 1990s, the USAF had started development of the software for a high-accuracy catalog using SP, and there were two papers on this subject. In response to the 1998 Russian paper by Boikov, Testov, and Khutorovsky on the Russian mathematical model for geosynchronous satellites, Dr. Alfriend presented a joint U.S./Russian paper on comparison of the United States and Russian models for geosynchronous satellites. Because the Russian theory was more precise, it was expected that it would result in a more accurate orbit prediction; the

Fig. 7.1 2000 workshop group picture at USNO.

question was how much more accurate it would be. The results presented generated an interesting exchange. Figure 7.2 compares the errors in the orbit determination from the two theories for a specific satellite, no. 2608, with the Russian theory in blue and the United States in red. The results show that the errors resulting from the Russian theory were less than half the errors from the U.S. theory. This selection of blue for Russia and red for the United States generated the comment from one of the Russian colleagues, "Exactly who is red and who is blue?" This generated some laughter and was indicative of how the atmosphere had changed since the first workshop in 1994.

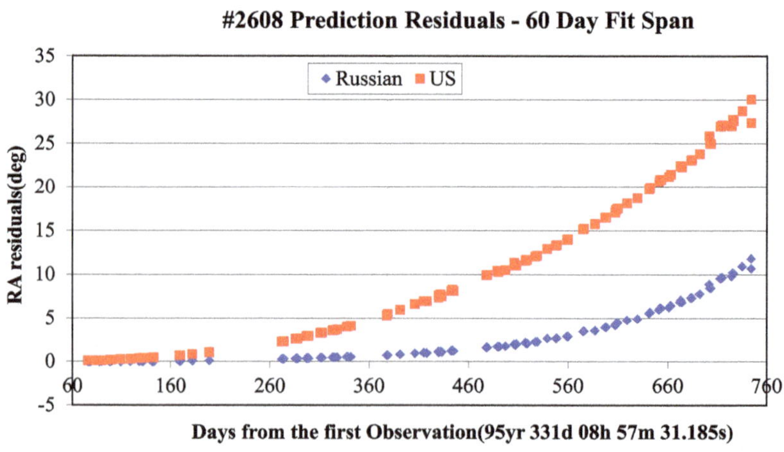

Fig. 7.2 Comparison of the right ascension residuals for Russian and U.S. orbit theories.

This fourth workshop saw an increase in the number of papers presented and the number of areas for joint research increased. The breadth of topics also widened to include papers on solar radiation pressure data, resonant orbit studies, upper atmosphere investigations, Russian algorithms for atmospheric corrections, and errors in covariance matrices. These drew the attention of participants from both countries.

After the first day of the conference, the U.S. Naval Observatory hosted a reception 1630–1900 hrs, which was followed by a tour of the observatory. On Thursday evening, there was a social hour at 1800 hrs followed by dinner at the Holiday Inn in Georgetown. Some Russians from the embassy joined the event. Dr. Hoots hosted a dinner at his home on Saturday evening to end the workshop. After the dinner, Rolan Kiladze and Stanislav Veniaminov entertained everyone with a piano concert.

It was felt by all that the workshop was a success; each workshop was improving in quality and should be continued. The Russians proposed that the next workshop be held in Russia, and all agreed this was a good idea.

V. FIFTH U.S.–RUSSIAN SPACE SURVEILLANCE WORKSHOP

The fifth workshop was held at the Pulkovo Astronomical Observatory in St. Petersburg, Russia, 24–27 September 2003 (Fig. 7.3). There were 62 attendees, 11 from the United States, 49 from Russia, and two from the European Space Agency (ESA). The 49 Russian attendees represented a breadth of space research organizations from all of Russia. ESA did not have a space surveillance system, but they had radars and optical telescopes that could be used for detecting and tracking objects in orbit to determine their

Fig. 7.3 Pulkovo Astronomical Observatory.

orbit and physical characteristics. In addition, there was an increasing interest in ESA in space surveillance. Thus, it had been decided after the fourth workshop to invite ESA to participate, and they sent two participants who were well known to the community, Dr. Heiner Klinkrad of ESA/ESOC and Dr. Thomas Schildknecht of the University of Bern. The increased attendance resulted in 30 papers presented during the workshop and indicated significant interest in the workshop in Russia. The USAF Space Command continued to support the workshop by sending several attendees. Figure 7.4 shows the group at the Pulkovo Observatory.

The 30 papers presented included a comparison of the U.S. and Russian catalogs, issues with small objects in LEO and GEO, radar and optical observing, atmospheric density effects on LEO objects and orbit estimation, essentially all the major issues in space surveillance. Many interesting discussions were generated by the papers. Having the workshop at the Pulkovo Observatory made evident the integral role of the Russian astronomical observatories in the Russian space surveillance system. In the United States, the tracking of high-altitude satellites is performed mostly by government telescopes dedicated to satellite tracking. Debris in high-altitude orbits, especially near the geosynchronous altitude, was a subject of interest. Much of the interest in the high-altitude debris objects was due to the difficulty in maintaining their orbits. Many of these debris objects are in highly perturbed

Fig. 7.4 U.S.–Russian space surveillance workshop attendees, Pulkovo Astronomical Observatory, 24–27 September 2003.

Fig. 7.5 Third U.S.–Russian space surveillance workshop banquet.

orbits because they have a HAMR. It is felt they may be Mylar insulation from geosynchronous satellites, which gives them the HAMR that is continually changing due to their orientation changing.

The attendees were given a tour of the Pulkovo Astronomical Observatory and entertained with a tour of the beautiful State Heritage Museum, one of the world's largest and most prestigious museums. The workshop ended with a banquet at the observatory (see Fig. 7.5).

The workshop was very successful. The Russian delegation suggested that the next workshop be held again at the Pulkovo Observatory, and the attendees agreed.

Some of the Russian delegation invited the U.S. and ESA attendees to visit Moscow, and those that went got a tour of Star City, Red Square, and the Kremlin, and a wonderful evening at the Bolshoi ballet. Some even visited the largest McDonald's in the world along the Arpad.

VI. SIXTH U.S.–RUSSIA SPACE SURVEILLANCE WORKSHOP

The sixth workshop was held again at the Pulkovo Astronomical Observatory, 22–26 August 2005. There were 11 U.S. attendees, 48 Russian attendees, and two from ESA. There were 30 papers presented, the same as the fifth workshop. This workshop covered the same broad range of space surveillance topics as the 2003 workshop. The ESA contributed two papers, and there were two joint research papers by Yurasov and Nazarenko from Russia and Alfriend and Cefola from the United States on issues related to the effects of atmospheric drag on LEO satellites. An earlier related paper [5] by these authors had been selected for the Best Paper Award at the 2004 AAS Space

Flight Mechanics Conference in Wailea, Hawaii. The research funding was obtained by Dr. Cefola for joint research, and the project was taken over by Dr. Alfriend in 2004 when Dr. Cefola changed jobs. Over the next few years, this project would result in numerous joint papers [6–21] between the United States and Russia that would demonstrate the positive impact of joint research by the two countries. Papers showing the use of various Russian astronomical observatories for maintaining the orbits of high-altitude objects, including the difficult-to-track HAMR objects, were presented. A new subject of growing interest was the relative motion of objects that are in close proximity to each other.

The workshop ended with another nice banquet at the Pulkovo Observatory.

VII. SEVENTH U.S.–RUSSIAN SPACE SURVEILLANCE WORKSHOP

After two workshops in Russia, it was decided to return to the United States, and because there had been three workshops at the Naval Observatory, it was decided to have the participants experience a different part of the United States. The Naval Postgraduate School in Monterey, California, was selected as the site of the seventh workshop. The workshop date was originally selected as the first week of October, but that conflicted with the 50th anniversary of the launch of Sputnik, 4 October 1957. The workshop was moved to 29 October–2 November 2007. The workshop started with a reception at the home of Dr. Alfriend. On arrival, the Russian team was honored with a cake (see Fig. 7.6) acknowledging their achievement of being first in space with the launch of Sputnik. Figure 7.7 shows the workshop participants with the cake. At the end of the reception Dr. Veniaminov again entertained the participants with a piano concert on the Alfriend's piano.

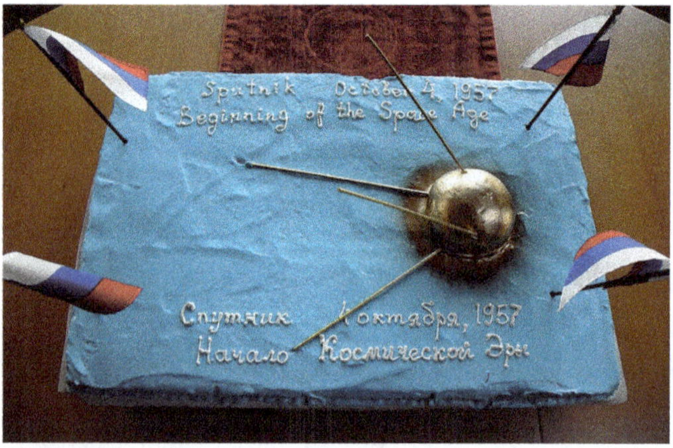

Fig. 7.6 Reception cake honoring Russia being first in space.

Fig. 7.7 Workshop attendees at reception at Alfriend home.

There were 29 participants, 19 from the United States, 8 from Russia, and 2 from ESA. A welcome addition were two professors from the Naval Postgraduate School, Dr. Oleg Yakimenko and Dr. Eugene Bournakov, whose native language was Russian, and they were interpreters. Being engineering professors, they did not struggle with the technical translation. The U.S. delegation was a balance of people from academia, industry, and government research labs.

The theme of the workshop was *Development of a More Complete Space Object Catalog*. In 2006, there were about 12,000 objects in the space object catalog with a minimum size of 10 cm in diameter. In January 2007, there was a Chinese ASAT test that added about 3000 objects to the catalog. A collision with an object more than 1 cm in diameter can cause catastrophic damage to a satellite. Also, it was estimated there were more than 100,000 objects greater than 5 cm in diameter in Earth orbit. Thus, the probability of a collision with an object that results in catastrophic damage is much higher with an uncataloged object than with a cataloged object. As a result, it was decided that the focus of this workshop would not be on the past and present, but on the future need of cataloging many more objects. The workshop focused on issues related to developing a catalog with objects as small as 5 cm in diameter, primarily the theories, technologies, and algorithms necessary to develop a satellite catalog more inclusive of smaller objects. Focus areas for the workshop were as follows:

1. Reduce orbit prediction errors due to atmospheric drag uncertainty, including covariance improvement.
2. Improve angles-only orbit determination for GEO, HEO, and LEO.

3. Correlate uncorrelated tracks, both radar and optical system tracks. Because every new object begins as an uncorrelated track, this is a major problem area for the development of the large new catalog.
4. Recognize the value of sharing observations.

The workshop consisted of 30 papers, with three of them having U.S. and Russian authors. There were three papers from the ESA dealing with the use of European assets for space surveillance and tracking small objects in orbit. In addition, the workshop was open to persons at the Naval Postgraduate School, and numerous faculty and students participated (Fig. 7.8).

Clint Eastwood is a resident of the Monterey Peninsula, at the time of the workshop a resident of Pebble Beach. He is a very popular actor in Russia. The Russian delegation jokingly said they wanted to meet him. We contacted his office, and during the time of the workshop, he was making a movie, but he agreed to autograph pictures for each member of the Russian delegation. Pictures of Clint from various movies were provided, and he wrote a private message to each Russian participant on pictures that were given to the participants at the reception. They were very grateful. As many know, Eastwood owns the Mission Ranch restaurant in Carmel. They typically do not take reservations, but they were contacted Monday afternoon 29 October, and they agreed to host a group of about 30 people. Monday evening at Mission Ranch is a piano bar, and it turned out that the pianist is Russian. He was very excited to be playing for a group of Russians, and when informed that Dr. Veniaminov was a concert pianist, he invited him to come play. Dr. Veniaminov then entertained the restaurant clientele for a few minutes (see Fig. 7.9). The Russian participants were very grateful and wrote Clint

Fig. 7.8 2007 workshop group picture at Naval Postgraduate School.

Fig. 7.9 Dr. Veniaminov playing piano at Mission Ranch restaurant.

Eastwood a beautiful letter thanking him for the pictures and the opportunity to eat in his restaurant.

The Russian technical chairman concluded that it may be necessary to unite the capabilities of all three space surveillance systems for obtaining the 100,000-object catalog. This may be complicated. He continued that

> We live in such an epoch when all the weather in the world is controlled by politicians. But we scientists and engineers, are capable to influence their decisions, just as the sly wife can steer her despot husband. Our weapon and tool is a convincing and persuasive substantiation of inevitable necessity of our proposals. We do already have our ideological and technical tool—Proceedings of our workshops. All revolutionary transformations in the past human history usually have been well-prepared and well-conditioned by a preliminary gradual and continuous pressure in the society brains. Perhaps it seems impossible or unreal now but one should begin someday.

The workshop ended with a banquet at the Hyatt Regency in Monterey, the host hotel for the workshop. The participants were entertained with a talk by Dan Bursch, a former astronaut and professor at the Naval Postgraduate School, who spent 196 days on the ISS in 2001–2002.

VIII. EIGHTH U.S.–RUSSIAN SPACE SURVEILLANCE WORKSHOP

In keeping with the previous decision to have the workshop in a different location in the United States, the eighth U.S./Russian Space Surveillance Workshop was held at the Wailea Marriott Resort in Wailea, Maui, Hawaii, 18–23 April 2010. Nearby Kihei, Hawaii, is the location of the AFRL Advanced Maui Optical and Supercomputing Center (AMOS) research facility, and the associated telescope systems that are at the top of Haleakala are used for observing satellites. Space surveillance research is performed at the facility, which made this an excellent workshop location.

Dr. Kenneth Seidelmann of the University of Virginia and Dr. Terry Alfriend of Texas A&M University continued as the general chair and U.S. technical chair, respectively. Dr. Stanislav Veniaminov who was the Russian chair at the previous workshops was unable to attend, and Dr. Sergei Kamensky of the Vympel Corp. and Dr. Vasiliy Yurasov of the Institute for Precision Instrument Engineering were the Russian chairs. Unfortunately, Dr. Kamensky passed away shortly after returning to Russia following the workshop. Following the approach of the 2007 workshop, this workshop focused on the future rather than the past. In the years before 2010, it had become feasible to maintain the entire space catalog with SP, and that was done on a routine basis. This experience revealed that, in the case of sparse tracking data, which often resulted in the loss of the satellite, SP models do not necessarily provide a significant accuracy improvement over GPs methods. This was the case for many highly eccentric and geosynchronous satellites, especially if they were small. It had been determined that there were numerous objects in near-geosynchronous orbit that were highly perturbed by solar radiation pressure, which meant they had a HAMR. These objects are called HAMR objects. It is surmised that many of them are sheets of insulation that have come from dead geosynchronous communication satellites. These objects limit the accuracy of collision risk assessment and maneuver planning, especially in the geosynchronous regime, and the quality of catalog maintenance deteriorates.

Other factors impacting the workshop were the 2007 Chinese ASAT test and the 2009 Iridium-Cosmos collision in February 2009. These events had created close to 6000 new objects in the space object catalog. Before these events, the USAF Space Command had performed conjunction assessments for the ISS and other high-value U.S. assets. For any close approaches, they would communicate the relevant data to the satellite operator who would calculate a probability of collision and decide whether to perform a collision avoidance maneuver. After the Iridium-Cosmos collision, it was decided that Space Command would perform this conjunction assessment on the operational satellites of all countries and inform the satellite operators of the situation and provide them the necessary data to compute the probability of collision. The satellite operator would then make the decision on whether to perform a collision avoidance maneuver. Initially, some countries, including Russia, had required that this information be passed through the communication channels in the embassy. It was finally realized that this procedure was too slow, and the communication needed to be directly to the operators. There were numerous discussions of this process during the workshop.

It was proposed that in this workshop, attention be focused on the deep space sparse data problem and perfecting the search methods. An enticing solution to mitigate weather and daylight constraints was satellite-based surveillance sensors. Other possible solutions for increasing the amount of data included cooperative networks of optical sensors, longer tracks from optical sensors, methods to optimize the mix of radar and optical tracking (including re-aiming sensors from one to another by imprecise measurement data), and so on. Thus, the suggestion was for a workshop in which presentations would report on existing and proposed theories, technologies, and algorithms necessary to mitigate the sparse and imprecise data environment for deep space satellites and, hence, to improve prediction accuracy of deep space satellites.

There were 42 attendees, 24 from the United States, 17 from Russia, and 1 from ESA. The workshop occurred shortly after the large volcano eruption in Iceland, and this impacted the travel for the Russians as they had to fly south around the ash cloud, and they missed their connection in New York. Thus, their trip took an extra day.

The workshop began with a reception at the home of Drs. Chris Sabol and Kim Luu that has a beautiful view of the Pacific Ocean and the west side of Maui. We were entertained with some Hawaiian dancer performances (Fig. 7.10). After the long trip from Europe, this provided a nice relaxing evening before the start of the workshop.

The workshop consisted of 28 papers of which there were 12 U.S.-authored papers, 11 Russian-authored papers, 3 joint U.S.–Russian-authored papers, and 2 ESA-authored papers. This was the most papers presented at any of the workshops to date. There were two papers on the new Russian International Scientific Optical Network (ISON) that is

Fig. 7.10 Hawaiian dancers at reception at Sabol-Luu home.

Fig. 7.11 Workshop group at the famous banyan tree in Lahaina, Hawaii, and at the luau.

managed by Dr. Vladimir Agapov of the Keldysh Institute of Mathematics of the Russian Academy of Sciences. ISON (http://lfvn.astronomer.ru/main/english.htm) is an international set of telescopes in many countries cooperating with the tracking of high-altitude space objects. It was originally established to obtain worldwide observations of the many objects created by explosions or breakups of geosynchronous satellites to confirm the theory of the evolution of these clouds of objects. It currently consists of about 30 telescopes in 10 countries, including the United States.

On Wednesday afternoon, the workshop ended early, and the participants went to Lahaina, Hawaii, to tour Lahaina and attend an original Hawaiian luau. It was a great evening (see Fig. 7.11).

The workshop ended at noon on Friday, and the participants were given a tour of the AMOS facility at the top of Haleakala. The workshop ended with a dinner on the rooftop of the Wailea Marriott (see Fig. 7.12).

IX. NINTH U.S.–RUSSIAN SPACE SURVEILLANCE WORKSHOP

After having the seventh and eighth workshops on the beautiful Monterey, California, peninsula and Maui, the ninth workshop returned to Russia, and

Fig. 7.12 U.S.–Russian eighth workshop final banquet.

Fig. 7.13 Lake Baikal.

they held it at beautiful Lake Baikal in Siberia, 27–30 August 2012. Lake Baikal (Fig. 7.13) is the deepest rift lake in the world and has the largest surface area of any lake in Europe. The workshop location was Krestovaya Pad, Listvyanka, Russia (see Fig. 7.14). The workshop was well attended, 24 attendees from Russia, 8 from the United States, and 2 from ESA. In addition, as seen from the pictures, there were 11 wives, 8 from Russia and 3 from the United States. Two new attendees from Russia were Dr. Boris Shustov, an astronomer with the Institute of Astronomy of the Russian Academy of Sciences, and Mr. Oleg Aksenov, a high-level manager at SRC "Kosmos." Dr. Shustov is the first member of the Russian Academy of Sciences (RAS) to participate in the workshops. Their attendance indicated high-level interest in the workshop. The conference hotel and facility and the attendees are provided in Fig. 7.14.

The basic theme of the past several conferences was *Development of a More Complete Catalog*, and it was decided to continue with this theme but to also add some focus on small satellites, especially "cubesats," many of which were beginning to be put in orbit. The small size of the cubesats can make it difficult to track them.

There were 29 papers presented at the workshop, 19 from Russia, 7 from the United States, 2 from ESA, and 1 joint Russia–U.S. paper. In the first

Fig. 7.14 Krestovaya Pad and workshop group.

session, Dr. Shustov gave a paper on the near-Earth object (NEO) threat problem and proposed integrating it and the space debris problem. It appeared that there might be agreement of this in Russia between ROSCOSMOS and the Space Council of the RAS. In the United States, the NEO problem is a concern to both NASA and DoD, but the debris tracking is the responsibility of the DoD, and NASA/JPL has primary responsibility for the NEO problem. Both Dr. Shustov and Mr. Aksenov participated in discussions at the workshop, and Mr. Aksenov seemed impressed with it and continuing the workshops.

Dr. Pavel Papushev of the Sayan Solar Observatory had participated in some of the workshops and was supposed to be the Russian technical chair, but he passed away suddenly a few months before the workshop. At the beginning, there was a moment of silence for him and Dr. Sergei Kamensky who passed away shortly after the 2010 workshop. In his opening remarks, Stanislav Veniaminov said he had a dream that at some workshop a U.S. participant would give his paper in Russian. When we went around the room for everyone to introduce themselves, Darren Garber introduced himself in Russian to the delight of Stanislav. Darren indicated at the next workshop he would give his paper in Russian.

On Tuesday, at the end of the day, there was an interesting excursion to the Baikal Institute of Solar and Terrestrial Physics facility (see Fig. 7.15). On Thursday, there was a very entertaining excursion to a museum and an old fort near Lake Baikal (see Fig. 7.16).

One issue raised was the sharing of space object catalogs. The exchange had started before the 1994 workshop but had ended about 2007. When the Russian catalog was sent to the United States, it was sent to NASA/JSC because of some agreement before the workshops. Because of their catalog including U.S. DoD satellites, which are not included in the U.S. catalog, NASA/JSC sent the Russian catalog to AFSpC A9, and they would not release it. Mr. Aksenov agreed that comparing the catalogs was

Fig. 7.15 Institute of Solar and Terrestrial Physics telescope.

Fig. 7.16 Old fort near Lake Baikal.

a good idea, and he was willing to restart the process. The discussion then centered on where to send it, and the conclusion was to try and resolve that issue.

Another item discussed was what should be the role of the future workshops? When the workshops started, it was two groups sitting on opposite sides of a table discussing through interpreters technical issues related to space surveillance. We had no idea that we would advance to having joint research projects and becoming colleagues and friends. There were often suggestions of areas of potential cooperation, but none of the participants were high-level administrators who had the authority to make such things happen. A question raised was "Should the workshop become more of a voice in seeking to develop international cooperation by jointly expressing our opinion and becoming involved from the technical level, rather than having the activities directed from above?"

Another discussion subject was the next workshop time and location. It was agreed that continuing with every other year was a good idea, and so 2014 was agreed for the next workshop. Mr. Aksenov proposed having it at Altay, a location with some of their sensors, and he stated by 2014 Russia expected to have some new sensors. It was agreed that the Altay optical and laser facility would be an interesting location.

The workshop ended with a very nice banquet at the hotel. All the U.S. attendees agreed that the Russian food was excellent. Figure 7.17 shows Mr. Oleg Aksenov giving a toast with excellent Russian vodka and making some final comments.

Dr. Alexander Samotokhin of the Keldysh Institute of Applied Mathematics had a friend that lived close to the hotel. They invited those in the U.S. delegation that were staying over the day following the workshop to a meal at the friend's home. For those who were there, it was an excellent meal and a wonderful ending to a great trip to Russia (see Fig. 7.18).

Fig. 7.17 Final banquet.

X. WORKSHOP SUMMARY

In 1994, two groups of scientists from the United States and Russia who performed research and development in space surveillance, and who had been raised during the Cold War, met at the U.S. Naval Observatory to give presentations on the algorithms, technology, and processes used by the two major space powers to track man-made space objects. This was hard to imagine as this subject had been sensitive or classified. From this 1994 meeting, there were eight more workshops, joint research projects, joint papers, and strong friendships. Probably no one there could have imagined that 18 years later they would be meeting at the beautiful Lake Baikal in Russia and discussing whether they should move beyond the exchange of information and joint research projects to be more proactive in attempts to help protect the world from the impact of space debris and allow space systems to better serve humanity. Although all presentations at the workshops had to be preapproved by their respective countries, the workshop participants and leaders decided on the focus of the workshops and the talks to be presented. This was somewhat unprecedented. Then in

Fig. 7.18 Lunch following workshop.

February 2014, the Russian Army invaded the Crimean peninsula of Ukraine, and things began to change. The atmosphere for cooperation slowly started to change. Having a workshop requires funding, and funding for cooperative activities with Russia was not viewed positively. We explored having the workshop at a neutral location such as the University of Bern in Switzerland, but approval for sufficient funding in both countries could not be obtained. Thus, was lost what might have been an excellent opportunity for the scientists from the United States, Russia, and Europe to help improve the space environment for all the world. It is interesting that in these times, which have continued to worsen from 2014 to now in 2025, there are groups of scientists who have demonstrated the willingness to work together and have the desire to continue to try and improve the space environment for all, but they cannot obtain the small amount of funding needed to meet and try to forge a path forward.

The contribution of the workshops to advancing science was recognized by the scientific community in 2005 when the American Association for the Advancement of Science (AAAS) awarded the 2005 AAAS International Scientific Cooperation Award to the leaders of the U.S. and Russian teams, Kyle T. Alfriend, Paul J. Cefola, Felix Hoots, and P. Kenneth Seidelmann of the United States, and Andrey I. Nazarenko, Stanislav S. Veniaminov, and Vasiliy S. Yurasov of Russia (see Fig. 7.19).

The attendees and titles of the papers presented at the first workshop are provided in Tables 7.2 and 7.3. The attendees and titles of the papers presented in workshops 2–9 are provided in Appendix B. All the papers presented at the workshops are accessible at https://ssworkshops.engr.tamu.edu.

From top left: Kyle T. Alfriend, Paul J. Cefola, Felix R. Hoots, P. Kenneth Seidelman

From top left: Andrew I. Nazarenko, Stanislav S. Veniaminov, Vasiliy S. Yurasov

Fig. 7.19 Recipients of the 2005 AAAS Award for International Scientific Cooperation.

REFERENCES

1 *Mathematical Foundation for SCC Astrodynamic Theory*, NORAD Publication, TP SCC 008, April 1982.

2 Khutorovsky, Z. N., Boikov, V. F., and Pylaev, L. N., "Catalog Maintenance of Low-Earth-Orbit Satellites," *AIAA J. of Guidance, Control and Dynamics*, Vol. 22, No. 6, Nov.–Dec. 1999.

3 Boikov, V., Testov, A., and Khutorovsky, Z., "Near-Geostationary Orbit Model Used in Satellite Catalog Maintenance," *AIAA Journal of Guidance, Control and Dynamics*, Vol. 25, No. 1, Jan.–Feb. 2002.

4 Khutorovsky, Z. N., "Monitoring of Geostationary Earth Orbit Satellites in Russian Space Surveillance Center," *AIAA Journal of Guidance, Control and Dynamics*, Vol. 25, No. 5, Sept.–Oct. 2002.

5 Yurasov, V., Nazarenko, A., Cefola, P., and Alfriend, K. T., "Results and Issues of Atmospheric Density Correction," *AAS Journal of the Astronautical Sciences*, Vol. 52, No. 3, July–Sept. 2004, pp. 281–300 (selected as conference best paper).

6 Yurasov, V., Nazarenko, A., Cefola, P., and Alfriend, K. T., "Density Corrections for the NRLMSIS-00 Atmosphere Model," AAS 05-168, 2005 AAS/AIAA Space Flight Mechanics Conference, Copper Mountain, CO, Jan. 2005.

7 Yurasov, V., Nazarenko, A., Alfriend, K. T., and Cefola, P., "Reentry Time Prediction Using Atmospheric Density Corrections," *AIAA Journal of Guidance, Control and Dynamics*, Vol. 31, No. 2, March–April 2008, pp. 282–289.

8 Yurasov, V., Nazarenko, A., Cefola, P. J., and Alfriend, K. T., "Application of the ARIMA Model to Analyze and Forecast Time Series of Density Corrections for NRLMSIS-00," Paper No. AAS 05-256, 2005 AAS/AIAA Astrodynamics Conference, Lake Tahoe, CA, Aug. 7–11, 2005.

9 Yurasov, V. S., Nazarenko, A. I., Cefola, P. J., and Alfriend, K. T., "Direct Density Correction Method: Review of Results," 57th IAF Congress, Valencia, Spain, Oct. 2006.

10 Nazarenko, A. I., Yurasov, V. S., Alfriend, K. T., and Cefola, P. J., "Optimal Measurement Filtering and Motion Prediction Taking into Account the Atmospheric Perturbations," Paper No. AAS 07-363, presented at the 2007 AAS/AIAA Astrodynamics Conference, Mackinac Island, MI, Aug. 2007.

11 Boikov, V. F., Khutorovsky, Z. N., and Alfriend, K. T., "Singular Value Decomposition and Least Squares Orbit Determination," Paper No. AAS 07-362, presented at the 2007 AAS/AIAA Astrodynamics Conference, Mackinac Island, MI, Aug. 2007.

12 Khutorovsky, Z. N., Kamensky, S., Sbytov, S., and Alfriend, K. T., "Orbit Determination of LEO Satellites for a Single Pass Through a Radar: Comparison of Methods," presented at the 2007 ISSFD Conference, Annapolis, MD, Sept. 2007.

13 Boikov, V. F., Khutorovsky, Z. N., and Alfriend, K. T., "Singular Value Decomposition and Least Squares Orbit Determination," presented at the Seventh U.S.–Russian Space Surveillance Workshop, Monterey, CA, 29 Oct.–2 Nov. 2007.

14 Khutorovsky, Z. N., Kamensky, S., Sbytov, S., and Alfriend, K. T., "Orbit Determination of LEO Satellites for a Single Pass Through a Radar: Comparison of Methods," presented at the Seventh U.S.–Russian Space Surveillance Workshop, Monterey, CA, 29 Oct.–2 Nov. 2007.

15 Nazarenko, A. I., Yurasov, V., Cefola, P. J., and Alfriend, K. T., "Direct Density Correction Method: Increasing the Accuracy of LEO Orbit Prediction," Paper No. AIAA 2008-6440, 2008 AIAA/AAS Astrodynamics Conference, Honolulu, HI, 17–21 Aug. 2008.

16 Kamensky, S., Stepanyants, V., Tuchin, A., and Alfriend, K. T., "Automatic Detection and Analysis of Non-Evolutionary Changes in the Orbital Motion of Geocentric Objects," Paper No. AAS 09-103, 2009 AAS/AIAA Space Flight Mechanics Conference, Savannah, GA, 8–12 Feb. 2009.

17 Khutorovsky, Z. N., Kamensky, S., Sbytoc, N., and Alfriend, K. T., "Comparison of Different Methods of LEO Satellites Orbit Determination for a Single Pass Through a Radar," Paper No. AAS 09-107, 2009 AAS/AIAA Space Flight Mechanics Conference, Savannah, GA, 8–12 Feb. 2009.

18 Agapov, V., Stepanyants, V., Kamensky, S., Khutorovsky, Z. N., and Alfriend, K. T., "Orbit Determination and Propagation of Objects with High AMR for Orbital Archive Maintenance," Paper No. AAS 09-232, 2009 AAS/AIAA Space Flight Mechanics Conference, Savannah, GA, 8–12 Feb. 2009.

19 Khutorovsky, Z. N., Sukhanov, S. A., Sbytov, N. N., Kamensky, S. Yu., and Alfriend, K. T., "Comparison of Different Algorithms of Orbit Determination Using Radar Measurements Acquired During One Penetration," 2009 AMOS Technologies Conference, Wailea, Maui, HI, 1–4 Sept. 2009.

20 Samotokhin, A. S., Khutorovsky, Z. N., Kamensky, S. Yu., and Alfriend, K. T., "Investigation of the Characteristics of the Algorithm of Primary Orbit Determination for LEO Satellites Based on Radar Measurements. Results of Simulation," 2009 AMOS Technologies Conference, Wailea, Maui, HI, 1–4 Sept. 2009.

21 Khutorovsky, Z. N., Samotokhin, A. S., Sukhanov, S. A., and Alfriend, K. T., "Guaranteed Approach for Orbit Determination with Limited Error Measurements," Paper No. AAS 13-706, presented at the 2013 AAS/AIAA Astrodynamics Conference, Hilton Head, SC, Aug. 2013.

Chapter 8

The U.S.–China Technical Interchanges

By 2008, the United States and Russia had held seven successful space surveillance workshops, had initiated joint research projects, and were having discussions on ways the two countries could cooperate in space surveillance to make space safer for all countries. We had exchanged space object catalogs and compared them. China was becoming a space power and on 17 January 2007 for some strange reason they shot down one of their own satellites with an antisatellite (ASAT) test. The satellite was in approximately an 800-km altitude orbit, which is the densest region in the space object population. The United States and Russia were both tracking over 3500 debris objects that had originated from this one ASAT test, which had increased the number of trackable space objects by more than 30%.

After some lengthy discussions, it was decided to see if China would be interested in initiating space surveillance workshops with the United States. This proved to be a much bigger challenge than with Russia. Obviously, they had to have some type of space surveillance system to track their own satellites, and even though they could have knowledge of the orbits of the satellites of the other space-faring nations from the public U.S. Space Object Catalog, they needed to do some tracking. Unlike the U.S. and Russia space surveillance scientists who were publishing papers, similar papers were not being published by Chinese scientists. Consequently, the challenge was with whom should we initiate discussions regarding such a technical interchange. Dr. Seidelmann, the U.S. general chair, was an astronomer and had been active in the IAU. He contacted four astronomers in China whom he knew from interactions through the IAU. One responded, Mdm Shuyua YE of the Shanghai Observatory, and she said she was coming to the National Radio Astronomy Observatory in Charlottesville, Virginia, and she would like to meet with him to discuss the workshop concept. Mdm YE was a past director of the Shanghai Observatory and a member of the Chinese Academy of

TABLE 8.1 U.S.–CHINA SPACE SURVEILLANCE TECHNICAL INTERCHANGE DATES AND LOCATIONS

Dates	Location
1–5 June 2009	Shanghai Astronomical Observatory, Shanghai, China
17–21 October 2011	National Astronomical Observatories, CAS Beijing, China
13–16 May 2013	Beijing Institute of Technology, Beijing, China

Sciences. During their meeting, she said she thought the workshop was a good idea, but it would need to be held in China due to the uncertainty related to the granting of visas for technical people from China coming to the United States. She suggested the Shanghai Observatory for the location. Dr. Seidelmann brought the concept back to the U.S. Planning Committee, and there was agreement with the proposed location. This was the first of three workshops held in China. The dates and locations of the three technical interchanges are provided in Table 8.1. The attendees and technical programs for each technical interchange are provided in Appendix C. As one can see from a review of Appendix C there was broad representation from space organizations in China.

I. FIRST U.S.–CHINA SPACE SURVEILLANCE TECHNICAL INTERCHANGE

The first U.S.–China Space Surveillance Technical Interchange was held 1–5 June 2009 at Shanghai Astronomical Observatory (SHAO). The U.S. general chair was Dr. Kenneth Seidelmann of the University of Virginia, and the U.S. technical chair was Dr. Terry Alfriend of Texas A&M University. Mdm YE and Professor YANG Fumin were the China general and technical chairs, respectively.

The first two days consisted of the technical presentations. There were 11 papers presented from the United States and 8 from China. Ten were planned, but two presenters had to cancel. The papers presented were interesting but mostly focused on debris issues and collision avoidance, not space surveillance issues. On the third day (Wednesday, 3 June) they arranged for a tour to the SHAO SLR laser ranging facility, the VLBI site, and some sightseeing in Shanghai. On Thursday morning, we took a train to Nanjing and visited Purple Mountain Observatory (PMO) and did some local sightseeing. We spent Thursday night in Nanjing. Friday morning we went to the PMO Xuyi facility (two-hour bus ride). After lunch, we returned to Nanjing and caught the train back to Shanghai.

The Chinese team were wonderful hosts. There was a fantastic opening banquet, the tours were very interesting and well planned, and they gave both teams the opportunity to have informal technical discussions on the issues of space surveillance. We had a very enjoyable final banquet. Final discussions indicated an interest in having another technical interchange in 2011. We agreed to go back to our respective organizations and communicate within several months. Both sides agreed that the next interchange should also be in China due to the continuing issues with visas for technical people from China.

II. SECOND U.S.–CHINA SPACE SURVEILLANCE TECHNICAL INTERCHANGE

The second U.S.–China Space Surveillance Technical Interchange was held at the National Astronomical Observatories, China Academy of Sciences, in Beijing, China, 17–20 October 2011. Dr. Seidelmann and Mdm YE were the U.S. and China cochairs, and Dr. Kyle Alfriend was the U.S. technical chair. Dr. ZHOU You of the National Astronomical Observatories was the China technical chair. The interchange was impacted with numerous issues that made it difficult to occur. First, the China leadership wanted U.S. presentations only with no presentations from China. It was made very clear to them that was unacceptable. Their proposal was not an interchange, and the U.S. team would not travel to China to just give presentations. The issue was resolved, and planning moved ahead. The NASA appropriations bill for FY2011 included a sentence preventing NASA employees from interacting with the Chinese. This prevented several NASA people from attending the technical interchange. Then more problems occurred in China. The first two days were held at the National Astronomical Observatories, and then the next two days were to be held at the Xi'an Satellite Control Center. The center was to financially support the U.S. travel to Xi'an. Then the week before the interchange, we and the Chinese team were informed that the visit to the Xi'an Satellite Control Center was canceled, and no reason was provided. This was obviously embarrassing to the Chinese leadership. We still traveled to Xi'an and held a one-day meeting at their National Timing Center and received a tour of the facility. We also received a tour of the Xinglong Observatory, which is about a 2.5-hour bus ride from Beijing. The primary telescopes there are the 6-m LAMOST and a 2.16-m telescope. The LAMOST is a unique telescope with a 6-m Schmidt corrector flat, with adaptive optics to correct for the atmosphere, in one building. In a separate building, it has a 4-m spherical mirror feeding a detector plate with 4000 3″ fibers for a spectrograph. The pointing is two hours on either side of the meridian and from the pole to the equator. They can reach 19.6 magnitude with one-hour exposures. The site is operational 260 days/year.

Having the interchange in Beijing resulted in better attendance from China, and the first day there were over 50 Chinese attendees with many of them students. There was still attendance from Shanghai and Nanjing as noted in Table C.3. At the opening banquet, there was representation from the Asia-Pacific Space Cooperation Organization (APSCO). China was participating in the International Scientific Optical Network (ISON) managed by the Keldysh Institute of Mathematics in Russia. Many other countries are also participating with ISON.

The presentations from China covered a wide range of subjects as can be seen from the program in Table C.4 in Appendix C. Issues with conjunction analysis and assessment is always a major topic. Another dominant subject was the use of optical sensors for tracking objects at high altitude. They also discussed the use of TLEs as they are publicly available. As with the first interchange, there was no participation with the military, even though there were several participants from the National University of Defense Technology.

In the final discussions, they indicated interest in continuing the interchanges, and we both agreed due to the continuing issues with obtaining visas to the United States that it was best to continue to hold it in China.

III. Third U.S.–China Space Surveillance Technical Interchange

The third technical interchange was held at Beijing Institute of Technology (BIT) 13–16 May 2013. This location increased attendance, as there were many students and faculty present. Unfortunately, there were no attendees from Shanghai, Nanjing, or Xi'an, but the National Defense University of Technology participated, and there was attendance from other entities in Beijing. Table C.5 provides the list of attendees, but we did not get a complete list of the Chinese participants. especially those from BIT.

The location of Beijing Institute of Technology indicated a possible interest in China in making this event more of an academic interchange. Academic participation enhances the breadth of the event, but space surveillance is the tracking of space objects, and a workshop requires the participation of the organizations involved, not just those doing academic research.

The 2012 sequestration and the GSA Conference scandal created issues with travel to China, which limited participation for the United States. Housing for the participants was the International Hotel for Education and Communication, basically student housing and not the quality we had for the first two workshops.

In this technical interchange, the papers were grouped into six categories:

1. Space Situational Awareness.
2. Space Debris Modeling and Prediction.
3. Observational and Measurement Techniques.
4. Satellite Catalog Maintenance.

5. Close Approach and Probability of Collision Computation.
6. Spacecraft Procedures in Response to Predicted Conjunctions.

These are valid subject areas, but what is missing is the detection and tracking. As with the second workshop, there was some discussion of the use of astronomical telescopes for high-altitude tracking indicating they are going the same route as Russia and the United States.

Final discussions indicated interest in continuing the interchanges, with the possibility of the next location being the National University of Defense Technology. There was also discussion about trying to develop some joint research projects at the academic level. Both sides agreed that with the issues of the Chinese attendees obtaining visas to the United States for technical events, any future workshops should be in China.

IV. SUMMARY

As with the U.S.–Russian Space Surveillance Workshops, the changing political environment resulted in not being able to get approval and funding for continuing the technical interchanges. In contrast to the U.S.–Russian workshops, where the participants felt the workshops were very successful and progress was being made, the lack of participation by China's military was making the interchanges more of an academic exchange, and progress on the major issues in space surveillance was lacking. There was no indication this would change.

The titles of the papers presented at the technical interchanges are provided in Appendix C. The papers are accessible at https://ssworkshops.engr.tamu.edu.

Epilogue: The Future of Space Surveillance

The newly organized U.S. Space Force is taking on the space surveillance mission just as international commercial activity in space is preparing to exceed government and military activity in space by almost every measure. We briefly highlight those elements of space surveillance and catalog maintenance whose character is undergoing sweeping changes.

In recent years, commercial providers of satellite tracking data have established worldwide networks of tracking sites. Data are provided from both optical and radar systems. These provide an abundance of additional data and offer frequent revisit of objects of interest. We discussed the U.S. and Canadian space-based sensors that have been invaluable in filling tracking gaps. The new S-band radar is already adding smaller objects to the catalog. Recent experiments with the European Union Space Surveillance and Tracking Partnership have demonstrated the benefits of international data sharing.

But new surveillance resources may bring with them a new set of challenges. The increased number-density of satellites will make it more difficult to associate observations with the proper cataloged satellite. New, more sophisticated algorithms for observation association, UCT processing, and catalog maintenance will likely be needed as the catalog grows.

New constellations such as Starlink and OneWEB bring a different challenge. The satellites in the constellations are autonomously controlled to preserve the constellation configuration. When another resident satellite is predicted to intrude on the constellation, the constellation members perform small maneuvers to allow safe passage of the intruder. This works fine for U.S.-sponsored constellations. But how will this be handled with the Chinese constellation now being deployed? There is certainly an emerging need for agreements and processes to address this situation.

In addition to proliferated launches, there are other events that add to the catalog. On 19 October 2024, there was a breakup of Intelsat 33E in geosynchronous orbit. The cause is unknown, but this adds some number of new debris objects yet to be cataloged.

We already discussed multiple nations demonstrating a capability to attack and destroy satellites. Any intentional creation of debris in space is unacceptable, especially at long-lived higher altitudes routinely used for operations.

Finally, space missions of national significance are currently being mounted throughout the Earth–Moon system, not merely in Earth orbit. The result is that some type of global situational awareness will soon have to be maintained throughout a volume of space that is orders of magnitude times the volume routinely monitored today. Moreover, many of the processes and tools used for catalog maintenance will have to be reinvented to accommodate a much more complex perturbation environment and reliance on a relative sparseness of tracking data in that much larger volume.

At the same time, the issue of space traffic coordination is becoming a tangible issue that must be addressed. There was a designation of the Department of Commerce, OSC, to provide space traffic management. Although this has historically been handled by the DoD, new legislation designated that DoD would manage defense-related satellites, and OSC would handle all other satellites.

All these efforts will depend critically on all parties having access to space surveillance information that is complete, timely, accurate, and precise, for every orbit regime, far into the future.

On 8 April 2024 at the Space Surveillance Conference in Colorado Springs, Colorado, the following information was publicly announced. This was presented by Richard DalBello, Office of Space Commerce Director, at the Space Traffic Management Policy workshop.

> *We envision that in the future there will be a global, coordinated system of SSA providers, with a series of national or regional hubs providing SSA information and services to spacecraft operators. These centers will be supported by networks of international partnerships, and their services will be augmented by a robust global commercial SSA sector.*

Taken collectively, these evolving challenges seem very daunting. But this is the very reason we love this business: there are always new problems just over the horizon. The dynamics of a real-world trajectory have yet to be perfectly modeled; the proper statistical model for the noise in radar or optical system measurements has yet to be formulated; the prediction of future solar activity and its effect on atmospheric density still confounds us; and how to maintain a catalog automatically without the occasional injection of human judgment is as elusive as the wind. We have made great strides over the 67 years of spaceflight, but there will always be plenty of challenging problems with which succeeding generations must grapple.

APPENDIX A

I. HYNEK'S 1958 MEMORANDUM TO THE SAO SATELLITE TRACKING PROGRAM STAFF ON THE OCCASION OF THE FIRST ANNIVERSARY OF THE LAUNCH OF SPUTNIK 1

A little-known internal memorandum from J. Allen Hynek, Deputy Director of the SAO, in charge of the STP, to his staff one year after the launch of Sputnik 1, provides a vivid and unusual account of the space surveillance mission at its very inception. Recall that the satellites in the Sputnik series did not use the IGY standard frequency of 108 MHz, so, at least initially, they could not be tracked by the American Minitrack radio interferometer system that was already deployed internationally. Moreover, none of the SAO's Baker–Nunn cameras were ready on 4 October 1957, as Hynek mentions, with the result that SAO's newly organized Project Moonwatch teams were among the first to make optical observations of both Sputnik 1 and its upper-stage rocket. The first "acquisition" alert that something new needed to be tracked came from the news media after the launch rather than from surveillance sensor systems, without any information about where to look for it. This particular scenario is actually close to the typical modern scenario in which new launches and new satellite deployments are usually not announced in advance. In this sense, Sputnik 1 really did open up, in a matter of several hours, the entire unprecedented space surveillance problem for the Americans. Hynek's memorandum records a number of unforgettable moments on that first evening and in the first few days of surveillance tracking.

The complete text of Hynek's five-page memorandum is recorded here in a font reminiscent of the original typewriter font, with the original spelling and punctuation, original paragraph breaks, and the original page numbers from the bottom margins just before the original page breaks. An electronic scan of a mimeograph copy that was distributed by Hynek to his staff at the SAO, referenced in Chapter 2 [12], is available online in the Smithsonian Institution Archive among the Fred Lawrence Whipple Papers. Included in that archive posting is a scan of Whipple's handwritten single-page listing of the first reports of sightings of Sputnik or its booster by various observers on the evening of 4 October 1957. That scan is reproduced here as Fig. A.1 after the body of the memorandum.

MEMORANDUM TO: STP Staff October 3, 1958

From: J. Allen Hynek

Subject: First Anniversary of Sputnik

Greetings on this first anniversary of the launching of man's first satellite to space. This gives me an opportunity to thank those of you who were with us on that historic day, (we had a total staff of 37 then -- 100 today) for the part you played in making the optical tracking program a success. Thanks go to, of course, those who joined the staff later, but I regret that all of you were not here last October 4 to have shared in as memorable an occasion as is likely to occur for centuries. Those of you who were here will recall well what happened, and will excuse me if I retell the saga of October 4, 1957, especially for the benefit of the "newcomers".

Our calendar being what it is, October 4 last year came on a Friday, and on that particular Friday by 6:30pm practically all of us had already gone home fully expecting to spend a quiet weekend. By some chance, only Ken Drummond and I were still around, in my office, leisurely discussing plans for the following week. And then the telephone rang.

To paraphrase Tom Lehrer, "I'm never forget the day Russians launch first satellite." Thinking perhaps that this was a wrong number or someone calling on a trivial matter, I nonchalantly picked up the phone. "Did I have any comments on the Russian satellite?" Still not taking it seriously I asked the reporter to read the full dispatch, and only then did I realize that this really was IT. As I hung up the receiver I just said to Ken in I'm afraid a rather weak voice, "They've done it. There is a Russian satellite up."

For a moment or two, I can tell you, both Ken and I felt extremely helpless--everybody gone and even Dr. Whipple not due to return from Washington for a few hours yet. In a few moments we realized that telephones were made to be used and we began calling both some Moonwatch team leaders and staff members. It was a little exasperating to have to convince some of the latter (but quite understandably) that this was not a gag. Some staff members didn't wait to be called but as soon as they heard it on the radio - popped right back. One of these was Mrs. Whidden (then Marty Holt) who as soon as she heard it on the radio hopped into a cab and was here in a matter of minutes. The entire staff reacted marvelously, without any question of time spent, disruption of personal plans, and without any hint or questions of overtime compensation. This was zero hour, not the one we had planned for, but here it was nonetheless.

-1-

It was not more than an hour before Kittredge Hall was besieged -- and I mean besieged -- by hoards of press, radio, and TV people. As reporters swarmed around, and reports began coming in, and Moonwatch teams all around were rushing to their posts, Dr. Whipple got in by plane from Washington. Kittredge Hall was a blaze of lights, and to that brightness was soon added the glare of TV and movie lights. My office was soon a shambles of extension cords, cigarette butts, and used up flash bulbs. Dr. Whipple who, of course, was the person almost entirely responsible for SAO's having the satellite tracking project in the first place (but that is another story) became the center of a storm of attention, thus relieving the tension from many of the rest of us.

During the first few nights of Sputnik I, Kittredge was alive with people completely around the 24 hours. Yes, even at three and four a.m. there was a buzz of activity. Our wives likewise rose to the occasion. They supplied coffee and sandwiches on an around-the-clock basis. Mrs. Whipple, on the very first night, bought up and distributed gobs of coffee and sandwiches, and this was continued by a regular brigade of wives, operating on an unarranged but nonetheless fairly regular schedule.

It is funny how these things stand out rather than the more technical aspects of the work. This went on, carried out by a level-headed staff that recognized that "first things come first" and, bothered as little as possible by the publicity and the whoop-la, carried on. Even those at SAO not on the regular STP staff became part of the team -- secretaries as well as scientific staff. Dr. Whitney and Dr. McCrosky, Dr. Sterne, Dr. Schilling, Dr. Jacchia, and Mr. Briggs -- all worked tremendously to assist in the push on getting out satellite orbits and predictions. Now all this is routinized, but then there had to be a lot of cut and try, and groping for ways of doing things. Those were stormy days on the technical sea, but both the ship and the crew came through -- but not unscathed. The months of quiet preparation on the part of Leon Campbell in Moonwatch, and of Dr. Henize in station preparation, all paid off. It would have been nicer if the Russians had given us a few more months of grace, but ----.

It is only in retrospect now that one can sense what really happened. The public reaction to Sputnik was, as you will recall, a strange mixture of awe, admiration and fear, the latter enhanced of course because there had been no warning. As my barber put it, "Most people realize that things in the sky were put there by God, and here was something that was brand new in the skies which although it wasn't put there by God, must have been put there by someone awfully close to God."

What we didn't realize until later was that since this was not a satellite of our own launching, and since Smithsonian Astrophysical Observatory had the task of tracking artificial earth satellites, we here at SAO became the information center for the western world on this new, frightening object. It was in many ways a terrible

-2-

responsibility to bear. The slightest word to the press carried tremendous weight. I remember well one instance during that period. After the orbit of Sputnik I was reasonably well known Life magazine wanted to know where to send planes to photograph it. Cost was no object. If I had said Tierra del Fuego, within minutes the Life crew would have been dispatched there. It seemed that any statement that Dr. Whipple and I, and others of the staff, made was front page stuff and this carried with it a tremendous responsibility to weigh words carefully and exactly. In helping us with press relations John White, our press officer of those days, helped us tremendously. Speaking only for myself, he kept me out of many embarrassing situations by cautioning me to rephrase statements which might otherwise be misinterpreted by the press. Even his watchfulness, however, did not prevent a few slips; perhaps the most glaring was the time the Boston Herald carried banner headlines "Mysterious Force Grips Sputnik", a situation which arose out of a misquoted explanatory statement from me concerning a release made by Washington that the orbit of Sputnik could not be explained solely by the laws of Newton.

By and large we got well, however, and I firmly believe that by patient explanation to press members (the lounge downstairs came to be known by many press members as their satellite "classroom" -- in fact, there was a sign on the door at one time "Astrophysical Press Association -- Members Only and all Others") we all aided in presenting the facts of satellites to the general public as scientifically and unspectacularly as possible.

The Russians may launch another Sputnik near this anniversary date but no matter how spectacular it is (even if it has a crew of ten and is equipped with a swimming pool) it could not possibly cause the excitement that October 4, 1957 saw.

Our STP had been geared to a 20-inch Vanguard satellite (which isn't up yet) and for an orbit of inclination not more than 35 degrees. Here we are faced with a high inclination orbit and an object much brighter than anyone had anticipated. In fact, when Dr. Gordon Little and his group in Alaska reported that they had seen an object as bright as Jupiter, none of us here would have at first believed that it was a satellite. It was, in fact, Dr. McCrosky who pointed out that what Dr. Little and his group had probably seen was the rocket carrier of the Russian satellite proper. Vanguard, of course, was due to have an orbiting rocket carrier also but we had become so used to thinking of satellites and carriers and all being at the very limit of naked eye visibility that this surprising object up there took us back. It is the brightness of the satellite, of course, that added immeasurably to the excitement on the part of the public. Here was something that they could get out and see plainly, and those of us who first saw Sputnik I sailing majestically across the sky I'm sure experienced a thrill we will never forget.

-3-

Despite the fact that our program was not ready to go on October 4, 1957, in the year that has passed since then the STP has taken gigantic strides to reach our present smooth-operating level. Whereas the first Baker-Nunn had just been assembled and tested in Pasadena by that date, all the satellite tracking stations have since come into full bloom. Worthy of mention is the fact that the first camera on October 4, having been completely disassembled for minor reworking, was immediately re-assembled by Dr. Henize, Mr. Stinnett and a few others to try for the first satellite photograph, only to learn that Sputnik I would not be visible for a while. Computations and analysis has attained a marvelous degree of organization in that preliminary and detailed orbits, and ephemerides of various kinds can be handled not for merely the expected one or two Vanguard satellites but for at least half a dozen objects up there simultaneously. Moonwatch, after October 4th, no longer had to have "fly-bys" which, for the benefit of the more recent members of the staff, were little one-tenth candlepower flashlight bulbs mounted in a plumber's friend (minus handle) and trailed behind an airplane by about one hundred feet. At a height of a mile, a one-tenth candlepower bulb shines with the brightness of a fifth magnitude star. Consequently, these devices that had been engineered largely through the efforts of Dr. Armand Spitz, and Leon Campbell, coordinator of Moonwatch, and flown by civilian air patrol pilots, were no longer necessary. Now there was a real satellite to practice on. Moonwatch deserves a special mention in that it was called upon to do something for which it was not originally intended. Before the Baker-Nunn cameras were ready, the Moonwatch network took on the task of being an interim tracking system, and it was through their observations telegraphed and mailed in that many of the early scientific results from satellites were made possible.

On October 4, 1957, there wasn't a single TWX machine installed in the communications room; in fact, there was no communications room. The problem of communication was being "studied". Suddenly there was a communication room. Chief Peterson pulled every string in his bow and late at night the first TWX machine was installed. To make things worse, partitions were being put in, and so there were boards and plaster all over what is now the communication room.

Before long, however, we had a world-wide nerve center focused in one room on the second floor of Kittredge.

It may interest you to know that during the past year some five million words have been handled by Chief Peterson and his relatively small gang from that one little room. Those of you who are interested in statistics might like to know that during the year there have been 535 Baker-Nunn photographs of satellites taken and 6784 Moonwatch observations made.

-4-

But that is all history now. As we face another October 4 and the second year of the satellite era we can all feel proud that the Smithsonian Satellite Tracking Program is now ready to handle just about anything that anyone puts up there. To use a phrase that has been used here since the inception of the STP: "if it moves we'll track it".

It would be proper to extend thanks on this anniversary, individually, to all who helped in those "pre-historic" days of Sputnik I, but this is frankly impossible. The names mentioned were used solely as part of the record of what happened -- "they were there when the shooting started", and not as a matter of thanks or acknowledgement.

May I therefore, now, on behalf of Dr. Whipple and myself, extend to everyone of you who was on the staff on October 4, 1957, and to those many more who have joined during the past year, our warm thanks for having met the very vital challenge that the world's first Sputnik launched into our personal skies.

-5-

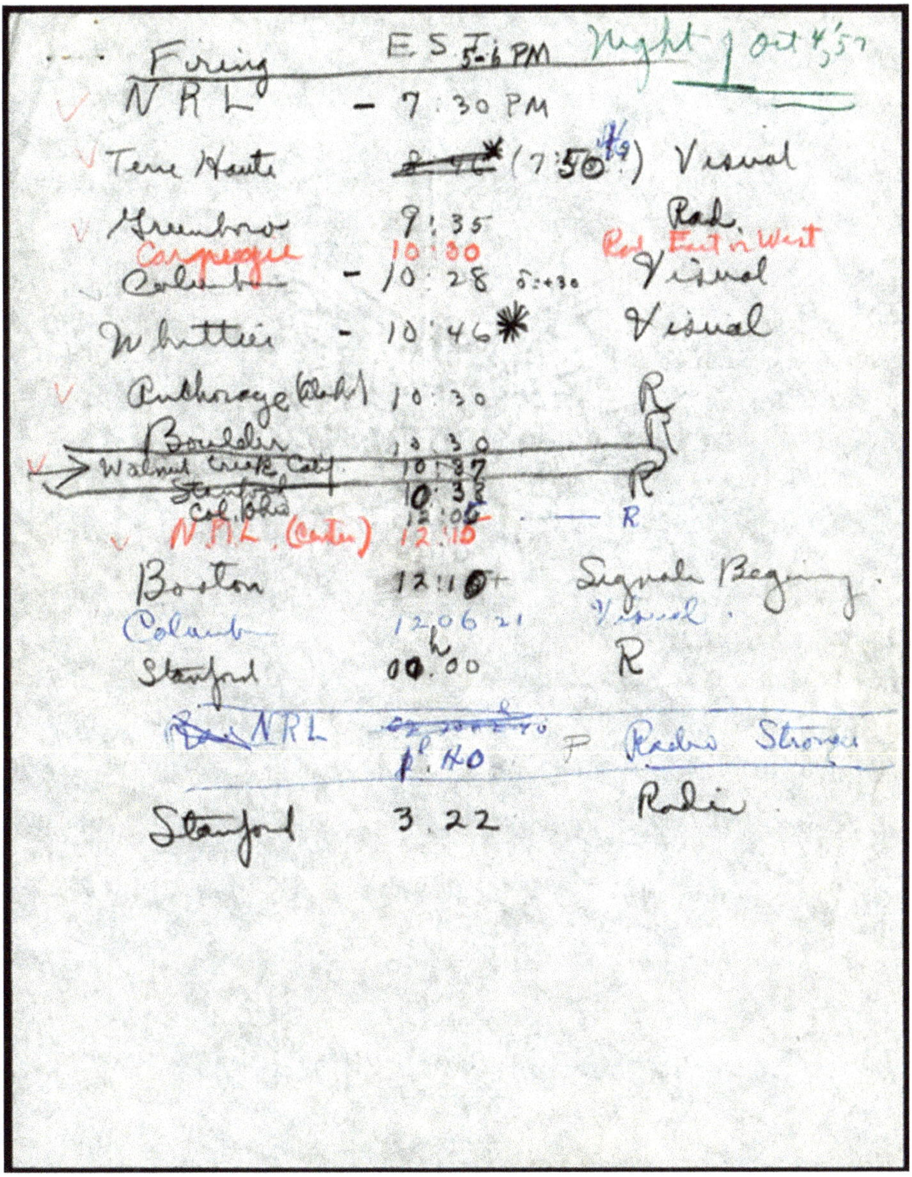
Firing E.S.T. 5-6 PM Night Oct 4, 57
NRL - 7:30 PM
Terre Haute (7:50) Visual
Greensboro 9:35 Rad
Carnegie 10:30 Rad East in West
Columbus - 10:28 Visual
Whittier - 10:46 Visual
Anchorage (Alaska) 10:30 R
Boulder 10:30 R
Walnut Creek, Calif. 10:37 R
Stanford 10:38
Cal. Ohio 12:06 R
NRL (Carter) 12:16
Boston 12:10 Signals Beginning.
Columbus 12:06 21 Visual
Stanford 00.00 R
NRL Radio Stronger
Stanford 3.22 Radio

Fig. A.1 Electronic scan of Whipple's handwritten list of Sputnik I sightings reported to SAO's Satellite Tracking Program in Kittredge Hall, Cambridge, Massachusetts, on 4 October 1957.

APPENDIX B

I. ATTENDEES AND PROGRAMS FOR U.S.–RUSSIAN SPACE SURVEILLANCE WORKSHOPS 2–9

TABLE B.1 SECOND U.S.–RUSSIAN SPACE SURVEILLANCE WORKSHOP ATTENDEES (POZNAN, POLAND)

U.S. Attendees	
Bernard Kaufman	Naval Research Laboratory
Paul Cefola	Draper Laboratory
Shannon Coffey	Naval Research Laboratory
Paul Schumacher	Naval Space Command
P. Kenneth Seidelmann	U.S. Naval Observatory
Scott Wallace	AFRL
Russian Attendees	
Vladimir Boikov	Vympel Corp
G. Chernyavskiy	Center for Program Studies
Valery Dicky	Space Research Center, COSMOS
Zakhary N. Khutorovsky	Vympel Corp.
Rolan I. Kiladze	Institute of Theoretical Astronomy
Andrey Nazarenko	Center for Program Studies
Stanislav Veniamonov	Space Research Center, COSMOS
E. Vovchyz	Lviv State University
Vasiliy Yurasov	Space Research Center COSMOS

TABLE B.2 SECOND U.S.–RUSSIAN SPACE SURVEILLANCE WORKSHOP PAPERS

Title	Presenting Author
Evaluation of the Accuracy of Forecasting Satellite Motion in the Atmosphere	A. Nazarenko and G. Chernyavskiy
Technology of Evaluation of Atmospheric Density Variation Based on the Space Surveillance System's Orbital Data	A. Nazarenko
Comparison of Atmospheric Density Models	T. Amelina, G. Batyr, V. Dicky, N. Tumolskaya, and V. Yurasov
The Features of Prediction Procedures, Used in Low Perigee Satellites' Catalog Maintenance	V. Boikov
Application of Parallel Processing to Astrodynamics	S. Coffey and H. Neal
Analytical Orbit Model for U.S. Naval Space Surveillance: An Overview	P. Schumacher and R. Glover
Techniques of Orbit Determination Using Minimum Data, Accounting of the Features of Russian Space Surveillance System	V. Boikov
Reentry Time Determination Analysis for Cosmos-398 and FSW-1-5	V. Andrewshchenko, G. Batyr, V. Bratchikov, V. Dicky, S. Veniaminov, and V. Yurasov
Geostationary Orbit Determination and Procedure	V. Yurasov and A. Moscovsky
On Orbital Evolution of Geostationary Satellites	R. Kiladze and A. Sochilina
Low Perigee Satellite Catalog Maintenance: Issues of Methodology	Z. Khutorovsky
Tracking Satellite Breakups	P . Schumacher
Universal Semianalytical Satellite Motion Propagation Method	V. Yurasov
Current Development of the Precision Semianalytical Satellite Theory	P. Cefola and R. Proulx

TABLE B.3 THIRD U.S.–RUSSIAN SPACE SURVEILLANCE WORKSHOP ATTENDEES (U.S. NAVAL OBSERVATORY, WASHINGTON, D.C.)

U.S. Attendees	
Terry Alfriend	Texas A&M University.
Gerald Banner	MIT Lincoln Laboratory
Paul Cefola	Draper Laboratory
Shannon Coffey	Naval Research Laboratory
Felix Hoots	General Research Corp.
Timothy P. Payne	SWC/AES
Eric C. Pearce	MIT Lincoln Laboratory
Paul W. Schumacher	Naval Space Command
P. Kenneth Seidelmann	U.S. Naval Observatory
R. Sridharan	MIT Lincoln Laboratory
Nicholas Johnson	Kaman Sciences
Jeff Huth	Kaman Sciences
Bernard Kaufman	NRL
Stephen Knowles	Naval Space Command
Mark Matney	Lockheed EFC
P. Kenneth Seidelmann	U.S. Naval Observatory
G. Stokes	MIT Lincoln Laboratory
David Vallado	AFRL
Russian Attendees	
Vladimir Boikov	Vympel Corp
Valery Dicky	Space Research Center "Kosmos"
Zakhary N. Khutorovsky	Vympel Corp.
Rolan I. Kiladze	Institute of Theoretical Astronomy
Alexander Menshikov	Vympel Corp.
Andrey Nazarenko	Center for Program Studies
Stanislav Veniaminov	Space Research Center "Kosmos"
Vasiliy Yurasov	Space Research Center "Kosmos"

TABLE B.4 THIRD U.S.–RUSSIAN SPACE SURVEILLANCE WORKSHOP PAPERS

Title	Presenting author
Deep Space Surveillance – Overview and Tracking Radar	G. P. Banner
Electro-Optical Deep Space Tracking	E. C. Pearce
Contributing Sensor Operations of the Space-Based Visible (SBV)	G. Stokes and R. Sridharan
Enhancing the Efficiency of Search for Deep-Space Objects by Using Available A'Priori Orbital Information—New Theory and Methods. Part I—New General Theoretical Approach	S. Veniaminov, P. Seidelmann, and P. Cefola
Enhancing the Efficiency of Search for Deep Space Objects by Using Available A'Priori Orbital Information—New Theory and Methods. Part II—Theory and Methods for Specific Practical Case	S. Veniaminov, V. Dicky, and N. Vasilyev
Generalization of the Notion of an Ephemeris for Acquisition of a Weak Intelligence Signal	S. Veniaminov and P. K. Seidelmann
Studies and Observations of Hazardous Objects in the Near-Earth Space	L. Rykhlova
Near-Geostationary Orbit Model, Used in Satellite Catalog Maintenance	V. Boikov, A. Testov, and Z. Khutorovsky
Using Hill-Brown Theory for Expanding Lunar Solid Spherical Harmonics	V. Boikov, A. Testov, and Z. Khutorovsky
Long-Term Prediction of Space Object Motion in GEO	V. Yurasov
A History of Analytical Orbit Modeling in the U.S. Space Surveillance System	F. Hoots
General Perturbations Orbit Model for U.S. Naval Space Surveillance Operations	P. Schumacher and R. Glover
Lunar–Solar Point-Mass Perturbations in the Precision Semi-analytical Satellite Theory	P. Cefola and R. Proulx
Geosynchronous Satellite Covariance Study	T. Alfriend and S. Paik
Satellite Observation Correlation Processing	T. Payne
Uncorrelated Observations Processing at Naval Space Command	S. Coffey, H. Neal, and M. Berry
A Future Look at Space Surveillance and Operations	D. Vallado and S. Alfano
Application of Special Perturbations and Parallel Processing to the Space Object Catalog	S. Coffey, H. Neal, and A. Segerman
Monitoring of GEO Satellites in Russian Space Surveillance Center	Z. Khutorovsky
Optical Measurements Processing Technology on Geostationary Satellites	P. Cefola and V. Yurasov
On Occasional Changes of Velocities of Geostationary Uncontrolled Objects	A. Sochilina, R. Kiladze, K. Grigoriev, and A. Vershkov
Analytical Estimates of GEO Determination Accuracy	V. Dicky, N. Vasilyev, and S. Veniaminov
History of Russian Space Surveillance System	A. Nazarenko

TABLE B.5 FOURTH U.S.–RUSSIAN SPACE SURVEILLANCE WORKSHOP ATTENDEES (U.S. NAVAL OBSERVATORY, WASHINGTON, D.C.)

U.S. Attendees	
Terry Alfriend	Texas A&M University.
Gerald Banner	MIT Lincoln Laboratory
Paul Cefola	Draper Laboratory
Shannon Coffey	Naval Research Laboratory
Felix Hoots	General Research Corp.
Nicholas L. Johnson	NASA Johnson Space Center
Richard L. Lambour	MIT Lincoln Laboratory
Eric C. Pearce	MIT Lincoln Laboratory
Paul W. Schumacher	Naval Space Command
John H. Seago	Honeywell Corporation
P. Kenneth Seidelmann	U.S. Naval Observatory
Craig V. Solodyna	MIT Lincoln Laboratory
Michael Stringer	1CACS
Russian Attendees	
Valery Dicky	Space Research Center "Kosmos"
Konstantin V. Grigoriev	Pulkovo Observatory
Zakhary N. Khutorovsky	Vympel Corp.
Rolan I. Kiladze	Institute of Theoretical Astronomy
Valerie Kolinko	Krasnogorskiy Mechanical Factory
Andrey Nazarenko	Center for Program Studies
Alla S. Sochilina	Pulkovo Observatory
Alexey Testov	Vympel Corp.
Yury N. Tretyakov	Space Research Center "Kosmos"
Stanislav Veniaminov	Space Research Center "Kosmos"
Vasiliy Yurasov	Space Research Center "Kosmos"

TABLE B.6 FOURTH U.S.–RUSSIAN SPACE SURVEILLANCE WORKSHOP PAPERS

Title	Presenting Author
Tracking site operations, detection and tracking methods and initial orbit development	
Narrowband and Wideband Radar Signatures in Support of the Space Catalog	C. Solodyna and G. Banner
Satellite Photometry with the GEODSS, MOSS, and SBV Systems	E. C. Pearce
Orbital Debris Size Estimation from Radar Cross Section Measurements	R. Lambour, T. Morgan, and N. Rajan
The New Approach to Revealing Latent Periodicities in Radar and Optical Signatures of Space Objects	S. Veniaminov, V. Dicky, and Yu. Tretyakov
Independent Radar Monitoring of the Earth Satellites in Emergency (The RORSAT Case)	A. Ladygin, S. Morozov, S. Serdyukov, Yu. Tretyakov, and S. Veniaminov
Use of Non-Coordinate Information in the Space Debris Problem	A. Bagrov, F. Masumi, and L. Rykhlova
Data correlation, initial orbit determination and satellite identification	
Correlation of Measurements in the Process of Satellite Catalog Maintenance	Z. Khutorovsky
Classification and Identification of Geostationary Space Objects by Using Coordinate and Photometric Observations	V. Yurasov, V. Vygon, and V. Shargorodskiy
Tools and Databases Used to Maintain the Space Catalog at 1CACS	M. Stringer and B. Teets
Identifying Satellite Launch Origins with Historical Examples	M. Stringer, B. Teets, and R. Thurston
General interest topics	
Explosions in Near-Earth Orbits and Their Relation with Solar Activity and with Meteor Streams	L. Rykhlova, T. Kasimenko, A. Mikisha, and M. Smirnov
Navy Calibration of the U.S. Space Surveillance Network Using Satellite Laser Ranging	J. Seago
Numerical-Analytical Algorithm for Prediction of Satellite Motion in High Eccentricity Orbits	A. Testov and V. Boikov
Transitioning from a General Perturbations to a Special Perturbations Space Object Catalog	M. Wilkins, T. Alfriend, S. Coffey, and A. Segerman
Secular Variations of the Semimajor Axis of Debris Particles Orbits in the Vicinity of GEO Caused by Solar Radiation Pressure	A. Mikisha, E. Novikova, L. Rykhlova, and M. Smirnov
On Ways of Modernization of Geostationary Ring Surveillance	R. Kiladze, A. Sochilina, K. Grigoriev, and A. Vershkov
Evolution of the NAVSPACE Catalog Processing System Using Special Perturbations	P. Schumacher and F. Hoots
On a Creation of Optronic Station of the Full Factory Availability for Service	S. Zdor, V. Grigroshenko, S. Morozov, and V. Kolinko

(Continued)

TABLE B.6 (*Continued*) FOURTH U.S.–RUSSIAN SPACE SURVEILLANCE WORKSHOP PAPERS

Title	Presenting Author
Cooperative research papers on subjects stimulated by the 1998 workshop	
Improvement of Orbits of Geostationary Satellites from Observations Over a Time Interval of 150–200 Days	K. Grigoriev and F. Hoots
Near-Real Time Monitoring of Variations of the Upper Atmosphere Density	P. Cefola, R. Proulx, G. Granholm, V. Yurasov, and A. Nazarenko
Comparison of the Russian and U.S. Algorithms for Catalog Maintenance for Geosynchronous Satellites	T. Alfriend, S. Paik, V. Boikov, Z. Khutorovsky, and A. Testov
Investigation of Satellite Catalog Maintenance Utilizing Sparse Tracking Data	F. Hoots, R. France, A. Sochilina, R. Kiladze, and K. Grigoriev
Revealing Latent Periodicities in Noisy Signals	S. Veniaminov, V. Dicky, Yu. Tretyakov, and P. Seidelmann
Comparative Analysis of the Russian and United States Space Object Catalogs	V. Dicky, Yu. Tretyakov, S. Veniaminov, V. Zavaliy, N. Johnson, and Z. Khutorovsky

TABLE B.7 FIFTH U.S.–RUSSIAN SPACE SURVEILLANCE WORKSHOP ATTENDEES (PULKOVO ASTRONOMICAL OBSERVATORY, ST. PETERSBURG, RUSSIA)

U.S. Attendees	
Jack Agee	AFOSR
Terry Alfriend	Texas A&M University.
Michael Carey	595th Space Group
Paul Cefola	MIT Lincoln Laboratory
Michael Hindley	Space Warfare Center.
Thomas S. Kelso	Air Force Space Command
Kent Miller	AFOSR
Timothy Payne	Air Force Space Command
P. Kenneth Seidelmann	University of Virginia
Mark Storz	Air Force Space Command
Scott Wallace	U.S. Air Force
Russian Attendees	
Vladimir M. Agapov	KIA Systems
Vadim A. Antonov	Central Astronomical Observatory at Pulkovo
Nail S. Bakhtigaraev	Institute of Astronomy of RAS
Vladimir F. Boikov	KIA Systems
Andrey N. Bukreev	KIA Systems
Yuriy V. Burtsev	Space Forces
Oleg P. Bykov	Central Astronomical Observatory at Pulkovo
Alexander V. Deviatkin	Central Astronomical Observatory at Pulkovo

(*Continued*)

TABLE B.7 (*Continued*) FIFTH U.S.–RUSSIAN SPACE SURVEILLANCE WORKSHOP ATTENDEES (PULKOVO ASTRONOMICAL OBSERVATORY, ST. PETERSBURG, RUSSIA)

Alexander V. Didenko	Astrophysical Institute, Kazakhstan Republic
Vladimir P. Efremov	Space Forces
Konstantin V. Grigoriev	Pulkovo Observatory
Georgy D. Gotovtsev	Lavochkin Association
Irina S. Gouseva	Central Astronomical Observatory at Pulkovo
Nikolay M. Ivanov	TSNIIMASH
Yuriy F. Kalyuka	TSNIIMASH
Sergey Kamensky	KIA Systems
Ivan I. Kanaev	Central Astronomical Observatory at Pulkovo
Gennady Yu	Kharlamov, Institute of Astronomy of RAS
Konstantin V. Kholshevnikov	Astronomical Institute of St. Petersburg University
Zakhary N. Khutorovsky	KIA Systems
Rolan I. Kiladze	Abastumani Astrophysical Observatory
Alexander F. Klichin	Lavochkin Association
Valery I. Kolinko	Krasnogorsky Zavod im. S. A. Zvereva
Valentin V. Koloda	Space Forces
Edward V. Kornilov	Central Astronomical Observatory at Pulkovo
Maxim N. Kretinin	Space Forces
Eduard D. Kuznetsov	Astronomical Observatory of Ural State University
Victor N. L'vov	Central Astronomical Observatory at Pulkovo
Liudmila Ya. Martynova	Rosoboronexport
Kirill L. Maslennikov	Central Astronomical Observatory at Pulkovo
Mikhail N. Meshkov	Special Design Bureau of Moscow Power Engineering Institute
Mikhail A. Mikhailov	TSNIIMASH
Igor E. Molotov	Central Astronomical Observatory at Pulkovo
Sergey A. Orlov	Astronomical Institute of St. Petersburg University
Olga A. Osokina	TSNIIMASH
Evelina Pedan	KIA Systems
Victor Shilin	Vympel Corp.
Sergey S. Smirnov	Central Astronomical Observatory at Pulkovo
Alla S. Sochilina	Central Astronomical Observatory at Pulkovo
Alexander V. Stepanov	Central Astronomical Observatory at Pulkovo
Yury N. Tretyakov	SRC "Kosmos"
Olga A. Tsiopa	Central Astronomical Observatory at Pulkovo
Vladimir S. Tyurin	Space Forces
Stanislav S. Veniaminov	SRC "Kosmos"
Andrey N. Vershkov	Central Astronomical Observatory at Pulkovo
Mikhail V. Yakovlev	TSNIIMASH
Vasily S. Yurasov	KIA Systems
Sergey I. Zdor	Krasnogorsky Zavod im. S. A. Zvereva
European Space Agency Attendees	
Heiner Klinkrad	ESA/ESOC
Thomas Schildknecht	University of Bern

TABLE B.8 FIFTH U.S.–RUSSIAN SPACE SURVEILLANCE WORKSHOP PAPERS

Title	Presenting Author
Comparison of space object catalogs	
The New Reference Systems	P. Seidelmann
Dynamics of the Principal Characteristics of the U.S./Russian Space Object Catalogs since 1995 to 2002	O. Aksenov, O. Andreyev, V. Dicky, N. Johnson, Yu. Tretyakov, S. Veniaminov, and V. Zavaliy
Space surveillance observational methods	
Experience of Formation of the GEO Orbital Information Archive Based on Different Data Sources	V. Agapov, Z. Khutorovsky, V. Boikov. N. Sbytov, and A. Samotokhin
Results from the 1999 Cobra Dane Debris Campaign	G. DeVere, J. Randolph, and T. Payne
Operational Test and Evaluation of Space Surveillance Systems	M. Carey and M. Hindley
Pulkovo Program for Research of the Near-Earth Space Objects	A. Devyatkin, I. Guseva, I. Kanaev, V. L'vov, I. Molotov, A. Sochilina, D. Gorshanov, V. Kupriyanov, M. Yu. Sidorov, R. Smekhacheva, and S. Tsekmejster
The Results of Photometric Observations of the Sun-Synchronous Satellites at the Zvenigorod Observatory	G. Harlamov, N. Bahtigaraev, S. Gavrish, V. Bogatyrev, and A. Chelyanov
Comparative Evaluation of Detection Performances of Radar "COBALT–RLS" with Foreign Radar Stations (RS) Intended for Monitoring the Man-Caused Contamination of Near-Earth Space	M. Meshkov and M. Suloviev
Space Surveillance Contributions to the STS 107 Accident Investigation	T. Kelso, R. Morris, G. DeVere, J. Randolph, B. Bowman, R. Racca, N. Ericson, and R. Thurston
Space Surveillance and Space Debris Mitigation in the GEO Region	V. Blagun, Yu. Levitskiy, S. Kulik, V. Lukyashchenko, S. Chekalin, N. Ivanov, Yu. Kalyuka, S. Loginov, M. Mikhailov, M. Yakovlev, and Yu. Bulinin
Methods in catalog formation	
Robust and Adaptive Techniques Used for Maintenance of the Catalog of LEO Satellites	Z. Khutorovsky
Performance of a Dynamic Algorithm for Processing Uncorrelated Tracks	T. Alfriend and J. Lim
High Accuracy Satellite Drag Model (HASDM)	M. Storz, B. Bowman, M. Branson, S. Casali, and K. Tobiska
On Investigation of Two- Line-Elements Precision of Geostationary Objects	A. Sochilina, R. Kiladze, I. Molotov, A. Vershkov, and E. Komilo

(Continued)

TABLE B.8 (*Continued*) FIFTH U.S.–RUSSIAN SPACE SURVEILLANCE WORKSHOP PAPERS

Title	Presenting Author
Catalog analysis	
Sigma Point Kalman Filters for Efficient Orbit Estimation in Space Surveillance	D. Lee and T. Alfriend
The Concept of an Exact Solution in Modem Celestial Mechanics	V. A. Antonov
Enveloping Surface of Trajectory Family of Isotropically Ejected Particles due to a Satellite Explosion	K. Kholshevnikov and S. Orlov
Estimation of Radar Orientation and Other Calibration Parameters Using Regularized Residuals	M. Storz
Near Real Time Corrections to the Atmospheric Density Model	V. Yurasov, A. Nazarenko, P. Cefola, and R. Proulx
Deep space observations	
Detection and Tracking of Space Objects and Debris in Europe	T. Schildknecht and W. Flury
An Optical Search for Small-Size Debris in GEO and GTO	T. Schildknecht, R. Musci, M. Ploner, W. Flury, J. Kuusela, J. de Leon Cruz and L. de Fatima Dominguez Palmero
New Deep Space Optical Search Strategies	T. P. Payne
Comparative Analysis of Different Methods for Revealing Latent Periodicities in Processes	S. Veniaminov, V. Dicky, and A. Shevyrev
Planning Optical Search of Celestial Bodies in Near-Earth Space	S. Zdor and V. Kolinko
First Results of the Space Debris Radar Observations Using Evpatoria RT-70 Transmitter and Low Frequency VLBI Network	I. Molotov, A. Konovalenko, B. Lipatov, G. Tuccari and V. Agapov
Deep space predictions and analysis	
Accounting for Precession and Nutation in Predicting Long-Term Motion in Quasi-Stationary Space Orbits Using a Combined Numerical-Analytical Algorithm	V. Boikov and A. Testov
Zone Catalog and Principles of Identification of Geostationary Satellites	A. Didenko, B. Demchenko, S. Nifontov, and A. Usoltseva
On the Evolution of Fragments of the Geostationary Objects	R. Kiladze and A. Sochilina
An Optical Search for Small-Size Debris in GEO and GTO	A. Brusnitsin, D. Kirichenko, and V. Semenikhi
The Long-Term Evolution of Geostationary Satellite Fragments After an Explosion	E. Kuznetsov

TABLE B.9 SIXTH U.S.–RUSSIAN SPACE SURVEILLANCE WORKSHOP ATTENDEES (PULKOVO ASTRONOMICAL OBSERVATORY, ST. PETERSBURG, RUSSIA)

U.S. Attendees	
Terry Alfriend	Texas A&M University.
Susan E. Andrews	MIT Lincoln Laboratory
Bruce R. Bowman	Air Force Space Command
Paul Cefola	MIT
Thomas S. Kelso	Analytical Graphics
Steven Knowles	Naval Space Command
Kim Luu	AFRL
Rich Lambour	MIT Lincoln Laboratory
Joseph Liu	Air Force Space Command
Chris Sabol	AFRL
David Vallado	Analytical Graphics
Russian Attendees	
Vladimir M. Agapov	KIA Systems
Vadim A. Antonov	Central Astronomical Observatory at Pulkovo
Nail S. Bakhtigaraev	Institute of Astronomy of RAS
Vladimir F. Boikov	KIA Systems
Andrey N. Bukreev	KIA Systems
Yuriy V. Burtsev	Space Forces
Oleg P. Bykov	Central Astronomical Observatory at Pulkovo
Alexander V. Deviatkin	Central Astronomical Observatory at Pulkovo
Alexander V. Didenko	Astrophysical Institute, Kazakhstan Republic
Vladimir P. Efremov	Space Forces
Konstantin V. Grigoriev	Pulkovo Observatory
Georgy D. Gotovtsev	Lavochkin Association
Irina S. Gouseva	Central Astronomical Observatory at Pulkovo
Nikolay M. Ivanov	TSNIIMASH
Yuriy F. Kalyuka	TSNIIMASH
Sergey Kamensky	KIA Systems
Ivan I. Kanaev	Central Astronomical Observatory at Pulkovo
Gennady Yu	Kharlamov, Institute of Astronomy of RAS
Konstantin V. Kholshevnikov	Astronomical Institute of St. Petersburg University
Zakhary N. Khutorovsky	KIA Systems
Rolan I. Kiladze	Abastumani Astrophysical Observatory
Alexander F. Klichin	Lavochkin Association
Valery I. Kolinko	Krasnogorsky Zavod im. S. A. Zvereva
Valentin V. Koloda	Space Forces
Edward V. Kornilov	Central Astronomical Observatory at Pulkovo
Maxim N. Kretinin	Space Forces
Eduard D. Kuznetsov	Astronomical Observatory of Ural State University
Victor N. L'vov	Central Astronomical Observatory at Pulkovo
Liudmila Ya. Martynova	Rosoboronexport
Kirill L. Maslennikov	Central Astronomical Observatory at Pulkovo
Mikhail N. Meshkov	Special Design Bureau of Moscow Power Engineering Institute
Mikhail A. Mikhailov	TSNIIMASH
Igor E. Molotov	Central Astronomical Observatory at Pulkovo
Sergey A. Orlov	Astronomical Institute of St. Petersburg Univ.

(Continued)

TABLE B.9 (*Continued*) SIXTH U.S.–RUSSIAN SPACE SURVEILLANCE WORKSHOP ATTENDEES (PULKOVO ASTRONOMICAL OBSERVATORY, ST. PETERSBURG, RUSSIA)

Olga A. Osokina	TSNIIMASH
Evelina Pedan	KIA Systems
Victor Shilin	Vympel Corp.
Sergey S. Smirnov	Central Astronomical Observatory at Pulkovo
Alla S. Sochilina	Central Astronomical Observatory at Pulkovo
Alexander V. Stepanov	Central Astronomical Observatory at Pulkovo
Yury N. Tretyakov	SRC "Kosmos"
Olga A. Tsiopa	Central Astronomical Observatory at Pulkovo
Vladimir S. Tyurin	Space Forces
Stanislav S. Veniaminov	SRC "Kosmos"
Andrey N. Vershkov	Central Astronomical Observatory at Pulkovo
Mikhail V. Yakovlev	TSNIIMASH
Vasily S. Yurasov	KIA Systems
Sergey I. Zdor	Krasnogorsky Zavod im. S. A. Zvereva
European Space Agency attendees	
Heiner Klinkrad	ESA/ESOC
Thomas Schildknecht	University of Bern

TABLE B.10 SIXTH U.S.–RUSSIAN SPACE SURVEILLANCE WORKSHOP PAPERS (PULKOVO OBSERVATORY, ST. PETERSBURG, RUSSIA)

Title	Presenting Author
Space surveillance systems and operations	
Optical Observations in a Proposed European Space Surveillance Network	T. Schildknecht, T. Flohrer, and R. Musci
Evolving Space Surveillance Capabilities – Mathematical Framework for Evaluating Future Surveillance Networks	S. Andrews
Orbital debris	
Problems of the Catalogue Maintenance for Faint GEO Objects	V. Agapov, V. Biryukov, S. Kamensky, I. Molotov, V. Rumyantsev, A. Sochilina, V. Titenko Sbytov, and A. Samotokhin
Satellite Orbital Conjunction Reports Assessing Threatening Encounters in Space (SOCRATES)	T. S. Kelso and S. Alfano
Re-Entry Prediction and On-Ground Risk Assessment	Heiner Klinkrad
A Study of the Surface Enveloping the Trajectory Family of Particles Isotropically Ejected by a Satellite Explosion with Account for the Orbit Node and Pericenter Motions	K. V. Kholshevnikov and S. A. Orlov
Space surveillance session A	
On the Theory and some Specific Methods for Searching for Space Objects by Use of Rough Orbital Information	S. Veniaminov, V. Zavaliy, Yu. Tretyakov, V. Pyrin, and V. Lapukhin

(*Continued*)

TABLE B.10 (*Continued*) SIXTH U.S.–RUSSIAN SPACE SURVEILLANCE WORKSHOP PAPERS (PULKOVO OBSERVATORY, ST. PETERSBURG, RUSSIA)

Title	Presenting Author
The High-Apogee Artificial Spacecraft Astrophotometrical and Astrometrical Observations in the Sayan Mountains Observatory of ITSP BB RAS and Their Applications to Checking the Spacecraft State of Operation	V. Goryashin, Yu. Karavaev, M. Mishina, M. Eselevich, T. Tsukker, V. Tergoev, O. Yurysheva, P. Papushev, R. Kopyatkevich, G. Mishin, P. Shaburov, and N. Sakva
Small Aperture Telescope Augmentation Study	R. L. Lambour, E. Pearce, S. Ferner, E. Rork, P. Trujillo, A. Decew, and P. Hopman
Atmospheric density	
The Semiannual Thermospheric Density Variation from 1970 to 2002 Between 200–1100 km	B. Bowman
Review of Results on Atmospheric Density Corrections	V. Yurasov, A. Nazarenko, T. Alfriend, and P. Cefola
Density Variations in the Upper Atmosphere During Several Solar Cycles Determined from Satellite Drag Measurements	S. Knowles
Application of the Arima Model to Analyze and Forecast the Time Series of Density Corrections for NRLMSIS-00	V. Yurasov, A. Nazarenko, T. Alfriend, and P. Cefola
Drag Coefficient Variability at 175–500 km from the Orbit Decay Analyses of Spheres	B. Bowman and K. Moe
Orbit determination	
An Algorithm for Preliminary HEO Orbital Parameter Determination Based on Three Optical Positional Measurements	V. Boikov, Z. Khutorovsky, and N. Sbytov
On a New Version of the Orbit Determination Method	A. Sochilina and R. Kiladze
Nonlinear Bayesian Filtering for State and Parameter Estimation	A. Sochilina, R. Kiladze, K. Grigoriev, and A. Vershkov
Optical systems and observations	
The Observations of the Near-Earth Objects with the ZA-320M Pulkovo Telescope	A. Devyatkin, E. Yu Aleshkin, K. Barshevich, G. Baturina, A. Bekhteva, D. Gorshanov, O. Krakosevich, V. Kupriyanov, V. L'vov, R. Smekhacheva, A. Sochilina, and S. Tsekmejster
New Wide Field Camera for Search for Fast Optical Transients	A. A. Pozanenko, S. Bondar, E. Ivanov, E. Katkova, G. Beskin, S. Karpov, A. Biryukov, I. Zolotukhin, and K. Hurley

(*Continued*)

TABLE B.10 (*Continued*) SIXTH U.S.–RUSSIAN SPACE SURVEILLANCE WORKSHOP PAPERS (PULKOVO OBSERVATORY, ST. PETERSBURG, RUSSIA)

Title	Presenting Author
The Pulkovo Cooperation for Radar and Optical Observations of Space Objects	I. Molotov, G. Tuccari, A. Konovalenko, M. Nechaeva, V. Abalakin, R. Kiladze, V. Titenko, Z. Khutorovsky, V. Agapov, A. Volvach, A. Deviatkin, A. Sochilina, I. Guseva, L. Xiang, and Yu. Gorshenkov
On Observation of Low-Mass and Small-Size Space Objects Using the Ground-Based Optoelectronic Means	S. Zdor, V. Kolinko, and V. Titenko
Space surveillance session B	
The Results of the Observations of the LEO Spacecrafts at the Experimental High-Potential Incoherent Scattered Radar	A. Medvedev, A. Potekhin, A. Zavorin, V. Lebedev, V. Khakhinov, and D. Kushnarev
Using EOP and Space Weather Data for Satellite Operations	D. Vallado and T. S. Kelso
A Concept of the Purposeful Research of the Dynamics of Man-Made Conditions on Geostationary Orbits (GEO) During the Spacecraft Development	A. Klishin and G. Gotovtsev
Indices of the Background Magnetic Field and the Polar Activity of the Sun	V. Makarov
Deep space	
Relative Orbit Determination of Geosynchronous Satellites Using the Cowpoke Equations	C. Sabol, K. Hill, C. McLaughlin, K. K. Luu, and M. Murai
On a New Theory of the Geostationary Satellite Motion and Its Applications	R. Kiladze
On Observations of Satellites at the Hissar Observatory	M. Gulyamov
Optical Facilities for Space Object Observations; Space Control and Analysis System (Ukraine)	V. Lopachenko
Observations of HEO at the Odessa Astronomical Observatory (Ukraine)	P. Sukhov

TABLE B.11 SEVENTH U.S.–RUSSIAN SPACE SURVEILLANCE WORKSHOP ATTENDEES (NAVAL POSTGRADUATE SCHOOL, MONTEREY, CALIFORNIA)

U.S. Attendees	
Terry Alfriend	Texas A&M University
Susan E. Andrews	MIT Lincoln Laboratory
Paul Cefola	Consultant
Suman Chakravorty	Texas A&M University
Felix Hoots	The Aerospace Corporation
Richard Hujsak	Analytical Graphics
Nicholas Johnson	NASA JSC
Thomas S. Kelso	Analytical Graphics
Paul Kervin	AFRL
Steven Knowles	SAIC
Kim Luu	AFRL
Eric Pearce	MIT Lincoln Laboratory
Chris Sabol	AFRL
Dan J. Scheeres	University of Colorado
Paul Schumacher	AFRL
John Seago	Analytical Graphics
Kenneth Seidelmann	University of Virginia
David Vallado	Analytical Graphics
Matt Wilkins	Schafer Corporation
Russian Attendees	
Vladimir Boikov	KIA Systems
Valery Dicky	Institute for Precision Instrument Engineering
Zakhary Khutorovsky	KIA Systems
Valery Kolinko	Federal Research and Production Center Krasnogorsk Plant
Andrey Nazarenko	Scientific Technological Center of Space Monitoring of Earth
Pavel Papushev	Institute of Solar Terrestrial Physics Siberian Branch of Russian Academy of Sciences
Yury Tretyakov	Scientific Research Center "Kosmos"
Stanislav Veniaminov	Scientific Research Center "Kosmos"
European Space Agency Attendees	
Heiner Klinkrad	ESA/ESOC
Thomas Schildknecht	University of Bern

TABLE B.12 SEVENTH U.S.–RUSSIAN SPACE SURVEILLANCE WORKSHOP PAPERS (NAVAL POSTGRADUATE SCHOOL, MONTEREY, CALIFORNIA)

Title	Presenting Author
Future space surveillance	
The U.S. Naval Space Surveillance Upgrade Program: 1999–2003	P. Schumacher
Investigations of the Feasibility of a European Space Surveillance System	T. Schildknecht, T. Flohrer, and R. Musci
Orbital debris	
Space Surveillance Network for a More Complete Catalog	E. Stansbery and N. Johnson
On the Probability of Collisions of Interplanetary Space Vehicles with Meteors	R. Kiladza
Detection of Small-Size Space Debris with the FGAN and EISCAT Radars	H. Krag, H. Klinkrad, R. Jehn, J. Markkanen, and R. Leushacke
On Some Difficulties of Planning the Search for a Space Object by Narrow-Beam Sensors	S. Veniaminov, V. Lapukhin ,and Yu. Tretyakov
Orbit determination I	
Technique for Operative Detection and Orbit Determination of Uncatalogued Space Objects	S. Veniaminov, V. Zavaliy, Yu. Tretyakov, V. Pyrin, and V. Lapukhin
Correlation of Optical Observations of Objects in Earth Orbit	D. Scheeres, J. Maruskin, and T. Alfriend
Orbit Determination of LEO Satellites for a Single Pass Through a Radar: Comparison of Methods	Z. Khutorovsky, S. Kamensky, N. Sbytov, and T. Alfriend
A Semianalytic Meshless Approach to the Transient Fokker-Planck Equation	M. Kumar, S. Chakravorty, and J. L. Junkins
Optical systems I	
A Survey of Wide-Field-of-View Optical Telescopes	P. Kervin
Astrometric Calibration for Wide-Area Space Surveillance Sensors	E. Pearce, R. Lambour, and J. Stuart
Catalog maintenance and development	
Cataloging with an Upgraded Space Surveillance Fence	F. Hoots, G. Pierce, L. Ford, and H. Hadley
The SVD and Least Squares Orbit Determination	V. Boikov, Z. Khutorovsky, and T. Alfriend
A Review of Data Correlation Methods for Space Surveillance	R. Hujsak
The Characteristics and Consequences of the Break-up of the Fengyun-C Spacecraft	N. Johnson, E. Stanbery, J.-C. Liou, M. Horstman, S. Stokely, and D. Whitlock
Orbit prediction accuracy	
An Analysis of State Vector Propagation Accuracy	D. Vallado
Accuracy of Determination and Prediction Orbits in LEO. Estimation Errors Depending on Accuracy and Amount of Measurements	A. Nazarenko
Discourse on Corrections to the NRLMSISE-OO Atmospheric Density Model	M. Wilkins, C. Sabol, P. Cefola, and T. Alfriend

(*Continued*)

TABLE B.12 (*Continued*) SEVENTH U.S.–RUSSIAN SPACE SURVEILLANCE WORKSHOP PAPERS (NAVAL POSTGRADUATE SCHOOL, MONTEREY, CALIFORNIA)

Title	Presenting Author
Orbit estimation	
Sensor Calibration as an Application of Optimal Sequential Estimation Toward Maintaining the Space Object Catalog	J. Seago and J. Woodburn
Optimum Measurement Filtration and Motion Prediction Taking into Account the Atmosphere Distributions	A. Nazarenko, V. Yurasov, T. Alfriend, and P. Cefola
Challenges Related to Discovery, Follow-up, and Study of Small High Area-to-Mass Ratio Objects at GEO	T. Schildknecht, R. Musci, T. Flohrer, and H. Klinkrad
Enhancing Multi-payload Launch Support with Netcentric Operations	S. Andrews, W. Bougas, T. Cott, S. Hunt, J. Kadish, and C. Solodyna
Characterization	
Improving the Hyperspectral Linear Unmixing Problem with Unsupervised Cluster and Covariance Estimates	E. Brevdo and K. Luu
Potentialities of Passive-RF Sensors for Characterization of Satellites	S. Veniaminov, O. Ryazanov, and Yu. Tretyakov
Photometric Calibration for Wide-Area Space Surveillance Sensors	J. Stuart, E. Pearce, and R. Lambour
Can Photometrical Data Help to Maintain a Catalogue of Small-sized Space Objects	M. Mishina, P. Papushev, V. Tergoev and T. Tsoukker
Reference stars	
Optical Reference Star Catalogs for Space Surveillance: Current Status and Future Plans	P. K. Seidelmann, R. Gaume, N. Zacharias, K. Johnston, and B. Dorland
Choosing Star Platforms when Measuring Angular Data of Space Objects with Relative Method	S. Zdor , V. Kol'inko, V. Titenko, and G. Yatskevich

TABLE B.13 EIGHTH U.S.–RUSSIAN SPACE SURVEILLANCE WORKSHOP ATTENDEES (WAILEA MARRIOTT RESORT, WAILEA, MAUI, HAWAII)

U.S. Attendees	
Terry Alfriend	Texas A&M University.
Susan E. Andrews	MIT Lincoln Laboratory
Duane Bird	USSTRATCOM/J31
Suman Chakravorty	Texas A&M University
Ed Fienga	JFCC Space
Ralph Gaume	U.S. Naval Observatory
Keric Hill	Pacific Defense Solutions
Marcus Holzinger	Univ of Colorado
Felix R. Hoots	Aerospace Corporation
Moriba Jah	AFRL
Thomas S. Kelso	SAIC
Paul W. Kervin	AFRL
Creon Levit	NASA Ames
Kim K. Luu	AFRL
Will Marshall	NASA Ames
Eric C. Pearce	MIT/LL
Chris Sabol	AFRL
Paul W. Schumacher	AFRL
Kenneth Seidelmann	University of Virginia
Charles Spillar	Air Force Space Command
E. Gene Stansbery	NASA JSC
Brad Thompson	AFOSR/EOARD
David Vallado	Analytical Graphics
Matt Wilkins	Schafer Corporation
Russian Attendees	
Nail S. Bakhtigaraev	Institute of Astronomy of RAS, Moscow
Vladimir Boikov	KIA Systems
Evgeniy A. Grishin	Vympel Corporation
Sergey Kamensky	Vympel Corporation
Zakhary Khutorovsky	KIA Systems
Valery Kolinko	Federal Research and Production Center Krasnogorsk Plant
Ivan Korebtsev	Institute of Solar Terrestrial Physics, Siberia
Vladimir Kouprianov	Central (Pulkovo) Observatory of RAS
Maxim E. Prokhorov	Vympel Corporation
Sergey Meshcheryakov	Central Research Institute of Machine Building
Igor E. Molotov	Keldysh Institute of Applied Mathematics
Pavel Papushev	Institute of Solar Terrestrial Physics Irkutsk
Igor V. Tarasenko	Research-and-Production Corporation "Precision Systems and Instruments"
Vasily S. Yurasov	Research-and-Production Corporation "Precision Systems and Instruments"
European Space Agency Attendees	
Holger Krag	ESA/ESOC

TABLE B.14 EIGHTH U.S.–RUSSIAN SPACE SURVEILLANCE WORKSHOP PAPERS (WAILEA MARRIOTT RESORT, WAILEA, MAUI, HAWAII)

Title	Presenting Author
Optical systems	
Simulation and Testing of Deep Space Search with the ETS 36-cm Telescope	R. Lambour, E. Pearce, R. Shah, and P. Trujillo
The IPIE Adaptive Optical System Application for LEO Observations	Eu. Grishin, V. Shargorodskiy, P. Inshin, V. Vygon, and M. Sadovnikov
Some Issues of Creation of Wide-Field Telescopes for Monitoring Satellites and Space Debris in Low and High Earth Orbits	I. Tarasenko, V. Shargorodskiy, S. Markelov, and V. Terebizh
ISON: Existing Structure, Tasks, Characteristics and Further Development	I. Molotov, V. Agapov, E. Akim, and S. Sukhanov
Orbit determination and prediction	
Radar Optical Observation Mix	F. Hoots
Software Tools for Determination of "Molniya" Type Orbits	V. Boikov and T. Alfriend
Improved Orbit Predictions Using Two-Line Elements	C. Levit and W. Marshall
Combined Determination of Orbital Parameters and Maneuver Impulse	V. Stepaniants, A. Tuchin, and S. Kamensky
Identification of Maneuvers Performed By Low-Thrust Engines	V. Tuchin, V. Stepaniants, V. Agapov, and A. Baronov
Approximation Approach for Motion Prediction for Satellites with High A/M Ratio	K. Sorokin and Z. Khutorovsky
Space surveillance systems and requirements	
Space Situational Awareness Sharing at the Operational Level	E. Fienga
Requirements for Space Surveillance Methods and Technologies for High-Elliptical, GEO and Quasi-Geostationary Orbits	V. Fateev. S. Sukhanov, V. Shilin, S. Kamensky, V. Agapov, and V. Burtzev
The European Space Surveillance System—Required Performance and Design Concepts	H. Krag, H. Klinkrad, T. Flohrer, E. Fletcher, and N. Bobrinsky
Commercial and Foreign Entities (CFE) Program Status Update and the Way-ahead	C. Spillar
Track correlation, estimation, and covariance	
Linearized Orbit Covariance Generation and Propagation Analysis	C. Sabol, T. Sukut, T. Alfriend, K. Hill, B. Wright, Y. Li, and P. Schumacher
Detection of LEO Satellites Break-Up Fragments Using Radar Measurements: Algorithms and Simulation Results	A. Samotokhin, Z. Khutorovsky, S. Kamensky, and T. Alfriend

(*Continued*)

TABLE B.14 (*Continued*) EIGHTH U.S.–RUSSIAN SPACE SURVEILLANCE WORKSHOP PAPERS (WAILEA MARRIOTT RESORT, WAILEA, MAUI, HAWAII)

Title	Presenting Author
Using Non-Coordinate Measurements for Evaluation of the Parameters of the Model of Solar Radiation Pressure for Prediction of Orbital Motion	R. Luchkin, M. Prokhorov, V. Shilin, and V. Lavrov
Object Correlation Using Control Effort Metrics with Boundary Condition Uncertainties	M. Holzinger and D. Scheeres
Evaluating Gooding Angles-only Orbit Determination of Space Based Space Surveillance Measurements	D. Vallado
Procedures for Recognition of Multi-Functional Satellites for the Case of Only One Learning Sample	M. Prokhorov, Yu. G. Ryndin, S. Sukhanov, and R. Luchkin
Guarantee Approach for Orbit Determination	Z. Khutorovsky and T. Alfriend
Space surveillance systems support	
Generalized Functions and Statistical Problems of Orbital Mechanics	S. Meshcheryakov
European Efforts to Survey, Track, and Characterize Small-Size Objects at High Altitudes	T. Schildknecht, A. Vananti, H. Krag, and H. Klinkrad
Analysis of the Characteristics of Satellite Collisions	V. Fateev, S. Sukhanov, Z. Khutorovsky, S. Kamensky, A. Samotokhin, V. Stepaniants, and Yu. Burtzev
Covariance-Based Network Tasking of Optical Sensors	K. Hill, P. Sydney, R. Corte, K. Luu, J. Houchar, D. Nahòolew, M. Jah, and P. Schumacher
Space surveillance support	
Two Years of International Cooperation on Conjunction Mitigation	T.S. Kelso
The UCAC3 Star Catalog: A New Catalog of Optical Reference Stars for Space Surveillance	R. R. Gaume, P. K. Seidelmann, N. Zacharias, and K. Johnston
Radar Open System Architecture for Lincoln Space Surveillance Activities	S. E. Andrews, P. Yoho, G. Banner, and T. Sangiolo

TABLE B.15 NINTH U.S.–RUSSIAN SPACE SURVEILLANCE WORKSHOP ATTENDEES (KRESTOVAYA PAD, LISTVYANKA, RUSSIA)

U.S. Attendees	
Kyle T. Alfriend	Texas A&M University.
Darren Garber	NXTRAC
Steve Knowles	SAIC
Tamara Payne	Applied Optimization
Ryan Russell	University of Texas
Paul W. Schumacher	AFRL
P. Kenneth Seidelmann	University of Virginia
David Walsh	Consultant
Russian Attendees	
Oleg Aksenov	SRC "Kosmos"
Vladimir Aleshin	RPC Precision Syst. and Instruments
Nail Bakhtigarayev	Institute of Astronomy, RAS
Andrey Chervonov	MCC TSNIIMASH
Denis Chirov	SRC "Kosmos"
Evgeny Grishin	RPC Precision Syst. and Instruments
Irina Guseva	Pulkovo Astronomical Observatory, RAS
Zakhary Khutorovsky	Vympel Corporation
Valery Kolin'ko	Zverev Krasnogorsky Plant
Ivan Korobtsev	ISTPh SB RAS
Sergey Kuzin	Lebedev Physical Institute, RAS
Vasilly Lavrov	Vympel Corporation
Sergey Meshcheryakov	MCC TSNIIMASH
Andrey Nazarenko	NTC "Kosmonit"
Igor Oleynikov	MCC TSNIIMASH
Vladimir Pirozhnik	SRC "Kosmos"
Andrey Rykin	Vympel Corporation
Alexander Samotokhin	KIAM
Boris Shustov	Institute of Astronomy, RAS
Vladimir Tergoyev	ISTPh SB RAS
Yury Tretyakov	SRC "Kosmos"
Stanislav Veniaminov	SRC "Kosmos"
Nikolay Yatskevich	Zverev Krasnogorsky Plant
Sergey Yazev	Irkutsk State University
ESA Participants	
Tim Florer	ESA/ESOC
Thomas Schildknecht	University of Bern

TABLE B.16 NINTH U.S.–RUSSIAN SPACE SURVEILLANCE WORKSHOP PAPERS (KRESTOVAYA PAD LISTVYANKA, RUSSIA)

Title	Presenting Author
Optical systems I	
The Results of Estimating the Performance of Electro-Optical Complex Using the Facility for Measuring the Sky Background and Astro-Climate Conditions	D. Chirov and V. Grigoroshenko
Optical Surveys for Space Debris in MEO	T. Schildknecht, A. Vananti, J. Herzog, A. Hinze, H. Krag, and T. Flohrer
On the Problems of Wide Field Image Processing	I. Guseva
Methods and Instruments of Solving the Asteroid/Comet Impact Hazard Problem: Necessity of Implementation in the (Inter)National Space Situation Awareness System	I. Molotov, V. Agapov, E. Akim, and S. Sukhanov
Orbit determination and prediction I	
Guaranteed Approach for Orbit Determination with Limited Error Measurements	Z. Khutorovsky, S. Sukhanov, A. Samotokhin, and T. Alfriend
Bounds on Range and Range Rate for Optical Tracking of Satellites	P. Schumacher
Comparison of Various Approaches for Reentry Prediction	A. Nazarenko and V. Yurasov
Radar Measurement Errors and Small Debris Density Effects on Orbit Prediction	D. Walsh
Space surveillance systems support 1	
On the Incorrectness of Application of the Fourier Harmonic Analysis for Revealing Latent Periodicities in Signals and Processes and on a Widespread Delusion While Analyzing Optical and Radar Signatures	O. Aksenov and S. Veniaminov
The Scope of Upper Atmosphere Density Variations During the Entire Space Age as Determined from Satellite Drag Measurements	S. Knowles
Real-Time Monitoring and Warning of Near-Earth Space Dangerous Events for Mission Control by the MCC of TSNIIMASH Federal State Unitary Enterprise	I. Oleynikov and V. Ivanov
Tracking of Phobos-Grunt Mission in Low Earth Orbit	A. Samotokhin, A. Tuchin, S. Sukhanov, Z. Khutorovsky, and A. Rykin
Optical systems 2	
Space Object Observation Using Optical Sensors Within the SPIRIT/CORONAS-F and THESIS/CORONAS-PHOTON Experiments	S. Kuzin, A. Ulyanov, S. Shestov, S. Bogache, and G. Karabadzhak
Analytical Analysis of Photometry of Three-Axis Stabilized Geosynchronous Satellites	T. Payne and N. Shine

(*Continued*)

TABLE B.16 (*Continued*) NINTH U.S.–RUSSIAN SPACE SURVEILLANCE WORKSHOP PAPERS (KRESTOVAYA PAD LISTVYANKA, RUSSIA)

Title	Presenting Author
Altay Optic-laser Center Capability to Satellites Emergencies Estimation	V. Aleshin, E. Grishin, V. Shargorodsky, and D. Novgorodtsev
Barrier Detection of Celestial Bodies	S. Zdor, V. Kolin'ko, and N. Yatskevich
Small objects and collision/conjunction assessment	
Parallel Sigma Point and Particle Filters for Orbit Determination	H. Shen, C. Karlgaar, R. Russell, V. Vittaldev, and E. Pellegrini
On Smaller Debris of Orbital Collisions	S. Meshcheryakov
The CubeSat Challenge: A Review of CubeSat Dependencies and Requirements for Space Tracking Systems	D. Garber
Model Study of the Possibilities of Space Debris Cataloging	A. Nazarenko
Methods and Algorithms of Contour Analysis of Optical and IR Images	V. Lavrov, R. Luchkin, O. Nemykin, M. Prokhorov, Yu. Ryndin, and S. Sukhanov
Observations of Near-Earth Space Objects at Sayan Observatory, Opportunities and Perspectives	V. Tergoev, Yu. Karavaev, E. Klunko, V. Goryashin, I. Korobtsev, M. Mishina, T. Tsukker, and M. Esclevich
Space surveillance systems support 2	
UTC Timekeeping in the Future	P. K. Seidelmann and J. H. Seago
Investigations of Radioactive Contamination of Near-Earth Space	N. Bakhtigaraev, M. Anikeeva, K. Boyarchuk, S. Kuzin, S. Ulin, and V. Chazov
Tailoring the Observation Scenarios and Data Processing Techniques for Supporting Conjunction Event Assessments	T. Flohrer, B. Bastida Virgili, H. Krag, H. Klinkrad, K. Merz, T. Schildknecht, and S. Lemmens
On the Problem of Predicting Development of Solar Cycle 24	S. Yazev
Orbit determination and prediction 2	
Once Again on International Cooperation	A. Rykin, A. Chervonov, and S. Veniaminov
Methods of Accounting for the Influence of Near-Earth Environment to Functioning of Space Surveillance Sensors and Facilities for Remote Sensing the Earth	Yu. Tretyakov and V. Trekin
Orbit Determination of Satellite Formations	T. Alfriend

APPENDIX C

I. PROGRAMS FOR U.S.–CHINA SPACE SURVEILLANCE TECHNICAL INTERCHANGES 1–3

TABLE C.1 FIRST U.S.–CHINA SPACE SURVEILLANCE TECHNICAL INTERCHANGE ATTENDEES

U.S. Attendees	
Kyle T. Alfriend	Texas A&M University
Paul Cefola	Consultant
Kenneth Chan	Aerospace Corporation
Stephen H. Knowles	Science Applications International Corp
Daniel J. Scheeres	University of Colorado
P. Kenneth. Seidelmann	University of Virginia
Gene Stansbery	NASA/Johnson Space Center
Matthew P. Wilkins	Schafer Corporation
China Attendees	
SHAO Meizhen	Center for Space Science and Applied Research, CAS
ZHANG Yao	Center for Space Science and Applied Research, CAS
HAN Zengyao	China Academy of Space Technology
WANG Xiuhong	Xi'an Satellite Control Center
ZHANG Rongzhi	Xi'an Satellite Control Center
ZHAO Changyin	Purple Mountain Observatory, CAS
WEI Dong	Purple Mountain Observatory, CAS
WANG Hongbo	Purple Mountain Observatory, CAS
CHAO Zhibin	Purple Mountain Observatory, CAS
MA Jingyuan	Purple Mountain Observatory, CAS
JIANG Xiaojun	National Astronomical Observatories, CAS
TANG Yijun	National Astronomical Observatories, CAS
ZHAO You	National Astronomical Observatories, CAS
GUO Xiaozhong	National Astronomical Observatories, CAS
TANG Datao	National Astronomical Observatories, CAS
SHEN Ming	National Astronomical Observatories, CAS
JWANG Jianfen	National Astronomical Observatories, CAS
MAO Yongna	National Astronomical Observatories, CAS
YE Shuhua	Shanghai Astronomical Observatory, CAS

(Continued)

TABLE C.1 (*Continued*) FIRST U.S.–CHINA SPACE SURVEILLANCE TECHNICAL INTERCHANGE ATTENDEES

HONG Xiaoyu	Shanghai Astronomical Observatory, CAS
YANG Fumin	Shanghai Astronomical Observatory, CAS
TANG Zhenghong	Shanghai Astronomical Observatory, CAS
HU Xiaogong	Shanghai Astronomical Observatory, CAS
ZIANG Zhongping	Shanghai Astronomical Observatory, CAS
MAO Yindun	Shanghai Astronomical Observatory, CAS
ZHOU Ruixian	Shanghai Astronomical Observatory, CAS
QU Weijing	Shanghai Astronomical Observatory, CAS
LIU Xiangyi	Xi'an JiaoTong University

TABLE C.2 FIRST U.S.–CHINA SPACE SURVEILLANCE TECHNICAL INTERCHANGE PAPERS

Title	Presenting Author
Monday, 1 June 2009	
Opening Remarks	Shuhua Y K. Seidelmann
Session 1: Conjunction and collision assessment	
The Development of Xi'An Satellite Control Center	ZHANG R.
Maintaining the Catalog of Earth Satellites	P. Schumacher (presented by T. Alfriend)
Covariance Based Track Association	K. T. Alfriend, K. Hill, C. Sabol, and P. W. Schumacher
Session 2: Orbital debris	
China Protection Activities Against Space Debris	HAN Z.
Orbital Debris Research at NASA	G. Stansbery
Laser Ranging Experiment to Space Debris at Shanghai Observatory	YANG F., ZHANG Z., CHEN J., CHEN W., WU Z., ZHANG H. and I. Prochazka
Session 3: High-altitude debris	
Monitoring Faint Space Debris with Rotating Drift-Scan CCD on a Small Telescope	TANG Z., MAO Y., LI Y, ZHANG, X., and YU, Y.
Optical Reference Stars for Space Surveillance: Latest Development	K. Seidelmann, R. Gaume, and N. Zacharias
The Study of Physical Characteristics of High Apogee Space Debris	JIANG, X.
Session 4: Conjunctions and collision avoidance	
An Analytic Method of Filtering Potential Dangerous Objects for Collision Avoidance	WANG X. and ZHANG R.
ISS Collision Probability	K. Chan
Analyzing Iri33-Cosmos2251 Satellite Collision Event and Effects Based on TLE Data	LIU, J. (canceled)
Tuesday, 2 June 2009	
Session 5: Atmospheric density modeling	
Evolution of Atmospheric Density Models	P. J. Cefola

(*Continued*)

TABLE C.2 (*Continued*) FIRST U.S.–CHINA SPACE SURVEILLANCE TECHNICAL INTERCHANGE PAPERS

Title	Presenting Author
Studies on the Characteristic of Thermospheric Density Variation During Geomagnetic Storms with CHAMP and GRACE Accelerometer Data	WANG H., ZHAO C., XIONG Y., and XIONG J.
Further Explorations in Dynamic Calibration of the Atmosphere Theory	M. J. Wilkins
Nonlinear Bayesian State Filtering for State and Parameter Estimation	K. T. Alfriend
Session 6: Orbit determination and prediction	
A History of Analytical Orbit Modeling in the United States Space Surveillance System	P. Schumacher (presented by T. Alfriend)
Current Status of Orbit Determination Method in PMO	ZHAO D., WEI D., XIONG Y., XIONG J., ZHANG W., and WANG X.
State Vector Propagation for the Space Catalog	P. J. Cefola
Orbit Determination for LEO/MEO/GEO and Lunar Orbiter at SHAO	HU X., WU B., HUANG Y., and PENG D.
Session 7: Catalog maintenance and development	
Survey of High Geocentric Orbit Space Debris by CNEOT	ZAO H. (canceled, presentation provided)
Uncorrelated Optical Observation: Properties and Implications for Object Correlation	D. J. Scheeres, K. Fujimoto, J. Maruskin, and K.T. Alfriend
Application of Image Restoration in GEO Space Debris Survey	PING Y. (canceled, presentation provided)
Closing Remarks	P. K. Seidelmann and S. YE

TABLE C.3 SECOND U.S.–CHINA SPACE SURVEILLANCE TECHNICAL INTERCHANGE ATTENDEES

U.S. Attendees	
Kyle T. Alfriend	Texas A&M University
Duane Bird	U.S. Strategic Command
Kenneth Chan	Aerospace Corporation
Ralph A. Gaume	U.S. Naval Observatory
Marcus Holzinger	Texas A&M University
Stephen H. Knowles	Science Applications International Corp.
Tamara E. Payne	Applied Optimization Inc.
P. Kenneth Seidelmann	University of Virginia
Ryan Paul Russell	University of Texas
David W. Walsh	AT&T/GSI
Matthew P. Wilkins	Schafer Corporation
China Attendees	
YE Shuyua	Member Chinese Academy of Sciences
HAO Jinxin	Dep. Director, National Astronomical Observatories, CAS
WEI Zhang	Secretary-General and CEO of the APSCO; China)
Maqbool Ahmad Chaudhry	Director General of APSCO (Pakistan)
YAN Huang	Department of Strategic Planning and Program Management, APSCO (China)
Tsoodol Nyamkhuu	Director General of APSCO (Mongolia)
YUAN Haibo	National Timing Service Center (NTSC), Shaanxi Astronomical Observatory
DONG Shaowu	NTSC, Shaanxi Astronomical Observatory, CAS (China)
XU Xiaoli	Purple Mountain Observatory
MA Chaowei	Beijing Institute of Tracking and Telecommunications Technology
ZHANG Shi-xue	National University of Defense Technology
XU Xiaoli	National University of Defense Technology
TANG Zhenghong	Shanghai Astronomical Observatory
LIU Zhonggui	Beijing Institute of Tracking and Telecommunications Technology
ZHANG Shi-xue	Changchun Institute of Optics, Fine Mechanics and Physics, CAS
ZHAO You	National Astronomical Observatories, CAS
LI Yuqiang	Yunnan Astronomical Observatory, CAS
GAO Pengqi	National Astronomical Observatories, CAS
GUO Xiaozhong	National Astronomical Observatories, CAS
SHEN Ming	National Astronomical Observatories, CAS
YU Huanhuan	National Astronomical Observatories, CAS
ZHAO Datao	National Astronomical Observatories, CAS
ZHAO Zhe	National Astronomical Observatories, CAS

TABLE C.4 SECOND U.S.–CHINA SPACE SURVEILLANCE TECHNICAL INTERCHANGE PAPERS

Title	Presenting Author
Monday, 17 October 2011	
Opening Remarks	YE S. K. Seidelmann
Session 1: Catalog maintenance and development	
Miss Distance Generalized Variance Non-Central Distribution	K. Chan
The Applicability Research of Nonlinear Collision Probability	Y. XIONG and X. XU
Fast Computation of All-pair Conjunction Analyses	N. Arora and R. Russell
Collision Risk Assessment for Spacecraft with CASS	C. MA and J. HUANG
Session 2: Orbital debris tracking	
The General Mission Analysis Tool (GMAT): A New Resource for Supporting Debris Orbit Determination, Tracking and Analysis	M.P. Wilkins
Periodicity Characterization of TLE Prediction Error and Poisson Series Fitting	BAI X., CHEN L., and TANG G.
ΔV Control Distance with Dynamics Parameter Uncertainty	M. Holzinger
Orbital Anomaly Detection and Application of Space Anomaly	CHEN L., BAI X. and ZHANG T.
Session 3: Orbit determination and prediction	
New Observation Results from a Rotating-drift-scan CCD System	TANG Z., MAO Y., LI Y., and YU Y.
Statistical Properties of Geosynchronous Satellite Photometry	T. Payne
Using Space Based Sensors to Catalog LEO Objects	LIU Z. and HAO S.
Tuesday, 18 October 2011	
Session 4: Catalog maintenance and development	
Sharing Space Situational Awareness Data	D. Bird
Analysis on Orbit Characteristics of a Space Object Based on a Low Thrust Orbit Transfer	ZHU K., XU D., GAO J., YANG. Y. and LI S.
Time Scales and the Future of UTC	K. Seidelmann and J. Seago
Scope of Upper Atmosphere Density Variations During the Entire Space Age Determined from Satellite Drag Measurements	S, Knowles
Session 5: Supporting technology and operations	
Signal Detection and Parameter Estimation of Fence -Type Space Surveillance Radar	HUANG J. and HU W.
Effect of Radar Measurement Errors on Small Debris Prediction	D. Walsh
Large FOV Mobile E-O Telescope for Searching and Tracking Low-Orbit Micro-satellites and Space Debris	WANG, J., GAO X., TANG J., HAN C., and ZHAO J.
Space Debris Measurements. Using Laser Ranging at Yunnan Observatory	LI Y., Li Z., LI R., FU H., ZHENG X., HE S. and ZHAI D.

TABLE C.5 THIRD U.S.–CHINA SPACE SURVEILLANCE TECHNICAL INTERCHANGE ATTENDEES

U.S. Attendees	
Kyle T. Alfriend	Texas A&M University
Kenneth Chan	Aerospace Corporation
Yang Cheng	Mississippi State University
Kohei Fujimoto	University of Colorado
Darren Garber	NXTRAC
Marcus Holzinger	Georgia Tech
Stephen Knowles	Integrity Applications
Tamara E. Payne	Applied Optimization Inc.
Chris Roscoe	Applied Defense Solutions
P. Kenneth Seidelmann	University of Virginia
David W. Walsh	AT&T/GSI
China Attendees	
WANG Shaokai	Beihang University
YIN Jianfeng	Beihang University
ZHANG Ran	Beihang University
CHEN Shenyan	Beihang University
RAO Yinrui	Beihang University
AN Haichao	Beihang University
SHAN Ming	National Astronomical Observatories, CAS
YU Huan	National Astronomical Observatories, CAS
GUO Xiaozhong	National Astronomical Observatories, CAS
MIAO Yongna	National Astronomical Observatories, CAS
JIANG Xiaojun	National Astronomical Observatories, CAS
ZHAO You	National Astronomical Observatories, CAS
YANG Hongwei	Tsinghua University
ZHANG Peng	Tsinghua University
LU Yao	Purple Mountain Observatory, CAS
LIN Houyuan	Purple Mountain Observatory, CAS
ZHU Tinglei	Purple Mountain Observatory, CAS
CHEN Ming	SINA
LI Cuilan	SINA
LIU Ye	SINA
BAI Xianzong	National University of Defense Technology
HU Weidong	National University of Defense Technology
LI Yuqiang	Yunnan Astronomical Observatory, CAS
LIU Zhonggui	Beijing Institute of Tracking and Telecommunications Technology
WANG Ronglan	National Space Science Center, CAS
SONG Qiang	China Academy of Launch Vehicles

TABLE C.6 THIRD U.S.–CHINA SPACE SURVEILLANCE WORKSHOP PAPERS

Title	Authors
Monday, 13 May 2013	
Opening Remarks	YE S. K. Seidelmann
Session 1: Probability of collision	
Miss Distance—Generalized Variance Non-Central Chi Distribution	K. Chan
Explicit Expression of Collision Probability in Terms of RSW and NTW Components of Relative Position	BAI X, CHEN L., and TANG G.
Uncertainty in Probability of Collision Calculation	D. Garber
Bayesian Group Tracking Method for Space Debris	HUANG J., HU W., and ZHANG L.
Session 2: Orbit uncertainty and propagation	
A Method to Improve the Precision of Predicted LEO Space Objects Ephemeris	LIU Z.
Analytical Nonlinear Uncertainty Propagation: Theory and Implementation	K. Fujimoto, D. J. Scheeres, and K. T. Alfriend
Maneuver Detection of Space Object for Space Surveillance	HUANG J., HU W., and ZHANG L.
Orbit Uncertainty Propagation of a Tumbling High Area-to-Mass-Ratio Space Object	Y. Cheng (FYI This is a US chinese)
Session 3: Tracking space debris 1	
Improving LEO Prediction With TLEs	Liu W., WANG R., Yan R., and GONG J.
An Automated Telescope for Orbital Debris	LU X.
GEO Cross-tag Resolution Using Color Signatures	T. Payne
Tuesday, 14 May 2013	
Session 4: Space situational awareness support	
Parallel Algorithm for Track Initiation for Optical Space Surveillance	P. W. Schumacher, M. P. Wilkins, and C. W. T. Roscoe
Maneuver Detection of Space Object for Space Surveillance	HUANG J., HU W., and ZHANG L
Autonomy Architecture for a Raven-Class Telescope for Space Situational Awareness	R. Coder and M. Holzinger
Session 5: Space debris tracking 2	
A Survey of Radars Capable of Providing Small Debris Measurements for Orbit Prediction	D. Walsh
Study on Optical Cross Section of Space Debris	MAO Y., WANG J., LU X., GE L., and JIANG X.
Long Term Behavior of the Thermosphere Determined from Satellite Drag Measurements	S. Knowles
Application of sCMOS in Space Debris Measurements	GE L.
Session 6: Space debris issues	
Space Debris Mitigation Status of China's Launch Vehicle	SONG Q.
Orbit Uncertainty Propagation of a Tumbling Area-to-Mass Ratio Space Object	Y. Cheng
Space Debris as an Epidemic	D. Finkleman (presented by Alfriend)

(*Continued*)

TABLE C.6 (*Continued*) THIRD U.S.–CHINA SPACE SURVEILLANCE WORKSHOP PAPERS

Title	Authors
Session 7	
Robust Formation Design for High-Eccentricity Satellite Missions Using a Stochastic Optimization Approach	C. Roscoe, S. Vadali, and K. Alfriend
Introduction of a New ISO project: Standard Calculation of Orbit Elements at Satellite–Launch Vehicle Separation	LIU J.
Mean Solar Time, UTI, UTC and Mean Solar Time	K. Seidelmann and J. Seago
Closing comments and discussion	YE S., ZHANG J., K. Seidelmann, and K. Alfriend

INDEX

Note: Page numbers are followed by f or t (indicating figures or tables).

www.ingramcontent.com/pod-product-compliance
Lightning Source LLC
Chambersburg PA
CBHW070840140925
32470CB00002B/2
* 9 7 8 1 6 2 4 1 0 7 4 3 6 *